Applied Video Processing in Surveillance and Monitoring Systems

Nilanjan Dey
Techno India College of Technology, Kolkata, India

Amira Ashour
Tanta University, Egypt

Suvojit Acharjee
National Institute of Technology Agartala, India

A volume in the Advances in Multimedia and
Interactive Technologies (AMIT) Book Series

www.igi-global.com

Published in the United States of America by
IGI Global
Information Science Reference (an imprint of IGI Global)
701 E. Chocolate Avenue
Hershey PA, USA 17033
Tel: 717-533-8845
Fax: 717-533-8661
E-mail: cust@igi-global.com
Web site: http://www.igi-global.com

Library of Congress Cataloging-in-Publication Data

Names: Dey, Nilanjan, 1984- editor. | Ashour, Amira, 1975- editor. |
   Acharjee, Suvojit, 1989- editor.
Title: Applied video processing in surveillance and monitoring systems /
   Nilanjan Dey, Amira Ashour, and Suvojit Acharjee, editors.
Description: Hershey PA : Information Science Reference, [2017] | Series:
   Advances in multimedia and interactive technologies | Includes
   bibliographical references and index.
Identifiers: LCCN 2016033139| ISBN 9781522510222 (hardcover) | ISBN
   9781522510239 (ebook)
Subjects: LCSH: Video surveillance. | Image processing--Digital techniques.
Classification: LCC TK6680.3 .A67 2017 | DDC 621.389/28--dc23 LC record available at https://lccn.loc.gov/2016033139

This book is published in the IGI Global book series Advances in Multimedia and Interactive Technologies (AMIT) (ISSN: 2327-929X; eISSN: 2327-9303)

British Cataloguing in Publication Data
A Cataloguing in Publication record for this book is available from the British Library.

All work contributed to this book is new, previously-unpublished material. The views expressed in this book are those of the authors, but not necessarily of the publisher.

For electronic access to this publication, please contact: eresources@igi-global.com.

Advances in Multimedia and Interactive Technologies (AMIT) Book Series

Joel J.P.C. Rodrigues
Instituto de Telecomunicações, University of Beira Interior, Portugal

ISSN: 2327-929X
EISSN: 2327-9303

Mission

Traditional forms of media communications are continuously being challenged. The emergence of user-friendly web-based applications such as social media and Web 2.0 has expanded into everyday society, providing an interactive structure to media content such as images, audio, video, and text.

The **Advances in Multimedia and Interactive Technologies (AMIT) Book Series** investigates the relationship between multimedia technology and the usability of web applications. This series aims to highlight evolving research on interactive communication systems, tools, applications, and techniques to provide researchers, practitioners, and students of information technology, communication science, media studies, and many more with a comprehensive examination of these multimedia technology trends.

Coverage

- Multimedia technology
- Digital Games
- Multimedia Streaming
- Gaming Media
- Digital Watermarking
- Mobile Learning
- Web Technologies
- Digital Images
- Digital Technology
- Multimedia Services

IGI Global is currently accepting manuscripts for publication within this series. To submit a proposal for a volume in this series, please contact our Acquisition Editors at Acquisitions@igi-global.com or visit: http://www.igi-global.com/publish/.

The Advances in Multimedia and Interactive Technologies (AMIT) Book Series (ISSN 2327-929X) is published by IGI Global, 701 E. Chocolate Avenue, Hershey, PA 17033-1240, USA, www.igi-global.com. This series is composed of titles available for purchase individually; each title is edited to be contextually exclusive from any other title within the series. For pricing and ordering information please visit http://www.igi-global.com/book-series/advances-multimedia-interactive-technologies/73683. Postmaster: Send all address changes to above address.

Titles in this Series

For a list of additional titles in this series, please visit: www.igi-global.com

Intelligent Analysis of Multimedia Information
Siddhartha Bhattacharyya (RCC Institute of Information Technology, India) Hrishikesh Bhaumik (RCC Institute of Information Technology, India) Sourav De (The University of Burdwan, India) and Goran Klepac (University College for Applied Computer Engineering Algebra, Croatia & Raiffeisenbank Austria, Croatia)
Information Science Reference • copyright 2017 • 520pp • H/C (ISBN: 9781522504986) • US $220.00 (our price)

Emerging Technologies and Applications for Cloud-Based Gaming
P. Venkata Krishna (VIT University, India)
Information Science Reference • copyright 2017 • 314pp • H/C (ISBN: 9781522505464) • US $195.00 (our price)

Digital Tools for Computer Music Production and Distribution
Dionysios Politis (Aristotle University of Thessaloniki, Greece) Miltiadis Tsalighopoulos (Aristotle University of Thessaloniki, Greece) and Ioannis Iglezakis (Aristotle University of Thessaloniki, Greece)
Information Science Reference • copyright 2016 • 291pp • H/C (ISBN: 9781522502647) • US $180.00 (our price)

Contemporary Research on Intertextuality in Video Games
Christophe Duret (Université de Sherbrooke, Canada) and Christian-Marie Pons (Université de Sherbrooke, Canada)
Information Science Reference • copyright 2016 • 363pp • H/C (ISBN: 9781522504771) • US $185.00 (our price)

Trends in Music Information Seeking, Behavior, and Retrieval for Creativity
Petros Kostagiolas (Ionian University, Greece) Konstantina Martzoukou (Robert Gordon University, UK) and Charilaos Lavranos (Ionian University, Greece)
Information Science Reference • copyright 2016 • 388pp • H/C (ISBN: 9781522502708) • US $195.00 (our price)

Emerging Perspectives on the Mobile Content Evolution
Juan Miguel Aguado (University of Murcia, Spain) Claudio Feijóo (Technical University of Madrid, Spain & Tongji University, China) and Inmaculada J. Martínez (University of Murcia, Spain)
Information Science Reference • copyright 2016 • 438pp • H/C (ISBN: 9781466688384) • US $210.00 (our price)

Emerging Research on Networked Multimedia Communication Systems
Dimitris Kanellopoulos (University of Patras, Greece)
Information Science Reference • copyright 2016 • 448pp • H/C (ISBN: 9781466688506) • US $200.00 (our price)

Emerging Research and Trends in Gamification
Harsha Gangadharbatla (University of Colorado Boulder, USA) and Donna Z. Davis (University of Oregon, USA)
Information Science Reference • copyright 2016 • 455pp • H/C (ISBN: 9781466686519) • US $215.00 (our price)

DISSEMINATOR OF KNOWLEDGE

www.igi-global.com

701 E. Chocolate Ave., Hershey, PA 17033
Order online at www.igi-global.com or call 717-533-8845 x100
To place a standing order for titles released in this series, contact: cust@igi-global.com
Mon-Fri 8:00 am - 5:00 pm (est) or fax 24 hours a day 717-533-8661

Table of Contents

Section 1
Introduction to Video Processing and Mining

Chapter 1

Abahan Sarkar, National Institute of Technology Silchar, India
Ram Kumar, National Institute of Technology Silchar, India

Chapter 2

Pushpajit A. Khaire, SRCOEM, India
Roshan R. Kotkondawar, GCOE, India

Chapter 3

Alaa M. AlShahrani, Taif University, Saudi Arabia
Manal A. Al-Abadi, Taif University, Saudi Arabia
Areej S. Al-Malki, Taif University, Saudi Arabia
Amira S. Ashour, Taif University, Saudi Arabia & Tanta University, Egypt
Nilanjan Dey, Techno India College of Technology, India

Chapter 4

S. Vasavi, V. R. Siddhartha Engineering College, India
T. Naga Jyothi, V. R. Siddhartha Engineering College, India
V. Srinivasa Rao, V. R. Siddhartha Engineering College, India

Detailed Table of Contents

Section 1
Introduction to Video Processing and Mining

Video processing has various applications in several domains. This section highlights the fundamental concepts and algorithms of video processing in the medical domain. The real time frames for the acquired images from different image modalities is used. Another context is introduced for data mining that has great benefit in data extraction various applications including educational and medical domains.

Chapter 1

    Abahan Sarkar, National Institute of Technology Silchar, India
    Ram Kumar, National Institute of Technology Silchar, India

In day-to-day life, new technologies are emerging in the field of Image processing, especially in the domain of segmentation. Image segmentation is the most important part in digital image processing. Segmentation is nothing but a portion of any image and object. In image segmentation, the digital image is divided into multiple set of pixels. Image segmentation is generally required to cut out region of interest (ROI) from an image. Currently there are many different algorithms available for image segmentation. This chapter presents a brief outline of some of the most common segmentation techniques (e.g. Segmentation based on thresholding, Model based segmentation, Segmentation based on edge detection, Segmentation based on clustering, etc.,) mentioning its advantages as well as the drawbacks. The Matlab simulated results of different available image segmentation techniques are also given for better understanding of image segmentation. Simply, different image segmentation algorithms with their prospects are reviewed in this chapter to reduce the time of literature survey of the future researchers.

Chapter 2

Pushpajit A. Khaire, SRCOEM, India

Roshan R. Kotkondawar, GCOE, India

Study on Video and Image segmentation is currently limited by the lack of evaluation metrics and benchmark datasets that covers the large variety of sub-problems appearing in image and video segmentation. Proposed chapter provides an analysis of Evaluation Metrics, Datasets for Image and Video Segmentation methods. Importance is on wide-ranging, Datasets robust Metrics which used for evaluation purposes without inducing any bias towards the evaluation results. Introductory Section discusses traditional image and video segmentation methods available, the importance and need of measures, metrics and dataset required to evaluate segmentation algorithms are discussed in next section. Main focus of the chapter explains the measures, metrics and dataset available for evaluation of segmentation techniques of both image and video. The goal is to provide details about a set of impartial datasets and evaluation metrics and to leave the final evaluation of the evaluation process to the understanding of the reader.

Chapter 3

Alaa M. AlShahrani, Taif University, Saudi Arabia

Manal A. Al-Abadi, Taif University, Saudi Arabia

Areej S. Al-Malki, Taif University, Saudi Arabia

Amira S. Ashour, Taif University, Saudi Arabia & Tanta University, Egypt

Nilanjan Dey, Techno India College of Technology, India

Marketing profit optimization and preventing the crops' infections are a critical issue. This requires crops recognition and classification based on their characteristics and different features. The current work proposed a recognition/classification system that applied to differentiate between fresh (healthy) from rotten crops as well as to identify each crop from the other based on their common feature vectors. Consequently, image processing is employed to perform the statistical measurements of each crop. ImageJ software was employed to analyze the desired crops to extract their features. These extracted features are used for further crops recognition and classification using the Least Mean Square Error (LMSE) algorithm in Matlab. Another classification method based on Bag of Features (BoF) technique is employed to classify crops into classes, namely healthy and rotten. The experimental results are applied of databases for orange, mango, tomato and potatoes. The achieved recognition (classification) rate by using the LMSE for all datasets (healthy and rotten) has 100%. However, after adding 10%, 20%, and 30% Gaussian noise, the obtained the average recognition rates were 85%, 70%, and 25%; respectively. Moreover, the classification (healthy and rotten) using BoF achieved accuracies of 100%, 88%, 94%, and 75% for potatoes, mango, orange, and tomato; respectively. Furthermore, the classification for all the healthy datasets achieved accuracy of 88%.

Chapter 4

S. Vasavi, V. R. Siddhartha Engineering College, India
T. Naga Jyothi, V. R. Siddhartha Engineering College, India
V. Srinivasa Rao, V. R. Siddhartha Engineering College, India

Now-a-day's monitoring objects in a video is a major issue in areas such as airports, banks, military installations. Object identification and recognition are the two important tasks in such areas. These require scanning the entire video which is a time consuming process and hence requires a Robust method to detect and classify the objects. Outdoor environments are more challenging because of occlusion and large distance between camera and moving objects. Existing classification methods have proven to have set of limitations under different conditions. In the proposed system, video is divided into frames and Color features using RGB, HSV histograms, Structure features using HoG, DHoG, Harris, Prewitt, LoG operators and Texture features using LBP, Fourier and Wavelet transforms are extracted. Additionally BoV is used for improving the classification performance. Test results proved that SVM classifier works better compared to Bagging, Boosting, J48 classifiers and works well in outdoor environments.

Chapter 5

Sayan Chakraborty, Bengal College of Engineering and Technology, India
Prasenjit Kumar Patra, Bengal College of Engineering and Technology, India
Prasenjit Maji, Bengal College of Engineering and Technology, India
Amira S. Ashour, Tanta University, Egypt
Nilanjan Dey, Techno India College of Technology, India

Image registration allude to transforming one image with reference to another (geometrically alignment of reference and sensed images) i.e. the process of overlaying images of the same scene, seized by assorted sensors, from different viewpoints at variant time. Virtually all large image evaluating or mining systems require image registration, as an intermediate step. Over the years, a broad range of techniques has been flourished for various types of data and problems. These approaches are classified according to their nature mainly as area-based and feature-based and on four basic tread of image registration procedure namely feature detection, feature matching, mapping function design, and image transformation and resampling. The current chapter highlights the cogitation effect of four different registration techniques, namely Affine transformation based registration, Rigid transformation based registration, B-splines registration, and Demons registration. It provides a comparative study among all of these registration techniques as well as different frameworks involved in registration process.

Section 2
Feature Detectors and Descriptors

Feature detectors and descriptors have a vital role in numerous applications including video camera calibrations, object recognition, biometrics, medical applications and image/video retrieval. Extract point correspondences "Interest points" between two similar scenes, objects, images/video shots is their main task. This section outlines the different feature detectors and descriptors types that involved in various applications.

Chapter 6

Al Hussien Seddik Saad, Minia University, Egypt
Abdelmgeid Amin Ali, Minia University, Egypt

Nowadays, due to the increasing need for providing secrecy in an open environment such as the internet, data hiding has been widely used. Steganography is one of the most important data hiding techniques which hides the existence of the secret message in cover files or carriers such as video, images, audio or text files. In this chapter; steganography will be introduced, some historical events will be listed, steganography system requirements, categories, classifications, cover files will be discussed focusing on image and video files, steganography system evaluation, attacks, applications will be explained in details and finally last section concludes the chapter.

Chapter 7

Sathish Shet, JSS Academy of Technological Education, India
A. R. Aswath, Dayananda Sagar College of Engineering, India
M. C. Hanumantharaju, BMS Institute of Technology and Management, India
Xiao-Zhi Gao, Aalto University School of Electrical Engineering, Finland

The most crucial task in real-time processing of image or video steganography algorithms is to reduce the computational delay and increase the throughput of a steganography embedding and extraction system. This problem is effectively addressed by implementing steganography hiding and extraction methods in reconfigurable hardware. This chapter presents a new high-speed reconfigurable architectures that have been designed for Least Significant Bit (LSB) and multi-bit based image steganography algorithm that suits Field Programmable Gate Arrays (FPGAs) or Application Specific Integrated Circuits (ASIC) implementation. Typical architectures of LSB steganography comprises secret message length finder, message hider, extractor, etc. The architectures may be realized either by using traditional hardware description languages (HDL) such as VHDL or Verilog. The designed architectures are synthesizable in FPGAs since the modules are RTL compliant. Before the FPGA/ASIC implementation, it is convenient to validate the steganography system in software to verify the concepts intended to implement.

Section 3
Segmentation, Classification and Registration based Image/Video Processing

Video and Image segmentation/classification as well as registration have numerous techniques. Performance metrics are used to evaluate and benchmark such techniques. This section elaborates an overview of these techniques with an analysis of the evaluation metrics for Image and Video Segmentation methods. Moreover, a classification process is included to classify different types of crops, which can be applied for further video applications.

Chapter 8

    Ammar Ladjailia, University of Souk Ahras, Algeria
    Imed Bouchrika, University of Souk Ahras, Algeria
    Nouzha Harrati, University of Souk Ahras, Algeria
    Zohra Mahfouf, University of Souk Ahras, Algeria

As computing becomes ubiquitous in our modern society, automated recognition of human activities emerges as a crucial topic where it can be applied to many real-life human-centric scenarios such as smart automated surveillance, human computer interaction and automated refereeing. Although the perception of activities is spontaneous for the human visual system, it has proven to be extraordinarily difficult to duplicate this capability into computer vision systems for automated understanding of human behavior. Motion pictures provide even richer and reliable information for the perception of the different biological, social and psychological characteristics of the person such as emotions, actions and personality traits of the subject. In spite of the fact that there is a considerable body of work devoted to human action recognition, most of the methods are evaluated on datasets recorded in simplified settings. More recent research has shifted focus to natural activity recognition in unconstrained scenes with more complex settings.

Chapter 9

    Shefali Gandhi, Dharmsinh Desai University, India
    Tushar V. Ratanpara, Dharmsinh Desai Univerisity, India

Video synopsis provides representation of the long surveillance video, while preserving the essential activities of the original video. The activity in the original video is covered into a shorter period by simultaneously displaying multiple activities, which originally occurred at different time segments. As activities are to be displayed in different time segments than original video, the process begins with extracting moving objects. Temporal median algorithm is used to model background and foreground objects are detected using background subtraction method. Each moving object is represented as a space-time activity tube in the video. The concept of genetic algorithm is used for optimized temporal shifting of activity tubes. The temporal arrangement of tubes which results in minimum collision and maintains chronological order of events is considered as the best solution. The time-lapse background video is generated next, which is used as background for the synopsis video. Finally, the activity tubes are stitched on the time-lapse background video using Poisson image editing.

Chapter 10

Claudio Urrea, Universidad de Santiago de Chile, Chile
Víctor Uren, Universidad de Santiago de Chile, Chile

A technical evaluation of the sensing, communication and software system for the development and implementation of a remote monitoring system for an electric golf cart is presented. According to the vehicle's characteristics and the user's needs, the technical and economic aspects are combined in the best possible way, thereby implementing its monitoring at a distance. The monitoring system is used in two important stages: teleoperation and the vehicle complete autonomy. This allows the acquisition of video images on the vehicle, which are sent wirelessly to the monitoring station, where they are presented through a user-friendly interface. With the purpose of complementing the information sent to the remote user of the vehicle, several important teleoperation variables of a land vehicle, such as voltage level, current and speed are sensed.

Chapter 11

Paromita Roy, Bengal College of Engineering and Technology, India
Nivedita Patra, Bengal College of Engineering and Technology, India
Amartya Mukherjee, IEM Kolkata, India
Amira S. Ashour, Tanta University, Egypt
Nilanjan Dey, Techno India College of Technology, India
Satya Priya Biswas, Bengal College of Engineering and Technology, India

Traffic congestion in cities is a major problem mainly in developing countries. In order to counter this, many models of traffic systems have been proposed by different scholars. Different ways have been proposed to make the traffic system smarter, reliable and robust. A model is proposed to develop an Intelligent Traffic Monitoring System (ITMS) which uses infrared proximity sensors and a centrally placed microcontroller and uses vehicular length along a lane to implement auto controlling of the traffic. The model also provides mean to control the traffic manually through a PC software and an Android application.

Chapter 12

Neethidevan Veerapathiran, Mepco Schlenk Engineering College, India
Anand S., Mepco Schlenk Engineering College, India

Computer vision techniques are mainly used now a days to detect the fire. There are also many challenges in trying whether the region detected as fire is actually a fire this is perhaps mainly because the color of fire can range from red yellow to almost white. So fire region cannot be detected only by a single feature and many other features (i.e.) color have to be taken into consideration. Early warning and instantaneous responses are the preventing ideas to avoid losses affecting environment as well as human causalities. Conventional fire detection systems use physical sensors to detect fire. Chemical properties of particles in the air are acquired by sensors and are used by conventional fire detection systems to raise an alarm. However, this can also cause false alarms. In order to reduce false alarms of conventional fire detection systems, system make use of vision based fire detection system. This chapter discuss about the fundamentals of videos, various issues in processing video signals, various algorithms for video processing using vision techniques.

Preface

Advancement in technology leads to wide spread use of mounting cameras to capture video imagery. Such surveillance cameras are predominant in commercial institutions through recording the cameras' output to tapes that stored in video archives. These recorded videos can be used for further monitoring and analysis to alert security officers in banks and institutes to avoid crime. Thus, automated video surveillance becomes a significant research area in several applications such as people/vehicles tracking and safety issues. Detected and tracked moving objects can be classified into several categories including human, vehicles and trucks using color and shape analysis as well as classifying the human activity into walking and running for example. This classification is applied to improve tracking using progressive reliable constraints. In order to achieve effective tracking and monitoring, advanced techniques for video analysis and processing are required.

In a digital video, the picture's information is digitized both spatially and temporally and the resultant pixel intensities are quantized. Generally, digital video and image processing are applied for improving the quality of the captured videos/images. Advancement in video technology leads to enormous amounts of data that accessible across the world in the form of images and video sequences through television and internet. Thus, there is a great potential for video-based applications in many areas. Moreover, the progression in multimedia acquirement and storage technology leads to a marvelous growth in multimedia databases.

The most essential task in video processing is to divide the long video sequences into a number of shots. Afterward, discover a key frame of each shot for supplementary video information retrieval tasks. Segmentation of the video track into smaller items facilitates the succeeding processing procedures on video shots, for instance semantic representation/ tracking of the selected video information, video indexing and recognizing the frames where a transition occurred from one shot to another. The low level features can be extracted from the segmented video. The video data must be manipulated appropriately for efficient information retrieval. The main challenging task is the retrieval of information from the video data. The majority task is to transform the unstructured data into structured one for video data processing. Prior to processing the video frames noise elimination and illumination changes should be removed.

Typically, video processing has diverse applications, such as in astronomy, medicine, image compression to reduce the memory requirement, sports to capture the motion of an athlete, rehabilitation to assess the locomotion abilities, motion pictures, surveillance to detect and track individuals and vehicles), production industries, robot control, TV productions, educational programs biometrics, and photo editing. Numerous of these applications rely on the same video and image processing methods. Accordingly, these basic methods will the focus of this book.

OBJECTIVE OF THE BOOK

This book deliberates the foremost techniques of video/image processing including noise elimination, segmentation, classification and encryption. It includes video steganography along with miscellaneous surveillance and monitoring systems for real-time applications. This book endeavors to endow with significant frameworks and the most recent empirical research findings in the area of video processing. It includes video processing fundamentals as well as advance topics to help readers building the initial concept and to carry out the research work on this particular field in appropriate manner. As well as it introduces variety of video processing applications in a wide range. It is written for professionals and researchers working in the field of video and imaging in various disciplines, e.g. Software/Hardware video security monitoring, medical devices engineering, researchers, academicians, advanced-level students, and technology developers.

ORGANIZATION OF THE BOOK

The book consists of and introductory chapter followed by 12 chapters that are organized into three sections as shown below. The first section of the book encloses five chapters that introduced the segmentation, classification and registration based image/video processing. The second section contains two chapters concerning video steganography. From Chapter 8 till Chapter 12, the third section introduces several surveillance and monitoring systems applications.

The introduction reported the main concept of the video processing concept and its relation to the image processing. It elaborated the video processing various applications that can be improved in new research aspects in the future.

Section 1: Segmentation-, Classification-, and Registration-Based Image/Video Processing (Chapters 1-5)

Segmentation, classification, and registration are significant image/video processes. Several metrics are used to evaluate and benchmark their performance. This section elaborated an overview of these techniques with an analysis of the evaluation metrics for Image and Video Segmentation methods. Moreover, a classification process was included to classify different types of crops, which can be applied for further video applications.

Chapter 1

This chapter presented a brief outline of the most common segmentation techniques such as thresholding based segmentation, model based segmentation, edge detection based segmentation, and clustering based segmentation. The Matlab simulated results for different image segmentation techniques were included for better understanding of image segmentation.

Chapter 2

Lack of evaluation metrics and benchmark datasets that covers the large variety of sub-problems appearing in image and video segmentation is a challenging issue. This chapter analyzed the evaluation metrics datasets for image and video segmentation methods. Furthermore, it discussed the traditional image and video segmentation methods and the required datasets to evaluate the segmentation algorithms.

Chapter 3

This chapter proposed a recognition/classification system for fresh (healthy) crops and rotten ones. Image analysis was employed to extract statistical features for each crop using ImageJ software. These features were used for crops' recognition using the Least Mean Square Error (LMSE) algorithm and classification based on Bag of Features (BoF) technique. The experimental results were applied on databases for orange, mango, tomato and potatoes.

Chapter 4

Typically, video analysis including object detection, classification and tracking require the entire video scanning, which in turn requires objects detection and classification. This chapter elaborated moving object classification based on the features extracted from the objects in the video sequence. Techniques such as edge detection using various filters, edge detection operators, CBIR (Content Based Image Retrieval) and Bag-of visual words (BoV) were used to classify videos into broad classes to assist searching and indexing using semantic keywords.

Chapter 5

Image registration refers to transforming one image with reference to other images from different viewpoints at variant time. This current chapter highlights the cogitation effect of four different registration techniques, namely the affine transformation based registration, rigid transformation based registration, B-splines registration, and Demons registration. It provided a comparative study among all of these registration techniques as well as different frameworks involved in registration process.

Section 2: Video Steganography (Chapters 6-7)

Recently, due to technology errorless with secured data transmission become a must during large multimedia transfer. Video Steganography known as the process of hiding secret information inside a video sequence without changing the pixel color is involved. In this section, steganography was introduced along with its system requirements, categories, and classifications with focus on image and video files. Furthermore, steganography system evaluation, attacks, and applications were explained.

Chapter 6

This chapter elaborated an overview of the steganography model including the components and the embedding and extraction processes with the evaluation process. Furthermore, the steganography system requirements were discussed as well as a comparison has been conducted among three categories of steganography.

Chapter 7

This chapter addressed the computation delay reduction and increase in the throughput for the LSB/multi-bit based image steganography. This was achieved by implementing the embedding and extraction schemes of LSB/multi-bit steganography in the reconfigurable device such as FPGA. Novel architectures were developed for real-time applications. The algorithm presented in this chapter was validated before the hardware implementation. The hardware simulation and synthesis established that the proposed technique achieved high speed compared to its software counterpart. Further, this scheme can be extended to complex steganography techniques and higher performance can be obtained.

Section 3: Surveillance and Monitoring Systems (Chapters 8-12)

Surveillance and monitoring systems are essential with the extensive dissemination of the human activates. In real time applications, each moving object is denoted as a space-time activity tube in the video. Automated recognition of human activities is a crucial topic that can be applied in several real time applications, including human computer interaction, and smart automated surveillance. Moreover, monitoring allows the detection of any interference that may occur during an action. Therefore, this section outlined and provided several applications and examples for surveillance and monitoring systems.

Chapter 8

Researchers are interested with the perception and recognition of human activities that involved in the automated visual surveillance. This chapter discussed different studies that devoted to human action recognition. Moreover, it addressed several public datasets that used to validate the automated activity recognition techniques.

Chapter 9

Video synopsis afforded long surveillance video representation, while preserving the essential activities of the original video. Since, activities should be displayed in different time segments than original video. Thus, the process begins with extracting moving objects. In this chapter, temporal median algorithm was used to model the background and foreground objects using background subtraction method for detection. The genetic algorithm was used to optimize the temporal shifting of the activity tubes. Finally, the activity tubes were stitched on the time-lapse background video using Poisson image editing.

Chapter 10

In this chapter, the development and implementation of a remote monitoring system for an electric golf cart was presented. The proposed monitoring system consisted of two phases: teleoperation and vehicle complete autonomy. The acquired video images on the vehicle were sent wirelessly to the monitoring station. Afterward, several significant teleoperation variables of a land vehicle, such as voltage level, current and speed were sensed.

Chapter 11

Traffic congestion in cities is a major problem mainly in the developing countries that led to develop several models for the traffic system. This chapter proposed a developed Intelligent Traffic Monitoring System using infrared proximity sensors and a centrally placed microcontroller. Vehicular length along a lane is used to implement auto controlling of the traffic. The proposed model provided a mean to control the traffic manually through PC software and an android application.

Chapter 12

Fire detection and warning are challenging problems as the fire color has a wide range from red yellow to almost white. Early warning and instantaneous responses are the preventing ideas to avoid losses affecting environment as well as human causalities. In order to reduce false alarms of conventional fire detection systems, vision systems were developed. This chapter was briefly described the fundamentals of videos, various issues in processing video signals, various algorithms for video processing using vision techniques.

Nilanjan Dey
Techno India College of Technology, India

Amira S. Ashour
Tanta University, Egypt & Taif University, Saudi Arabia

Suvojit Acharjee
National Institute of Technology Agartala, India

Acknowledgment

The best minute you spend is the one you invest in people. - Kenneth H. Blanchard

We are greatly thankful to our parents for endless support, guidance and love through all our life. We are grateful to our beloved family members for standing beside us throughout our career and through editing this book. Our thanks are also directed to our students, who have put in their time and effort to support and contribute in some manner. We dedicate this book to all of them.

We would like to express our gratitude to the all people support, share, talked things over, read, wrote, offered comments, allowed us to quote their remarks and assisted in editing, proofreading and design; through the book journey.

We believe that the team of authors provides the perfect blend of knowledge and skills that went into authoring this book. We thank each of the authors for devoting their time, patience, perseverance and effort towards this book.

Special thanks to the IGI-publisher team, who showed us the ropes to start and continue.

Last, but definitely not least, we'd like to thank our readers, who gave us their thrust and hope our work inspired and guide them.

The best minute you spend is the one you invest in people. - Kenneth H. Blanchard

We are greatly thankful to our parents for endless support, guidance and love through all our life. We are grateful to our beloved family members for standing beside us throughout our career and through editing this book. Our thanks are also directed to our students, who have put in their time and effort to support and contribute in some manner. We dedicate this book to all of them.

We would like to express our gratitude to the all people support, share, talked things over, read, wrote, offered comments, allowed us to quote their remarks and assisted in editing, proofreading and design; through the book journey.

We believe that the team of authors provides the perfect blend of knowledge and skills that went into authoring this book. We thank each of the authors for devoting their time, patience, perseverance and effort towards this book.

Special thanks to the IGI-publisher team, who showed us the ropes to start and continue.

Last, but definitely not least, we'd like to thank our readers, who gave us their thrust and hope our work inspired and guide them.

Acknowledgment

Nilanjan Dey
Techno India College of Technology, India

Suvojit Acharjee
National Institute of Technology, India

Amira S. Ashour
Tanta University, Egypt & Taif University, Saudi Arabia

Introduction

INTRODUCTION

Video Processing Perception

Nowadays, due to the great advancement in technology, digital images and videos become ubiquitously. Applications in medical, industrial, art, astronomical and military domains depend mainly on image/video processing. Digital image and video processing empower the revolution of any multimedia technology for communications and entertainment. Images/videos allocate a widespread range of the electromagnetic spectrum. Originally, videos were captured and transmitted in analog form that developed to digital video due to the advent of the digital integrated circuits. Videos are considered to be a sequence of successive images that captured on film streaming using a camera (Zabih *et al.*, 1998). For example, movies are played through light flashes to illuminate each frame on the moving film at sufficient rates that provide continues motion. Moreover, other examples for image/video processing include the blur removal from a radar picture of a fast moving car, and the compression and efficient transmission of images and videos.

Generally, videos can be represented in digital form starting from pixels, where a row of pixels from a line at which a collection of lines can represent a video frame. Thus, videos are originated from images that consist mainly of a combination of pixels. Digital video system is considered to be a video recording system that uses a digital video signals, which is a sequence of time varying images. Since, images are a spatial distribution of different intensities that stay constant with time, while videos are time varying images. Thus, video signal handles frames which are as a series of images. Moreover, a continuous real time video is attained by changing the frames in a faster way known as frame rate. In order to acquire such images which are the milestone of any video, a variety of physical devices based on the application under concern are employed. Such devices include microscopes, endoscopy, x-ray, ultrasound, video cameras, and radar. These devices are employed to capture images/videos that analyzed and processed to extract useful information about the scene being imaged. Since, digital images/ videos pictures and movies; respectively that converted into a binary format. Hence, digital video bitstream manipulation referred to as digital video processing that used for data compression, enhancement, restoration, motion analysis and other processes to obtain superior images quality images and to extract significant information (Winkler, 2005).

The main concern of the current work is to introduce the concept of video processing. In addition, it included several video processing techniques with an emphasis on the some applications that drive digital video technology.

Video Processing

Video refers to visual information included in:

1. Sequence of still image, which is a constant spatial distribution of intensity with respect to time, and
2. Time-varying images that represents a spatio-temporal intensity pattern (Tekalp, 1995).

Video coding converts the video sequences into an efficient bitstream. The digital video bitstream requires digital video processing for several applications, where the still images are use in commercial, military, industrial, medical, satellite system and surveillance imaging. One of the main differences between the still images' processing and the digital video processing is that the later contains a significant amount of temporal correlation between the frames. Processing multi-frames enables the development of effective algorithms, such as motion-compensated prediction and motion compensated filtering. Video processing is known as the content analysis of the video to obtain information of the scene that it describes. Video processing includes object recognition, video compression, video indexing, video segmentation and video tracking.

Video Compression

Rapid progress in the media captures, storage and transmission technologies lead to the significant role of the video compression process. Recently, video data storage becomes important for several applications such as in military, forensic, medical, and the multimedia applications. Thus, video compression has a vast number of applications ranging from video on demand and video conferencing to video phones. Basically, compression refers to the data reduction in images/videos, to reduce the media overheads for distributing their sequences (Le Gall, 1991). Several approaches can be used for data reductions in images, including:

1. Color resolution reduction with respect to the dominant light intensity,
2. Color nuances reduction,
3. Small and invisible parts removal, and
4. Adjacent images comparison with the removal of unchanged details.

The same approaches can be applied on the videos' frames, where a video sequence consists of a series of frames. Efficient video compression can be realized by minimizing both temporal and spatial redundancy. Consequently, temporal redundancy among adjacent frames should be exploited to achieve efficient storage and transmission. Temporal redundancy implies similar adjacent frames, while spatial redundancy implies similar neighboring pixels.

During video coding a removal of redundant information from video sequence is achieved. The exclusion of spatial redundancy is known as intraframe coding, while the exclusion of temporal redundancy is called interframe coding. When both interframe and itraframe coding are used, this is referred to hybrid coding. Video compression algorithms can be categorized into:

1. Lossless video compression, and
2. Lossy video compression (Richardson, 2004).

Generally, lossless and lossy compressions refer to whether or not all original data can be recovered during the uncompressing process. In lossless compression, every original data bit remains after the uncompressing process, i.e. all the information is completely restored and recovered. Conversely, lossy compression permanently eliminates certain information, mainly the redundant information, thus only a part of the original information is restored after the compression process.

Video Segmentation

Video segmentation reduces the video data through extracting significant basic parts that have robust correlation with the real information in the video data. It can be described as the process that separates the foreground/background in a video. It can be spatial, temporal, or spatio-temporal (Tarabalka *et al.*, 2014). In the spatial domain, segmenting a video frame is similar to in a static image (Inigo, & Suresh, 2012), while temporal segmentation refers to the segmenting a video frames sequence in temporal domain (Petersohn, 2009). In video processing, spatial segmentation determines the object's boundaries if original objects have a different visual appearance from the background. Whereas, temporal segmentation is used to determine the shots boundary that identify the moving objects from the background. Consequently, several segmentation algorithms use temporal segmentation to identify the moving objects and spatial segmented the object's boundaries, then integrate both results to enhance the accuracy.

Typically, segmentation methods are often preceded by a pre-processing step to discard the unnecessary information using low-pass filtering for example, and a feature extraction step. Region-based image/video description, interactive region-based annotation schemes, indexing/retrieval, video summarization, objects detection, and region-based coding are some of the video segmentation applications. The techniques involved in the video segmentation process can be categorized based on the attributes (features) presented with the input video data, the employed model and the involved strategies as illustrated in Figure 1.

Object Tracking in Video Sequences

Object/motion tracking in video sequences is a significant issue in the computer vision and various research domains (Bovik, 2009). It derives the trajectory of the moving object over time in video sequences. Object tracking has several domains including surveillance, education, security, clinical applications, entertainment, human robot interaction, and biomechanical applications. Object extraction, object recognition and tracking are the main steps for objects tracking. There are several challenges for this process,

Figure 1. Video segmentation techniques

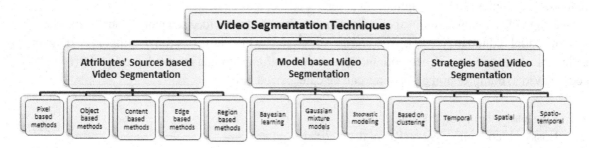

Figure 2. Object tracking techniques

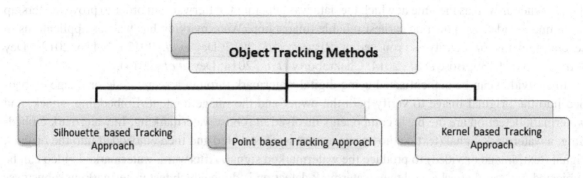

such as the luminance and intensity changes, presence of noise and blur in the video, partial/full object occlusion, the object's abrupt motion, and the variation that may occur in the shape/size of the object from frame to frame. The procedure used for video objects tracking is based on algorithms to analyze the sequential video frames and produces the target movement between the frames as an output. Such techniques for object tracking in videos can be categorized as illustrated in Figure 2.

After determining the region of interest in each frame using object detection algorithm, a tracking process corresponding to the objects across frames is performed using one of the methods (Yilmaz *et al.*, 2006) in Figure 2. Thus, based on object extraction, object recognition and tracking the object under concern within the video sequence is tracked. Kernel based/contour based tracking necessitates the detection step when the object first appears in the scene. It uses various estimating approaches to determine the corresponding region to the target object. However, the instance point tracking method includes the detection in every frame. Based on the representation type, the silhouette tracking can form the motion models.

Video Security

Secure image/video processing is an emerging technology to qualify the processing tasks in a secure way. It has attracted researchers to enrich functionalities for private data storage and transmission. Secure communication and digital data transfer requirements are potentially increased with the progress of multimedia systems. Secure video processing requires effective solutions for the trade-off between the security and complexity. Authenticated video is required in various situations such as in

1. Video-on-demand (VOD) systems to prevent unauthorized viewing with illegal copying, and for
2. Digital video rentals, which requires a reliable third-party time source to offer time stamping.

A new possibility is to rent digital video for a fixed number of viewings over a longer time period. Thus, cryptography for digital video is necessary to achieve video security (Macq, & Quisquater, 1995).

For image/video protection from unauthorized viewing, distribution, copying, and modification, several techniques can be used (Puech *et al.*, 2012). These include authentication, encryption, and time stamps (Menezes *et al.*, 1996). Encryption covers the digital media content, therefore only users who possess the decryption "key" are able to convert the encrypted data into its original form. Without the key, it is computationally impossible to extract and recover the original data. However, authentication proves who created the document without hiding the data content. Whereas, time stamps identifies the data's owner as well as the time at which the data was generated. Afterward, in order to prove ownership of an image/video is to own its earliest reliable time stamp. Watermarking has various applications in several domains for security as proposed in (Biswas *et al.*, 2012; Dey *et al.*, 2012a; 2012b; 2012c; Dey *et al.*, 2013a; 2013b; Bose *et al.*, 2014; Chakraborty *et al.*, 2014; Dey *et al.*, 2014).

Image/video can be authenticated using digital watermark through a secret code or image embedded into the original image to verify both the owner and the image content. Embedding, attack, and watermark detection are the basic components involved in robust watermarking. In watermark embedding, a watermark signal (text, audio or image etc.) is constructed and then embedded into the original signal (text, image or video) to produce the watermarked signal. Afterward, watermarked video can be subjected to several attacks, and then watermark detection is done which test watermarking robustness against attacks. A verification algorithm authenticates the image to determine both the owner as well as the integrity of the image/video. Video watermarking process embeds data in the video for copyright, annotation and identification purposes. Generally, image watermarking techniques can be extended to videos watermarking, while the later faces other challenges than that in image watermarking techniques. Thus, the following extensions should be considered during video watermarking:

1. The huge amount of redundant data that exist between the frames,
2. There must be a strong balance between the motionless and the motion regions, and
3. Real time and streaming video applications. Basically, the watermarking technologies are classified in various domains like the spatial domain or temporal domain (Neeta *et al.*, 2006) and the frequency domain (Hong *et al.*, 2001).

Various techniques can be employed to maintain original video fidelity with robust watermark. Some aspects for the video watermarking systems design are:

1. Robustness of the watermark to common signal processing manipulations,
2. Imperceptibility to measure that the watermark embedding did not cause large degradation to the host video,
3. Security to ensure that the embedded information is secured against tampering, and
4. Capacity, where the embedded information amount must be sufficient to uniquely identify the video's owner.

Video Processing Applications

Video processing is involved in various applications including:

Video Surveillance

Video surveillance is used to monitor the activities, behavior, or any information changing using usually video cameras to observe the area of interest. Video surveillance systems can be included to monitor public safety/security, airports/seaports/railways, retail stores, financial institutions, and hospitals. The system framework starts with the video images acquisition using video cameras, and then a pre-processing to enhance the frames quality in the sequences. Afterward, motion segmentation is performed to separate the foreground images from background images followed by objects classification, and tracking. Finally, the activity analysis is processed from the obtained information.

Wijnhoven, & De Peter (2006) proposed model-based entity detection for traffic surveillance. The object orientation was sensed using a gradient direction histogram within identified regions-of-interest (ROIs) of the moving objects in the scene. Leibe *et al.* (2008) presented a novel method using Minimum Description Length hypothesis for multi-object tracking. This allowed the system to overcome mismatches and provisionally lost tracks. The proposed system executed multi-view object recognition in order to detect vehicles in the input images. Nan Lu *et al.* (2008) proposed a new real time motion detection procedure that integrated the temporal differencing technique, optical flow technique, double background filtering (DBF) technique, and morphological processing techniques to obtain fine performance. The proposed system achieved high ability of anti-interference and maintains high accurate rate detection. Furthermore, it reduced computation time than other real-time surveillance methods.

Mobile Communications

Telecommunications industry progression depends mainly on the mobile telecommunications. The infrastructures of the mobile network are deployed with the bandwidth and capabilities to lastly deliver captivating video services that can support the various mobile devices, such as video conferencing, video mail, and interactive voice and video response services (Maharajh *et al.*, 2007). Video analysis development leads to more efficient use of the bandwidth and power in the handheld devices with users sharing videos and with cloud computing servers. Advanced video codecs enable improved video compression by providing the high quality of the video at significantly lower bit rates. In addition, standardized codecs allow greater flexibility using their ability to work in a wide range of network environments.

Vision-Based Robotics

Robots equipped with video cameras are working alongside humans. They can help in routing through complex environments and areas. This robotics system requires advancement in different video processing aspects including recognition, tracking, and distributed processing. Naik *et al.* (2016) discussed the design constraints of the mobile robot and processing the real time video received from the mobile robot based on computer vision. The discrete wavelet transform (DWT) algorithm was applied on the real time video to reduce the required memory size for storage. The mobile engaged robot that fed in the video live steaming of the surrounding area was of low quality. Memory controller was employed as an interface between the memory and the processing element. Alias *et al.* (2015) proposed relocation method for a mobile robot using video processing technique. This method was capable of detecting the mapped line to follow by the mobile robot. The proposed system consisted of line detection and path recovery techniques in order to relocate the mobile robot automatically. The captured images from the webcam were used to detect the line trajectory. Using line detection technique, the error pathway followed by mobile robot generated the path's error data. Based on path's error data, speed and time period supplied to the mobile robot's wheels were measured. High definition webcam and Arduino board were used to capture the video and to relocate the mobile robot; respectively. This video was framed and formed into to-dimensional sequential images.

CONCLUSION

Digital images and videos universally exist in several domains including bio-medical, astronomical and industrial applications. Therefore, the capability to process such resources of image/video signals is an incredible process and task. Typically, both image and video processes are similar, however the major difference between the image signal and video signal is that a video signal consists of temporal information. This temporal information introduces the concept of object and camera motion. Thus, videos have both spatial (static) nature and temporal nature.

Image/video processing have several purposes such as image enhancement to reduce the noise, pattern recognition for automatic detection of objects, data reduction to easily handled and interpreted information, and data compression to reduce the images size and speed up image transmission across a network.

Video processing has several steps including compression, segmentation, classification and objects/motion tracking. It involved in several applications as the preceding addressed ones. Furthermore, video processing can assist medical practitioners in their diagnoses. In today's environment, educational institution, virtually every municipality, agency, mass transportation centers, financial institution, and medical centers/hospitals must plan to protect the images/video information. In order to protect video transmission, watermarking and security techniques are employed.

Nilanjan Dey
Techno India College of Technology, India

Amira S. Ashour
Tanta University, Egypt & Taif University, Saudi Arabia

REFERENCES

Alias, M. F., Mansor, M. A., Ahmad, M. R., & Othman, N. (2015). Straight-Line Mobile Robot Using Video Processing Technique. *Journal of Global Research in Computer Science*, *6*(11).

Biswas, S., Roy, A. B., Ghosh, K., & Dey, N. (2012). A Biometric Authentication Based Secured ATM Banking System. *International Journal of Advanced Research in Computer Science and Software Engineering*.

Bose, S., Chowdhury, S. R., Sen, C., Chakraborty, S., Redha, T., & Dey, N. (2014, November). Multi-thread video watermarking: A biomedical application.*2014 International Conference on Circuits, Communication, Control and Computing (I4C)*, (pp. 242-246). IEEE. doi:10.1109/CIMCA.2014.7057798

Bovik, A. C. (2009). *The essential guide to video processing*. Academic Press.

Chakraborty, S., Maji, P., Pal, A. K., Biswas, D., & Dey, N. (2014, January). Reversible color image watermarking using trigonometric functions.*2014 International Conference on Electronic Systems, Signal Processing and Computing Technologies (ICESC)*, (pp. 105-110). IEEE. doi:10.1109/ICESC.2014.23

/

Dey, M., Dey, N., Mahata, S. K., Chakraborty, S., Acharjee, S., & Das, A. (2014, January). Electrocardiogram Feature based Inter-human Biometric Authentication System.*2014 International Conference on Electronic Systems, Signal Processing and Computing Technologies (ICESC)*, (pp. 300-304). IEEE. doi:10.1109/ICESC.2014.57

Dey, N., Das, P., Chaudhuri, S. S., & Das, A. (2012c, October). Feature analysis for the blind-watermarked electroencephalogram signal in wireless telemonitoring using Alattar's method.*Proceedings of the Fifth International Conference on Security of Information and Networks* (pp. 87-94). ACM. doi:10.1145/2388576.2388588

Dey, N., Das, P., Roy, A. B., Das, A., & Chaudhuri, S. S. (2012a, October). DWT-DCT-SVD based intravascular ultrasound video watermarking.*2012 World Congress on Information and Communication Technologies (WICT)*, (pp. 224-229). IEEE. doi:10.1109/WICT.2012.6409079

Dey, N., Maji, P., Das, P., Biswas, S., Das, A., & Chaudhuri, S. S. (2013b, January). An edge based blind watermarking technique of medical images without devalorizing diagnostic parameters.*2013 International Conference on Advances in Technology and Engineering (ICATE)*, (pp. 1-5). IEEE. doi:10.1109/ICAdTE.2013.6524732

Dey, N., Mukhopadhyay, S., Das, A., & Chaudhuri, S. S. (2012b). Analysis of P-QRS-T components modified by blind watermarking technique within the electrocardiogram signal for authentication in wireless telecardiology using DWT. *International Journal of Image, Graphics and Signal Processing*, 4(7), 33–46. doi:10.5815/ijigsp.2012.07.04

Dey, N., Samanta, S., Yang, X. S., Das, A., & Chaudhuri, S. S. (2013a). Optimisation of scaling factors in electrocardiogram signal watermarking using cuckoo search. *International Journal of Bio-inspired Computation*, 5(5), 315–326. doi:10.1504/IJBIC.2013.057193

Hong, I., Kim, I., & Han, S. S. (2001). A blind watermarking technique using wavelet transform. *IEEE International Symposium on Industrial Electronics* (Vol. 3, pp. 1946-1950). IEEE. doi:10.1109/ISIE.2001.932010

Inigo, S. A., & Suresh, P. (2012). General study on moving object segmentation methods for video. *Int. J. Adv. Res. Comput. Eng. Technol*, 1, 265–270.

Le Gall, D. (1991). MPEG: A video compression standard for multimedia applications. *Communications of the ACM*, 34(4), 46–58. doi:10.1145/103085.103090

Leibe, B., Schindler, K., Cornelis, N., & Van Gool, L. (2008). Coupled object detection and tracking from static cameras and moving vehicles. *Pattern Analysis and Machine Intelligence. IEEE Transactions on*, 30(10), 1683–1698.

Lu, N., Wang, J., Wu, Q. H., & Yang, L. (2008). An improved motion detection method for real-time surveillance. *IAENG International Journal of Computer Science*, 35(1), 1–10.

Macq, B. M., & Quisquater, J. J. (1995). Cryptology for digital TV broadcasting. *Proceedings of the IEEE*, 83(6), 944–957. doi:10.1109/5.387094

Maharajh, K., MacNeil, B., & Walker, T. (2007). *U.S. Patent Application No. 11/956,184*. Washington, DC: US Patent Office.

Menezes, A. J., Van Oorschot, P. C., & Vanstone, S. A. (1996). *Handbook of applied cryptography*. CRC Press. doi:10.1201/9781439821916

Naik, P. K., Tigadi, A., Guhilot, H., & Vyasaraj, T. (2016). Design and Analysis of Real Time Video Processing Based on DWT Architecture for Mobile Robots. *Procedia Computer Science, 78*, 544–549. doi:10.1016/j.procs.2016.02.100

Neeta, D., Snehal, K., & Jacobs, D. (2006, December). Implementation of LSB steganography and its evaluation for various bits. *2006 1st International Conference on Digital Information Management,* (pp. 173-178). IEEE.

Petersohn, C. (2009, November). Temporal video structuring for preservation and annotation of video content. *2009 16th IEEE International Conference on Image Processing (ICIP),* (pp. 93-96). IEEE.

Puech, W., Erkin, Z., Barni, M., Rane, S., & Lagendijk, R. L. (2012, September). Emerging cryptographic challenges in image and video processing. *2012 19th IEEE International Conference on Image Processing (ICIP),* (pp. 2629-2632). IEEE.

Richardson, I. E. (2004). *H. 264 and MPEG-4 video compression: video coding for next-generation multimedia*. John Wiley & Sons.

Tarabalka, Y., Charpiat, G., Brucker, L., & Menze, B. H. (2014). Spatio-temporal video segmentation with shape growth or shrinkage constraint. *IEEE Transactions on Image Processing, 23*(9), 3829–3840. doi:10.1109/TIP.2014.2336544 PMID:25020092

Tekalp, A. M. (1995). *Digital video processing*. Prentice-Hall, Inc.

Wijnhoven, R., & De Peter, H. N. (2006). 3D wire-frame object modeling experiments for video surveillance. In *Proc. 27th Int. Symp. Inform. Theory in Benelux*.

Winkler, S. (2005). *Digital video quality: vision models and metrics*. John Wiley & Sons. doi:10.1002/9780470024065

Yilmaz, A., Javed, O., & Shah, M. (2006). Object tracking: A survey. *ACM Computing Surveys, 38*(4), 13.

Zabih, R., Miller, J. F., & Mai, K. W. (1998). *U.S. Patent No. 5,767,922*. Washington, DC: U.S. Patent and Trademark Office.

KEY TERMS AND DEFINITIONS

Digital Watermarking: Is a marker secretly embedded in a noise-tolerant signal, such as an audio, image, or video for more security and authorization.

Feature Extraction: Is the transforming the existing features into a lower dimensional space.

Features: Are inherent properties of data, independent of coordinate frames.

Image Analysis: Is the extraction of meaningful information from images; mainly from digital images by means of digital image processing techniques.

Image Processing: The analysis and manipulation of a digitized image, especially in order to improve its quality.

Image Segmentation: Is the process of partitioning a digital image into multiple segments as a set of pixels or super-pixels. It assists to locate objects and boundaries in images.

Video Processing: Is the manipulation of video content to resize, clarify or compress it for further information extraction.

Video Segmentation: Is the process that separates the foreground/background in a video.

Section 1
Introduction to Video Processing and Mining

Video processing has various applications in several domains. This section highlights the fundamental concepts and algorithms of video processing in the medical domain. The real time frames for the acquired images from different image modalities is used. Another context is introduced for data mining that has great benefit in data extraction various applications including educational and medical domains.

Chapter 1
Study of Various Image Segmentation Methodologies:
An Overview

Abahan Sarkar
National Institute of Technology Silchar, India

Ram Kumar
National Institute of Technology Silchar, India

ABSTRACT

In day-to-day life, new technologies are emerging in the field of Image processing, especially in the domain of segmentation. Image segmentation is the most important part in digital image processing. Segmentation is nothing but a portion of any image and object. In image segmentation, the digital image is divided into multiple set of pixels. Image segmentation is generally required to cut out region of interest (ROI) from an image. Currently there are many different algorithms available for image segmentation. This chapter presents a brief outline of some of the most common segmentation techniques (e.g. Segmentation based on thresholding, Model based segmentation, Segmentation based on edge detection, Segmentation based on clustering, etc.,) mentioning its advantages as well as the drawbacks. The Matlab simulated results of different available image segmentation techniques are also given for better understanding of image segmentation. Simply, different image segmentation algorithms with their prospects are reviewed in this chapter to reduce the time of literature survey of the future researchers.

OBJECTIVE

Segmentation is the most important part in image processing. Fence off an entire image into several parts which is something more meaningful and easier to further process. These several parts that are rejoined will cover the entire image. Segmentation may also depend on the various features that are contained in the image. It may be either color or texture. Before denoising an image, it is segmented to recover the original image. The main motto of segmentation is to reduce the information for easy analysis. Segmentation is also useful in Image Analysis and Image Compression.

DOI: 10.4018/978-1-5225-1022-2.ch001

INTRODUCTION

Images are primarily used in the field of computer vision for tasks such as navigation of robots, disease identifications from MR images, identification of number plates of moving vehicles, etc. (Somasundaram & Alli, 2012). The primary objective of computer vision and digital image processing is to automate different tasks and image segmentation is an important step in it.

Image Segmentation is a process of dividing an image into different parts. This helps to simplify or change the overall presentation of an image into such data which is more meaningful and easier for a system to analyze. In segmentation, value is assigned to every pixel of an image in such a way that the pixels which share certain characteristics, such as color, intensity or texture in a particular region are grouped together. Adjacent regions which are not grouped together must be significantly different with respect to the same characteristics. Purpose of dividing an image is to further analyze each of these subparts or sub-images so that some high level information can be extracted. Sometimes image denoising is done before the segmentation to enhance the image and improve the quality of the segmentation process (Dass & Devi, 2012). A novel method is proposed to find the contours of Hippocampus brain cell using microscopic image analysis (Hore, Chakroborty, Ashour et al., 2015). An algorithm is proposed for segmentation of blood vessels (Dey, Roy, Pal et al., 2012). Rough-set based image segmentation methods are reviewed in (Roy, Goswami, Chakraborty et al., 2014) and classification is done. A new algorithm is proposed to detect the shot boundary by using the minimum ratio similarity measurement between the characteristic features of two consecutive frames (Pal, Acharjee, Rudrapaul et al., 2015).

The chapter is organized as follows:

- Introduction is presented in Section 1.
- Section 2 discusses the background to the proposed study.
- Section 3 deals with available different segmentation methods.
- Matlab implemented results are discussed in Section 4 for better understanding.
- Finally, Section 5 concludes the chapter.

BACKGROUNDS

The details literature survey associated with different available image segmentation techniques are briefly discuss.

In region based technique pixels that are related to an object are grouped for segmentation (Kaganami & Beiji, 2009). The thresholding technique is bound with region based segmentation. The area that is detected for segmentation should be closed. Region based segmentation is also termed as "Similarity Based Segmentation" (Canny, 1986). There won't be any gap due to missing edge pixels in this region based segmentation (Chen & Chen, 2009). The boundaries are identified for segmentation. In each and every step, at least one pixel is related to the region and is taken into consideration (Shafait, Keysers & Breuel, 2008). After identifying the change in the color and texture, the edge flow is converted into a vector. From this the edges are detected for further segmentation (Ma & Manjunath, 1997).

Segmentation can also be done by using edge detection techniques. In this technique the boundary is identified to segment. Edges are detected to identify the discontinuities in the image. The edges of the region are traced by identifying the pixel value and it is compared with the neighboring pixels. For this classification they use both fixed and adaptive features of Support Vector Machine (SVM) (Gómez-Moreno, Maldonado-Bascón, & López-Ferreras, 2001). The technique that is used for segmenting the remote sensing image has high spatial resolution. The two step procedure for segmentation are extracting the edge information from the edge detector and then the pixels are labeled. The advantage of this technique is retrieving information from the weak boundary too (Saha, Ray, Greiner et al., 2012). Based on the edge flow, the image is segmented. It identifies the direction of the change in color and texture of a pixel in an image to segment (Ma & Manjunath, 1997). Segmentation can also be done through edges. There will be some gap between the edges as it is not closed. So, the gap is filled with edge linking. The broken edges are extended in the direction of the slope for the link to get the connectivity for segmentation (Shih & Cheng, 2004).

Thresholding is the easiest way of segmentation. It is done through that threshold value which are obtained from the histogram of those edges of the original image (Karthikeyan, Vaithiyanathan, Venkatraman et al., 2012). The threshold values are obtained from the edge detected image. So, if the edge detections are accurate, then the threshold too. Segmentation through thresholding has fewer computations compared to other techniques. For a particular segment there may be set of pixels which is termed as "his ton". Roughness measure is followed by a thresholding method for image segmentation (Mushrif & Ray, 2008). Segmentation is done through adaptive thresholding. The gray level points where the gradient is high, is then added to the thresholding surface for segmentation (Yanowitz & Bruckstein, 1988).

Segmentation is also done through Clustering. They followed a different procedure, where most of them apply the technique directly to the image, but here the image is converted into histogram and then clustering is done on it (Comaniciu & Meer, 1997). Pixels of the color image are clustered for segmentation using an unsupervised technique Fuzzy C. This is applied to ordinary images. If it is a noisy image, it results to fragmentation (Lim & Lee, 1990).

A basic clustering algorithm, i.e., K-means is used for segmentation in textured images. It clusters the related pixels to segment the image (Jain & Dubes, 1988). Segmentation is done through a feature clustering and there it will be changed according to the color components (Khan, Bhuiyan, & Adhami, 2011). Segmentation is also purely depending on the characteristics of the image (Wang & Wang, 2010). Features are taken into account for segmentation. The difference in the intensity and color values are used for segmentation (Felzenszwalb & Huttenlocher, 2004).

Markov Random Field (MRF) based segmentation is known as Model based segmentation (Lehmann, 2011). Components of the color pixel tuples are considered as independent random variables for further processing. MRF is combined with edge detection for identifying the edges accurately (Luo, Gray & Lee, 1998).

The segmentation may also be done by using Gaussian Markov Random Field (GMRF) where the spatial dependencies between pixels are considered for the process (Kang, Yang, & Liang, 2009). Gaussian Markov Model (GMM) based segmentation was used for region growing. The extension of the Gaussian Markov Model (GMM) that detects the region as well as edge cues within the within the GMM framework.

IMAGE SEGMENTATION AND ITS CLASSIFICATION

Segmentation is the most important part in image processing. Fence off an entire image into several parts which is something more meaningful and easier to further process. These several parts that are rejoined will cover the entire image. Segmentation may also depend on the various features that are contained in the image. It may be either color or texture. Before denoising an image, it is segmented to recover the original image. The main motto of segmentation is to reduce the information for easy analysis. Segmentation is also useful in Image Analysis and Image Compression.

3.1 Segmentation Techniques Categories

In recent years, a lot of research is done in the field of the image segmentation process. There are currently thousands of algorithms, each doing the segmentation process slightly different from another, but still there is no particular algorithm that is applicable for all types of digital image, fulfilling every objective. Thus, the algorithm developed for a group of images may not always apply to images of another class (Dass & Devi, 2012, pp. 7-8).

Segmentation can be classified as follows:

- Region Based,
- Edge Based,
- Threshold,
- Feature Based Clustering,
- Model Based.

The classification is specified in Figure 1.

Figure 1. Various types of image segmentation

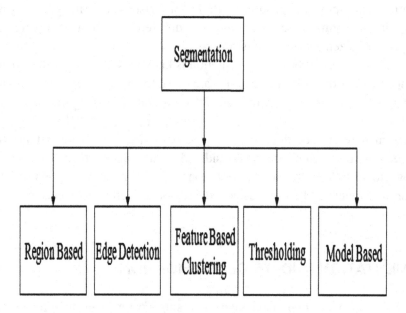

3.1.1 Region Based Segmentation

Segmentation methods based on regions are relatively simple and are more immune to noise. Contrary to edge based Segmentation techniques who segmenting image based on the abrupt changes in the intensities of neighboring pixel, region based segmentation algorithms segment an image into regions that are similar according to a set of predefined criteria (Wang & Wang, 2010). Region based segmentation include: Region Growing, Region Splitting and Merging.

3.1.1.1 Region Growing

This method group pixel in an entire image into sub regions or large regions based on predefined criterion. In other words, the basic idea is to group a collection of pixels with similar properties to form a region (Langote & Chaudhari, 2012). Region growing can be processed into four steps:

1. Select a group of seed particles in the original image.
2. Select a set of criteria for determining similar seeds based on properties such as gray level intensity or color and then set up a stopping rule.
3. Grow the region by adding to each seed those neighboring pixels that have predefined properties similar to the seed pixel.
4. Stop the region's growth when there are no more pixels that match the criterion for inclusion in that region.

3.1.1.2 Region Splitting and Merging

Previous mentioned techniques; region grows by selecting a set of seed points. However, in this technique, the image is subdivided into a set of arbitrary unconnected regions and merge/split the region according to the condition of the segmentation. This particular splitting technique is usually implemented with a theory based on quad tree data. Quad tree is a tree in which each node has exactly four branches (Langote & Chaudhari, 2012). This includes the following steps:

1. Start splitting the region into four branches.
2. Merge any region when no further splitting is possible.
3. Stop when no further merging is possible.

3.1.2 Segmentation Based on Edge Detection

Edge detection is a very important step in digital image processing and computer vision. In an image, the edges represent object boundaries and thus help in detection and segmentation of objects in an image (Nadernejad, Sharifzadeh, & Hassanpour, 2008; Khalifa, 2010). Edge detection refers to algorithms which try to identify points in a digital image where there is an abrupt change in image brightness or there is a difference in intensities. These points are then linked together to form closed object boundaries (Khalifa, 2010). The result of segmentation using edge detection is a binary image.

There are many different ways to perform edge detection; however, two most prominently used algorithms are mentioned here:

3.1.2.1 Gray Histogram Techniques

In this technique, segmentation depends upon the separation of foreground from background by selecting a threshold value T. The difficulty arises in selecting the threshold values since gray threshold is uneven due to the presence of noise. Thus, we substitute the curves of the object and the background with two conic Gaussian curves (Kang, Yang, & Liang, 2009), whose intersection is chosen as the value of threshold T.

3.1.2.2 Gradient Based Methods

The gradient is the first derivative for image f(x, y), when there is an abrupt change in the intensity near edge. Another noise, gradient based method (Kang, Yang, & Liang, 2009) involves convolving gradient operators with the image. High value of gradient magnitude can be pointed with abrupt change between intensities of the two regions. These points are called edge pixels and can be linked together to form closed boundaries.

Normally the Sobel operator, canny operator, Laplace operator, Laplacian of Gaussian (LOG) operator, etc. is used as an operator with gradient based method. Usually canny operator is used, but it takes more time as compared to Sobel operator. In practice edge detection algorithms require a balance between detecting edges accurately and reducing the level of noise. If the level of accuracy is too high, noise will create a detection of numerous additional and fake edges. On the other hand, if we try to reduce the level of noise too greatly, we might reduce the accuracy of the edges and many of the useful edges might not be detected. Thus, edge detection algorithms are usually suitable for images that are simple and noise free. The various edge detection techniques are shown in Figure 2.

Figure 2. Various types of edge detection techniques

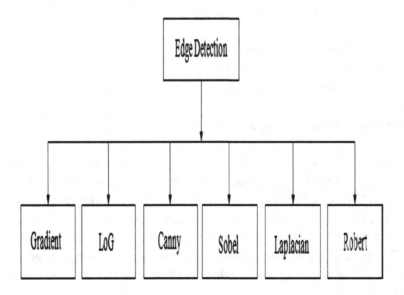

Various edge detection techniques are discussed below:

1. The edges are detected by calculating the minimum and maximum of first derivative in gradient edge detector.
2. Zero Crossing is found in second derivative to identify the edges in Laplacian edge detector.
3. The main objective is to determine the differences between adjacent pixels, one way to find an edge is to explicitly use {+1,-1} that calculates the difference between adjacent pixels. Mathematically, these are called forward differences. The Robert kernels are, in practice, too small to reliably find edges in the presence of noise. The simplest way to implement the first order partial derivative is by using the Roberts cross-gradient operator. The Roberts operator masks are given by

$$G_X = \begin{bmatrix} -1 & 0 \\ 0 & 1 \end{bmatrix}$$

$$G_Y = \begin{bmatrix} 0 & -1 \\ 1 & 0 \end{bmatrix}$$

These filters have the shortest support, thus the position of the edges is more accurate, but the problem with the short support of the filters is its vulnerability to noise.

4. The Prewitt kernels are named after Judy Prewitt. Prewitt kernels are based on the idea of central difference. The Prewitt edge detector is a much better operator than the Roberts operator. Consider the arrangement of pixels about the central pixel [i, j] as shown below:

The partial derivatives of the Prewitt operator are calculated as

$$G_X = \left(a_2 + ca_3 + a_4 \right) - \left(a_0 + ca_7 + a_6 \right)$$

and

$$G_Y = \left(a_6 + ca_5 + a_4 \right) - \left(a_0 + ca_1 + a_2 \right)$$

The constant c in the above expressions implies the emphasis given to pixels closer to the center of the mask. G_X and G_Y are the approximations at [i, j].
Setting c=1, the Prewitt operator mask is obtained as

$$G_X = \begin{bmatrix} -1 & -1 & -1 \\ 0 & 0 & 0 \\ 1 & 1 & 1 \end{bmatrix}$$

and

$$G_Y = \begin{vmatrix} -1 & 0 & 1 \\ -1 & 0 & 1 \\ -1 & 0 & 1 \end{vmatrix}$$

The Prewitt masks have longer support. The Prewitt mask differentiates in one direction and averages in another direction; so the edge detector is less vulnerable to noise.

5. Sobel Edge Detector uses Convolution Kernel to detect the edges. The Sobel kernel relies on central differences, but gives greater weight to the central pixels when averaging. The Sobel kernels can be thought of as 3*3 approximations to first derivatives of Gaussian kernels. The partial derivatives of the Sobel operator are calculated as

$$G_X = \left(a_2 + 2a_3 + a_4 \right) - \left(a_0 + 2a_7 + a_6 \right)$$

and

$$G_Y = \left(a_6 + 2a_5 + a_4 \right) - \left(a_0 + 2a_1 + a_2 \right)$$

So, the Sobel masks in matrix form are given as

$$G_X = \begin{bmatrix} -1 & -2 & -1 \\ 0 & 0 & 0 \\ 1 & 2 & 1 \end{bmatrix}$$

and

$$G_Y = \begin{vmatrix} -1 & 0 & 1 \\ -2 & 0 & 2 \\ -1 & 0 & 1 \end{vmatrix}$$

The noise –suppression characteristics of a Sobel mask is better than that of a Prewitt mask.

6. Canny Edge Detector also uses high spatial gradient, but it takes more computation than Sobel and Robert's Edge Detector. In that Canny edge detector has some step by step procedure for segmentation is mentioned in Figure 3, which is as follows:

a. To reduce the effect of noise, the surface of the image is smoothened by using Gaussian Convolution.
b. Sobel operator is applied to the image to detect the edge strength and edge directions.
c. The edge directions are taken into considerations for non-maximal suppression i.e., the pixels that are not related to the edges are detected and then, they are minimized.
d. Final step is removing the broken edges, i.e., the threshold value of an image is calculated and then the pixel value is compared with the threshold that is obtained. If the pixel value is high than the threshold, then, it is considered as an edge or else it is rejected (Canny, 1986).

3.1.3 Theory Based Segmentation

This type of image segmentation algorithm includes derivatives from different fields and are very important for segmentation approach. They include genetic algorithms, wavelet based algorithms, fuzzy based algorithms, and neural network based algorithms, clustering based algorithms and so on (Wang & Wang, 2010).

3.1.3.1 Clustering Techniques

Clustering is an unsupervised learning task, where one needs to identify a finite set of categories known as clusters to classify pixels (Dehariya, Shrivastava, & Jain, 2010). A similarity criteria are defined between the pixels and then similar pixels are grouped together to form clusters. Similarity criteria include attribute of an image such as size, color, texture, etc. The quality of a cluster depends on both the quality of similarity criteria used and how it is implemented. Clustering methods are classified as hard clustering, k-means clustering, fuzzy clustering, etc.

Figure 3. Procedure of Canny edge detector

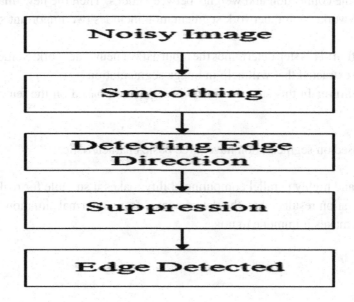

1. **Hard Clustering:** Hard clustering assumes that a pixel can only belong to a single cluster and also that there exists sharp boundaries between clusters. One of the most popular and well used hard clustering algorithms is K -means clustering algorithm (Dehariya, Shrivastava, & Jain, 2010). K-mean clustering is a clustering technique group n pixels of an image into a K number of clusters, where K < n and K is a positive integer. Initially the centroids of the predefined clusters are initialized randomly. Clusters are formed on the basis of some similar features like gray level intensity of pixels and distance of pixel intensities. The process is as follows:
   a. Randomly choose number of clusters K.
   b. Randomly choose K pixels of different intensities as Centroids.
   c. Centroids are finding out by calculating mean of the pixel values in a region. Place Centroids as far away from each other as possible.
   d. Now, compare a pixel to every Centroid and assign pixel to the closest Centroid to form a cluster. When all the pixels have been assigned, initial clustering has been completed
   e. Recalculate the mean of each cluster and recalculate the position of Centroids in K clusters.
   f. Repeat steps 4 & 5 until the Centroids no longer move.

2. **Fuzzy Clustering:** Fuzzy clustering can be used in situations when there are no defined boundaries between different objects in an image. Fuzzy clustering divides the input pixels into clusters or groups on the basis of some similarity criterion. Similarity criterion can be distance, connectivity, intensity etc. Fuzzy clustering algorithms include FCM (fuzzy C means) algorithm, GK (Gustafson-Kessel), GMD (Gaussian Mixture Decomposition), FCV (Fuzzy C varieties) etc. Fuzzy Clustering Mean algorithm (Khalifa, 2010) is most accepted since it can preserve much more information than other approaches. In this technique, a dataset is grouped into N clusters with every data point in the dataset belonging to every cluster to a certain degree.

3.1.3.2 Neural Network-Based Segmentation

In this algorithm, an image is firstly mapped into a neural network where every neuron represents a pixel (Pal & Pal, 1993; Kang, Yang & Liang, 2009). The neural network is trained with training sample set in order to determine the connection and weights between nodes. Then the new images are segmented with trained neural network. Neural network segmentation includes two important steps:

1. **Feature Extraction:** This step determines the input data of neural network. Some important features from images are extracted that will help in image segmentation
2. **Image Segmentation:** In this step the image is segmented based on the features extracted from the images

Neural network based on segmentation have three basic characteristics:

1. Fast computing and highly parallel computing ability makes it suitable for real time application.
2. Improve segmentation results when the data deviated from a normal situation.
3. High robustness makes it immune to noise.

3.1.4 Threshold Methods

Image segmentation by using the threshold method is quite simple but very powerful approach for segmenting images based on image-space region i.e. characteristics of the image (Kang, Yang, & Liang, 2009). This method is usually used for images having the light object on darker background or vice versa. Thresholding algorithm will choose a proper threshold value T to divide image's pixels into several classes and separate objects from the background. Any pixel (x, y) for which f(x, y) >=T is considered to be foreground while any pixel (x, y) which has value f(x, y) <T is considered to be background. The drawback of this segmentation technique is that it is not suitable for complex images. Based on the selection of a threshold value, there are two types of thresholding method that are in existence. These are Global Thresholding and Local Thresholding.

3.1.4.1 Global Thresholding

Global (single) thresholding method is used when there the intensity distribution between the objects of foreground and background are very distinct. When the difference between foreground and background objects is very distinct, a single value of threshold can simply be used to differentiate both objects apart. Thus, in this type of thresholding, the value of threshold T depends solely on the property of the pixel and the grey level value of the image. Some most common used global thresholding methods are Otsu method, entropy based thresholding, etc. (Wang & Wang, 2010).

3.1.4.2 Local Thresholding

This method divides an image into several sub regions and then choose various thresholds for each sub region respectively. Thus, threshold depends on both f (x, y) and p (x, y). Some common used Local thresholding techniques are simple statistical thresholding, 2-D entropy-based thresholding histogram transformation thresholding etc. (Kang, Yang, & Liang, 2009).

Thresholding is often used as an initial step in a sequence of image processing operations. The main limitation of thresholding techniques is that in its simplest form, only two classes are generated and it cannot be applied to multi-channel images.

3.1.5 Model Based Segmentation

The human eyes have the ability to recognize objects, even if they are not completely visible. All the algorithms mentioned above utilize only local information. In this case, we require specific knowledge about the geometrical shape of the object, which can then be compared with the local information to recreate the object. This segmentation technique is applicable only if we know the exact shape of the objects contained in the image.

3.1.6 Watershed Transformation

The watershed transformation is a powerful tool for image segmentation. A grey-level image may be seen as a topographic relief, where the grey level of a pixel is interpreted as its altitude in the relief. A drop of water falling on a topographic relief flows along a path to finally reach a local minimum. Intuitively, the watershed of a relief corresponds to the limits of the adjacent catchment basins of the drops of water.

In image processing, different types of watershed lines may be computed. In graphs, watershed lines may be defined on the nodes, on the edges, or hybrid lines on both nodes and edges. Watersheds may also be defined in the continuous domain. There are also many different algorithms to compute watersheds. Watershed algorithm is used in image processing primarily for segmentation purposes.

The drawbacks of watershed algorithm based image segmentation are:

1. Over Segmentation,
2. Sensitivity to noise,
3. Poor performance in the area of low contrast boundaries, and
4. Poor detection of thin structures.

4. EXPERIMENTAL RESULTS

The Matlab simulated results of different available image segmentation techniques are given for better understanding of image segmentation.

4.1 Region Growing

This function performs "region growing" in an image from a specified seed point (x, y). The region is iteratively grown by comparing all unallocated neighboring pixels to the region. The difference between a pixel's intensity value and the region's mean is used as a measure of similarity. The Input image, the logical output image of the region and combined output are shown in Figure 4, Figure 5, and Figure 6 respectively.

The pixel with the smallest difference measured this way is allocated to the respective region. This process stops when the intensity difference between region mean and new pixel become larger than a certain threshold (t). Segmentation is the most important part in image processing. Fence off an entire

Figure 4. Input image for region growing

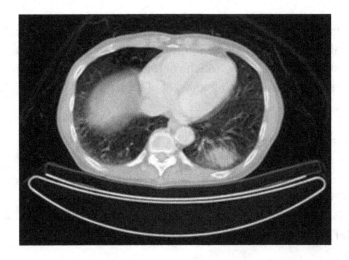

Figure 5. Logical output image of region

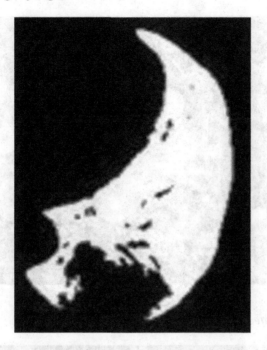

Figure 6. Result after region growing

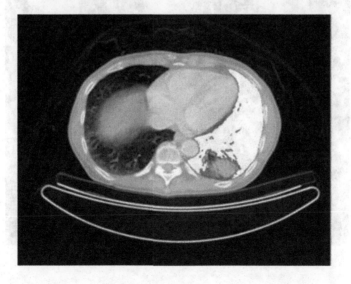

image into several parts which is something more meaningful and easier for further process. These several parts that are rejoined will cover the entire image which is shown in Figure 6.

4.2 Region Splitting and Merging

Figure 7, Figure 8, and Figure 9 show the original image, 9 small segmented image and resultant image after merging respectively.

Figure 7. Original image for region splitting and merging

Figure 8. Nine small segmented images

4.3 Differentiation of Gaussian Function

The Matlab result for differentiation of Gaussian function with Sigma value 1, 1.5 and 2 are shown in Figures 10, 11, and 12.

Figure 9. Segmented image after merging

Figure 10. Matlab result for DOG with Sigma value 1

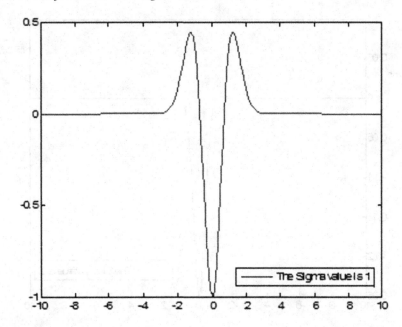

4.4 Different Edge Detector

EDGE takes intensity or a binary image I as its input, and returns a binary image BW of the same size as I, with 1's where the function finds edges in I and 0's elsewhere. EDGE supports six different edge-finding methods. The Sobel, Prewitt and Robert method find edges using the Sobel, Prewitt and Robert approximation to the derivative. It returns edges at those points where the gradient of I is maximum. The

Figure 11. Matlab result for DOG with Sigma value 1.5

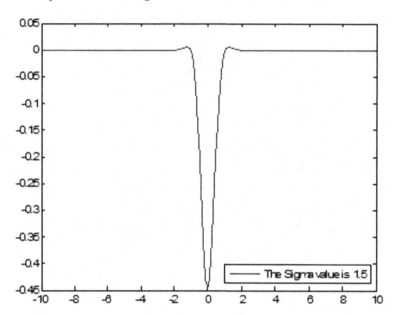

Figure 12. Matlab result for DOG with Sigma value 2

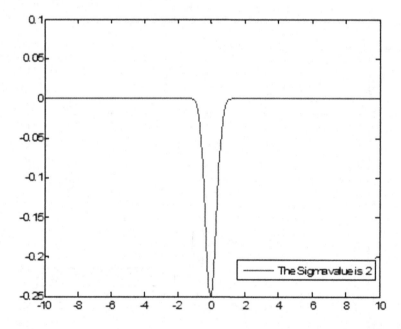

Laplacian of Gaussian method finds edges by looking for zero crossings after filtering I with a Laplacian of Gaussian filter. The Canny method finds edges by looking for local maxima of the gradient of I. The gradient is calculated using the derivative of a Gaussian filter. The method uses two thresholds, to detect strong and weak edges, and includes the weak edges in the output only if they are connected to strong edges. This method is therefore less likely than the others to be "fooled" by noise, and more likely to detect true weak edges. Simulated results are shown in Figures 13-18.

Figure 13. Original image for edge detection

Figure 14. Segmented image using Roberts kernel

Figure 15. Segmented image using Sobel kernel

Figure 16. Segmented image using Prewitt kernel

Figure 17. Segmented image using LOG

Figure 18. Segmented image using Canny edge detector

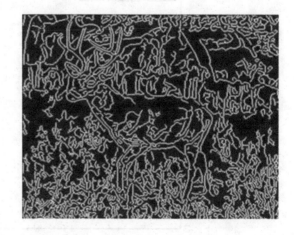

4.5 Segmentation Using Clustering

The randomly generated data and cluster assignments with centroid are shown in Figure 19 and Figure 20 respectively.

Figure 19. Randomly generated data

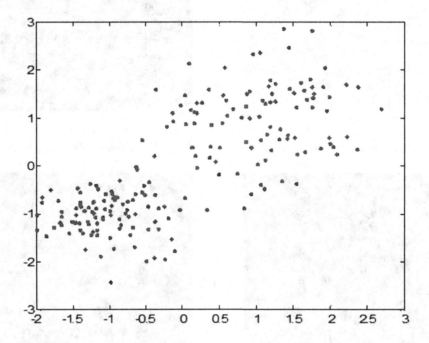

Figure 20. Randomly generated data after K-means clustering

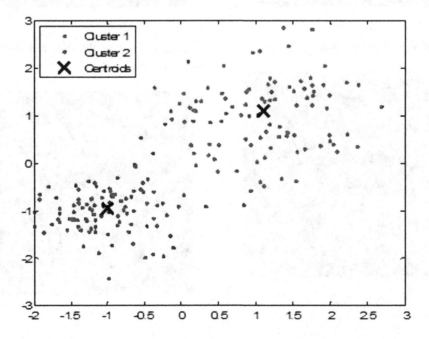

4.6 Segmentation Using Thresholding

Mean Gray Level Algorithm is simply applied by summing up all the pixel values in the image and then taking the mean of it to obtain the threshold. It is employed on a face image. The original image and its histogram are shown in Figure 21 and Figure 22 respectively. The output of Mean Gray Level is also shown in Figure 23.

Method of Two Peaks is employed by finding two local maximum points in the histogram and defining a threshold separating them.

In Edge pixel, Laplacian is calculated for each pixel and then histogram of pixels with large Laplacian is created. Using this new histogram a threshold can be detected using any of the previous methods.

Figure 21. Original face image

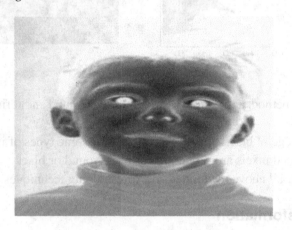

Figure 22. Histogram of original face image

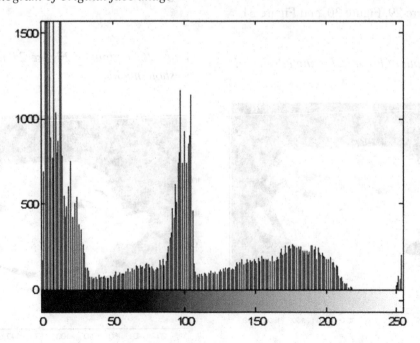

Figure 23. Output of Figure 21 using mean gray level thresholding

Figure 24. Output of Figure 21 using method of two peaks

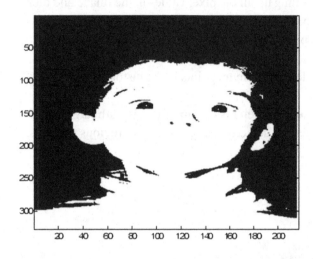

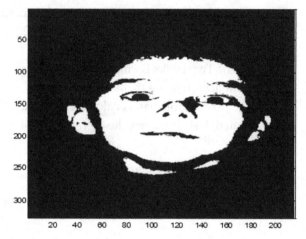

In Iterative Threshold method, a threshold is iteratively calculated and refined by consecutive passes through the image.

Assuming that percentage of black pixels is a constant for some types of images, lower pixel values up to the number of assumed pixels are segmented as background or black.

The threshold value for all above mentioned five thresholding techniques are tabulated in Table 1.

4.7 Watershed Transformation

Matlab simulated results of original image and noisy image with watershed transformation are given Figure 28, Figure 29, Figure 30, and Figure 31.

Figure 25. Output of Figure 21 using edge pixel

Figure 26. Output of Figure 21 using iterative threshold method

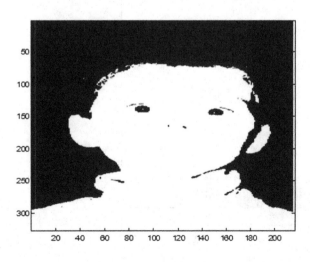

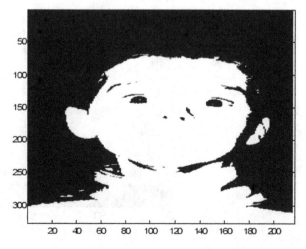

Figure 27. Output of Figure 21 using percentage of black pixels

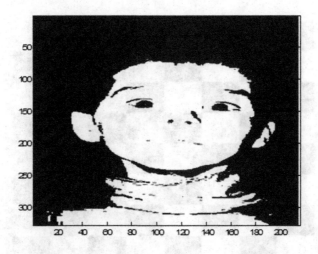

Table 1. Threshold value with different techniques

Name of the Techniques to Find the Threshold Value	Threshold Value
Mean Gray Level Algorithm	72.7906
Method of Two Peaks	130.00
Edge pixel	44.4347
Iterative Threshold method	77.4457
Percentage of Black Pixels	95 with 60% of Black Pixels

Figure 28. Original image of chess board

Figure 29. Image of chess board with noise

Figure 30. Watershed transformation of chess board image

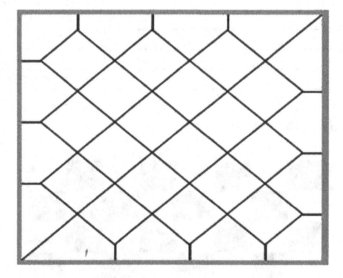

CONCLUSION

In this chapter, we discuss available main image segmentation algorithms, and implemented in Matlab for better visualization. Based on this, we now discuss the prospect of image segmentation. As the basic technique of image processing and computer vision, image segmentation has a promising future and the universal segmentation algorithm has become the focus of contemporary research.

Although there are a myriad of segmentation algorithms designed day in and day out, none of them can apply to all types of images and actual segmentation techniques usually aim at certain application. The result of image segmentation is affected by lots of factors, such as: homogeneity of images, spatial

Figure 31. Watershed transformation of noisy image

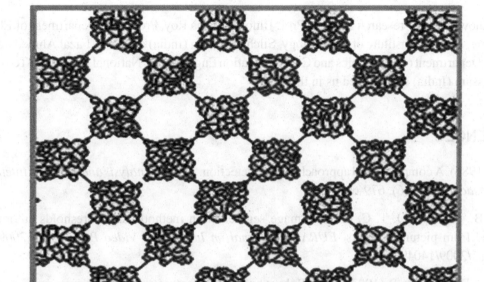

structure character of the image, continuity, texture, image content, and physical visual character and so on. Good image segmentation algorithm should take all-sided consideration on those factors.Due to all above factors, image segmentation remains a challenging problem in image processing and computer vision and is still a pending problem in the world. Based on the aforementioned statements, we can foresee the trend of image segmentation as follows:

1. **Combination of Multi-Algorithms:** For example we can integrate the advantages of edge detection and regionbased segmentation by combining those algorithms together and merging the segmentation results according to certain criteria.
2. **The Application of Artificial Intelligence:** Nowadays, although there are many existing image segmentation algorithms, almost each of them aims at a specific single application and only uses a fraction of image information, which limits their use to a great extent. Fortunately, this problem can be solved by introducing artificial intelligence into image segmentation.
3. **The Rise of Manual Alternating Segmentation:** It is effortless for a human to partition and detect an image. The efficiency and effectiveness of human eyes on image processing is far beyond the level of any computer. The reason is clear and simple: we human use a lot of synthetical knowledge when we are observing an image. Based on this reality, it is optimistic that manual alternating segmentation can realize better segmentation results.

ACKNOWLEDGMENT

It is to acknowledge our research advisers Prof. Binoy Krishna Roy, Professor, Department of Electrical Engineering, National Institute of Technology, Silchar, Assam (India) and Prof. Fazal Ahmed Talukdar, Professor, Department of Electronics and Communication Engineering, National Institute of Technology, Silchar, Assam (India) who trained us in this area.

REFERENCES

Canny, J. (1986). A computational approach to edge detection. *Pattern Analysis and Machine Intelligence, IEEE Transactions on*, (6), 679-698.

Chen, Y. B., & Chen, O. T. C. (2009). Image segmentation method using thresholds automatically determined from picture contents. *EURASIP Journal on Image and Video Processing, 2009*, 1–15. doi:10.1155/2009/140492

Comaniciu, D., & Meer, P. (1997, June). Robust analysis of feature spaces: color image segmentation. In *Computer Vision and Pattern Recognition, 1997. Proceedings., 1997 IEEE Computer Society Conference on* (pp. 750-755). IEEE. doi:10.1109/CVPR.1997.609410

Dass, R., & Devi, P. S. (2012). Image Segmentation Techniques. *International Journal of Electronics & Communication Technology, 3*(1), 66–70.

Dehariya, V. K., Shrivastava, S. K., & Jain, R. C. (2010, November). Clustering of image data set using K-means and fuzzy K-means algorithms. In *Computational Intelligence and Communication Networks (CICN), 2010 International Conference on* (pp. 386-391). IEEE.

Dey, N., Roy, A. B., Pal, M., & Das, A. (2012). *FCM based blood vessel segmentation method for retinal images*. arXiv preprint arXiv: 1209.1181.

Felzenszwalb, P. F., & Huttenlocher, D. P. (2004). Graph-based image segmentation. *International Journal of Computer Vision, 59*(2), 167–181. doi:10.1023/B:VISI.0000022288.19776.77

Gómez-Moreno, H., Maldonado-Bascón, S., & López-Ferreras, F. (2001). Edge detection in noisy images using the support vector machines. In Connectionist Models of Neurons, Learning Processes, and Artificial Intelligence (pp. 685-692). Springer Berlin Heidelberg. doi:10.1007/3-540-45720-8_82

Hore, S., Chakroborty, S., Ashour, A. S., Dey, N., Ashour, A. S., Sifaki-Pistolla, D., & Chowdhury, S. R. (2015). Finding Contours of Hippocampus Brain Cell Using Microscopic Image Analysis. *Journal of Advanced Microscopy Research, 10*(2), 93–103. doi:10.1166/jamr.2015.1245

Jain, A. K., & Dubes, R. C. (1988). *Algorithms for clustering data*. Prentice-Hall, Inc.

Kaganami, H. G., & Beiji, Z. (2009, September). Region-based segmentation versus edge detection. In *Intelligent Information Hiding and Multimedia Signal Processing, 2009. IIH-MSP'09. Fifth International Conference on* (pp. 1217-1221). IEEE. doi:10.1109/IIH-MSP.2009.13

Kang, W. X., Yang, Q. Q., & Liang, R. P. (2009, March). The comparative research on image segmentation algorithms. In *2009 First International Workshop on Education Technology and Computer Science* (pp. 703-707). IEEE. doi:10.1109/ETCS.2009.417

Karthikeyan, B., Vaithiyanathan, V., Venkatraman, B., & Menaka, M. (2012). Analysis of image segmentation for radiographic images. *Indian Journal of Science and Technology*, 5(11), 3660–3664.

Khalifa, A. R. (2010). Evaluating the effectiveness of region growing and edge detection segmentation algorithms. *J. Am. Sci*, 6(10).

Khan, J. F., Bhuiyan, S., & Adhami, R. R. (2011). Image segmentation and shape analysis for road-sign detection. *Intelligent Transportation Systems. IEEE Transactions on*, 12(1), 83–96.

Langote, V. B., & Chaudhari, D. D. (2012). Segmentation Techniques for Image Analysis. *International Journal of Advanced Engineering Research and Studies*, 1.

Lehmann, F. (2011). Turbo segmentation of textured images. Pattern Analysis and Machine Intelligence. *IEEE Transactions on*, 33(1), 16–29.

Lim, Y. W., & Lee, S. U. (1990). On the color image segmentation algorithm based on the thresholding and the fuzzy c-means techniques. *Pattern Recognition*, 23(9), 935–952. doi:10.1016/0031-3203(90)90103-R

Luo, J., Gray, R. T., & Lee, H. C. (1998, October). Incorporation of derivative priors in adaptive Bayesian color image segmentation. In *Image Processing, 1998. ICIP 98. Proceedings. 1998 International Conference on* (pp. 780-784). IEEE.

Ma, W. Y., & Manjunath, B. S. (1997, June). Edge flow: a framework of boundary detection and image segmentation. In *Computer Vision and Pattern Recognition, 1997. Proceedings., 1997 IEEE Computer Society Conference on* (pp. 744-749). IEEE. doi:10.1109/CVPR.1997.609409

Mushrif, M. M., & Ray, A. K. (2008). Color image segmentation: Rough-set theoretic approach. *Pattern Recognition Letters*, 29(4), 483–493. doi:10.1016/j.patrec.2007.10.026

Nadernejad, E., Sharifzadeh, S., & Hassanpour, H. (2008). Edge detection techniques: Evaluations and comparison. *Applied Mathematical Sciences*, 2(31), 1507–1520.

Naz, S., Majeed, H., & Irshad, H. (2010, October). Image segmentation using fuzzy clustering: A survey. In *Emerging Technologies (ICET), 2010 6th International Conference on* (pp. 181-186). IEEE. doi:10.1109/ICET.2010.5638492

Pal, G., Acharjee, S., Rudrapaul, D., Ashour, A. S., & Dey, N. (2015). Video segmentation using minimum ratio similarity measurement. *International Journal of Image Mining*, 1(1), 87–110. doi:10.1504/IJIM.2015.070027

Pal, N. R., & Pal, S. K. (1993). A review on image segmentation techniques. *Pattern Recognition*, 26(9), 1277–1294. doi:10.1016/0031-3203(93)90135-J

Rotem, O., Greenspan, H., & Goldberger, J. (2007, June). Combining region and edge cues for image segmentation in a probabilistic gaussian mixture framework. In *Computer Vision and Pattern Recognition, 2007. CVPR'07. IEEE Conference on* (pp. 1-8). IEEE. doi:10.1109/CVPR.2007.383232

Roy, P., Goswami, S., Chakraborty, S., Azar, A. T., & Dey, N. (2014). Image segmentation using rough set theory: A review. *International Journal of Rough Sets and Data Analysis*, *1*(2), 62–74. doi:10.4018/ijrsda.2014070105

Saha, B. N., Ray, N., Greiner, R., Murtha, A., & Zhang, H. (2012). Quick detection of brain tumors and edemas: A bounding box method using symmetry. *Computerized Medical Imaging and Graphics*, *36*(2), 95–107. doi:10.1016/j.compmedimag.2011.06.001 PMID:21719256

Sapna Varshney, S., Rajpa, N., & Purwar, R. (2009, December). Comparative study of image segmentation techniques and object matching using segmentation. In *Methods and Models in Computer Science, 2009. ICM2CS 2009.Proceeding of International Conference on* (pp. 1-6). IEEE. doi:10.1109/ICM2CS.2009.5397985

Shafait, F., Keysers, D., & Breuel, T. M. (2008, January). Efficient implementation of local adaptive thresholding techniques using integral images. In *Electronic Imaging 2008* (p. 681510–681510). International Society for Optics and Photonics. doi:10.1117/12.767755

Shih, F. Y., & Cheng, S. (2004). Adaptive mathematical morphology for edge linking. *Information Sciences*, *167*(1), 9–21. doi:10.1016/j.ins.2003.07.020

Somasundaram, S. K., & Alli, P. (2012). A Review on Recent Research and Implementation Methodologies on Medical Image Segmentation. *Journal of Computer Science*, *8*(1), 170–174. doi:10.3844/jcssp.2012.170.174

Wang, Y., & Wang, Q. (2010, October). Image segmentation based on multi-scale local feature. In *Image and Signal Processing (CISP), 2010 3rd International Congress on* (Vol. 3, pp. 1406-1409). IEEE. doi:10.1109/CISP.2010.5648301

Yanowitz, S. D., & Bruckstein, A. M. (1988, November). A new method for image segmentation. In *Pattern Recognition, 1988., 9th International Conference on* (pp. 270-275). IEEE.

Zhang, Y. J. (1997). Evaluation and comparison of different segmentation algorithms. *Pattern Recognition Letters*, *18*(10), 963–974. doi:10.1016/S0167-8655(97)00083-4

KEY TERMS AND DEFINITIONS

Clustering: The clustering technique attempt to access the relationships among patterns of the data set by organizing the patterns into groups or clusters such that patterns within a cluster are more similar to each other than patterns belonging to different clusters.

Image Processing: The analysis and manipulation of a digitized image, especially in order to improve its quality.

Image Segmentation: In computer vision, image segmentation is the process of partitioning a digital image into multiple segments (sets of pixels, also known as superpixels). The goal of segmentation is to simplify and/or change the representation of an image into something that is more meaningful and easier to analyze.

Region Growing: An approach to image segmentation in which neighbouring pixels are examined and added to a region class if no edges are detected.

Region of Interest: A region of interest (often abbreviated ROI), is a selected subset of samples within a dataset identified for a particular purpose.

Region Splitting: A top-down approach. It begins with a whole image and divides it up such that the segregated parts are more homogenous than the whole.

Thresholding: Thresholding is the simplest method of image segmentation. From a grayscale image, thresholding can be used to create binary images.

Chapter 2
Measures of Image and Video Segmentation

Pushpajit A. Khaire
SRCOEM, India

Roshan R. Kotkondawar
GCOE, India

ABSTRACT

Study on Video and Image segmentation is currently limited by the lack of evaluation metrics and benchmark datasets that covers the large variety of sub-problems appearing in image and video segmentation. Proposed chapter provides an analysis of Evaluation Metrics, Datasets for Image and Video Segmentation methods. Importance is on wide-ranging, Datasets robust Metrics which used for evaluation purposes without inducing any bias towards the evaluation results. Introductory Section discusses traditional image and video segmentation methods available, the importance and need of measures, metrics and dataset required to evaluate segmentation algorithms are discussed in next section. Main focus of the chapter explains the measures, metrics and dataset available for evaluation of segmentation techniques of both image and video. The goal is to provide details about a set of impartial datasets and evaluation metrics and to leave the final evaluation of the evaluation process to the understanding of the reader.

INTRODUCTION

This introductory section will express the notions of digital image and digital video, with the importance of both in today's age, it will also describe the importance of datasets and objective measures for evaluation of image segmentation and video segmentation techniques. According to (Jain, 1989) the term Digital Image Processing generally refers to processing of a two dimensional picture by a digital computer and in a broader sense, it implies digital processing of any two-dimensional data. A digital image is an array of real or complex numbers represented by a finite number of bits. (Pratt, 2007) describes segmentation of an image as it entails the division or separation of the image into regions of similar attribute and the most basic attribute for segmentation is image luminance amplitude for a monochrome image and color

DOI: 10.4018/978-1-5225-1022-2.ch002

components for a color image. Important features called edges and other attribute like texture are also useful for segmentation.

Digital images and videos are, respectively, defined as pictures and movies that have been converted into a computer-readable binary format consisting of logical 0s and 1s. Theoretically an image is understood by a still picture that does not change with time, and a video is created by number of frames or images that changes with time, mostly containing moving and/or changing objects (Bovik, 2000). Digital images or video are usually obtained by converting continuous signals into digital format, although "direct digital" systems are becoming more prevalent. Likewise, digital visual signals are viewed by using diverse display media, including digital printers, computer monitors, and digital projection devices. Due to rapid increasing of multimedia in the form of text and moreover in the form of images and video this information is transmitted, stored, processed, and displayed in a digital visual format, and thus the design of engineering methods for efficiently transmitting, maintaining, and even improving the visual integrity of this information is of utmost interest (Bovik, 2000).

Applications of Image and Video Segmentation

Image Segmentation has several applications some of them are discussed here in brief.

1.   Target Detection and Recognition of missile or aircraft can be done by applying image segmentation techniques on radar and other images, addition to this it can be used in remote sensing, weather forecasting and space analysis by applying segmentation techniques to satellite images etc.
2.   In robotics and industrial automation, objects can be detected, recognized and can be moved from one place to another using image segmentation method.
3.   Segmentation operations on images can be used in automated colorization of objects in scenes of movies and cartoons etc. Moreover, the segmentation operation can be applied to the conversion of 2D scene to 3D scene and creation of visual effects.
4.   In medical applications it can used for detection of tumors in patients with processing and segmenting of X-Ray, Magnetic Resonance (MRI) and other medical images.

There are number of applications of video segmentation including:

1.   Video surveillance, where the segmentation result is used to allow the identification of an intruder or of an anomalous situation. (Li et al., 2007)
2.   Another applications of video segmentations are video summarization, in the field of Computer vision; it can be used for video matting, video toning, and rendering, further the construction of segmented objects from 2D video into 3D scenes.
3.   Segmentation can achieve a better quality in Videoconferencing and video-phony applications by coding most relevant objects at higher quality. (Li et al., 2007)
4.   There are several other applications of video segmentation including application in medical science, Space Research, Industrial automation, Robotics, environmental monitoring, climate predictions, military target detection and aviation systems, etc.

For image segmentation and video segmentation, many algorithms have been proposed from early 70's till date, due to lack of segmentation dataset and respective ground truth, it was difficult for researchers in the field of computer vision, image processing, to evaluate their performance in the absence of ground-truth images. Ground-truth for a still image defines the support of the objects, which corresponds to the location, size, and the shape of the object. Table 1 shows the comparison of recent proposed methods of one of the technique called edge detection used for image segmentation. Comparison is carried out mainly on evaluation parameters, ground truth and datasets used; from the table it is clear that

1. Very few researches evaluate their result with objective evaluation parameter.
2. Ground truth of image is not being used by most of the authors to evaluate the performance of segmentation method.
3. Very few images are being used to compare the results of proposed method with another technique or method.

Due to unawareness of the availability of the dataset and objective quality metrics, researchers published the results of their proposed methods without using dataset, ground truth and proper evaluation metrics. The table presented is not to point any method or author's work, but to understand the use and importance of dataset, ground truth and objective evaluation for fruitful advancement of research in image and video segmentation techniques. Following paragraph discuss the need of appropriate image and video segmentation dataset and metrics and presents the reviews of various authors about the same.

The main purpose of studying Computer vision is to create artificial systems. (Price, 1986) stated that the creating and studying tend to be separate and unrelated efforts-one group "creates" another group "studies," but they do not use the same model of what to do. This will not help by the fact that created methods hardly ever survive long enough for any extensive study, even by their creator. Research in computer vision has suffered from a lack of building on past work. Only through real effort programs can be

Table 1. Comparison of recently proposed edge detection (segmentation) algorithms

Source	Comparison with Ground Truth	Methods Compared	Number of Images Used for Comparison	Evaluation Parameter Used for Comparison
(Biswas et al., 2012)	No	×	Lena, Cameraman (4)	×
(Bhardwaj et al., 2012)	No	Canny	Horse image (1)	×
(Patel et al., 2011)	No	LOG, Sobel	Experimental images without noise (2)	×
(Ou et al., 2011)	No	Sobel, VOS	Lena (color) (1)	×
(Mendhurwar et al., 2011)	No	Canny	Barbara, (3 gray, 3 color images) with Gaussian noise	PSNR
(Setayesh et al., 2011)	No	Canny	Lena, egg, coffee maker, with Gaussian noise (5)	PFOM (Pratt's figure of merit)
(Zhang et al., 2011)	No	Canny	Building (1)	×
(Yang et al., 2011)	No	Sobel, Prewitt, Robert	Building (1)	×
(Gao et al., 2010)	No	Canny, Prewitt, Sobel	Lena with Gaussian Noise (1)	×
(Alshennawy et al., 2009)	No	Sobel	Experimental images (3)	×

shared among a large group of researchers. Researchers should make the effort to obtain implementations of other researchers' systems so that we can better understand the limitations of our own work. (Fram et al., 1975) focuses on one of the fundamental problems in the area of image and video processing i.e. Segmentation of objects within an image or video, while a considerable number of segmentation schemes have been devised towards this end, no comprehensive attempt has been made to compare the various schemes available. Such a comparison is necessary in order to establish the performance limitations of each scheme under the various conditions encountered. Present lots of results on image and video segmentation lacks for standard dataset and quality measures. Results are often displayed with very few images and without objective evaluation. It is very difficult for a researcher to evaluate and compare his algorithm or method, given the results with subjective evaluation. (Price, 1986) stated that a thesis requires results on at least six different images. Statistical studies use training sets and test sets to control this problem. We cannot produce 100 test samples when each sample requires hours of computation for generating the results. It is hard for six "natural" images to all have the same obscure property that makes everything work right. When a large number of images is tried, the limits of the program become clearer and parameters are harder to tune for specific images (general solutions are used). Beware of any research that has only one image in the results. There is a reason-it is difficult to replicate and difficult to run. Any realistic (commercial) system must run on thousands of images before it is acceptable. What is more interesting is that we are willing to develop one more edge detector or another segmentation method, but we do not want to develop objective and quantitative methods to evaluate the performance of segmentation techniques. About four decades of research on image and video segmentation has produced number of segmentation techniques without a solid basis to evaluate the performance. In most disciplines, researchers evaluate the performance of a technique by a controlled set of experiments and specify the performance in clear objective terms. In segmentation, practically no efforts were even made to define objective measures. We still evaluate the performance of a segmentation method by looking at the results (Jain & Binford, 1991).

(Pal & Pal, 1993) discussed the same problem of qualitative evaluation of segmentation of images. A quantitative measure would be quite useful for vision applications where automatic decision is required and would help to justify the algorithm. Unfortunately, the human being is the best judge to evaluate the output of the segmentation algorithm. According to (Heath et. al, 1998), because of the difficulty of obtaining ground truth for real images, the traditional technique for comparing low-level vision algorithms is to present image results, side by side, and to let the reader subjectively judge the quality. This is not a scientifically satisfactory strategy. However, human rating experiments can be done in a more rigorous manner to provide useful quantitative conclusions. They present a paradigm based on experimental psychology and statistics, in which humans rate the output of low level vision algorithms. The use of human judges to rate image outputs must be approached systematically. Experiments must be designed and conducted carefully, and results must be interpreted with the appropriate statistical tools. The use of statistical analysis in vision system performance characterization has been rare. Another possible strategy would be to measure the performance enhancement of a complete general vision system with different segmentation methods. Thus, it has become an accepted practice to compare segmentation methods by presenting visual results side-by-side for the reader's subjective evaluation. That is, we resort to asking the only known object recognition system with proven competence—the human. But this practice raises many questions. How variable is the subjective judgment about a segmentation method across images? How well do different people agree in their subjective judgments of an image? To what extent are the conclusions affected by choice of images? (Heath et al., 1998) provide an experimental framework which allows to make quantitative comparisons using subjective ratings made by people. This approach avoids the issue of pixel-level ground truth. As a result, it does not allow making statements about the frequency of false positive and false negative errors at the pixel level. Instead, using experimental design and statistical techniques borrowed from psychology, it makes statements about whether the outputs of one edge detector are rated statistically significantly higher than the outputs of another (Heath et al., 1998).

Finding the requirement of dataset for productive research in image segmentation (Martin et. al 2001) presents a database containing 'ground truth' segmentations produced by humans for images of a wide variety of natural scenes. They define an error measure which quantifies the consistency between segmentations of differing granularities and find that different human segmentations of the same image are highly consistent. Use of this dataset is demonstrated in two applications:

1. Evaluating the performance of segmentation algorithms, and
2. Measuring probability distributions associated with Gestalt grouping factors as well as statistics of image region properties (Martin et al., 2001).

As discussed by (Unnikrishnan, Pantofaru, & Hebert 2007), most presentations of segmentation algorithms contain superficial evaluations which merely display images of the segmentation results and appeal to the reader's intuition for evaluation. There is a consistent lack of numerical results, thus it is difficult to know which segmentation algorithms present useful results and in which situations they do so. Appealing to human intuition is convenient, however if the algorithm is going to be used in an automated system then objective results on large datasets are to be desired.

On video segmentation (Goyette et. al, 2014) discussed that lack of a comprehensive dataset has a number of negative implications. Firstly, it makes it difficult to ascertain with confidence which algorithms would perform robustly when the assumptions they are built upon are violated. Many algorithms tend to over-fit specific scenarios. A method tuned to be robust to shadows may not be as robust to background motion. A dataset that contains many different scenarios and applies a variety of performance measures would go a long way towards providing an objective assessment. Further they (Goyette et al., 2014) added that Video segmentation research is currently limited by the lack of a benchmark dataset that covers the large variety of sub-problems appearing in video segmentation and that is large enough to avoid over-fitting. Consequently, there is little analysis of video segmentation which generalizes across subtasks, and it is not yet clear which and how video segmentation should leverage the information from the still-frames, as previously studied in image segmentation, alongside video specific information, such as temporal volume, motion and occlusion. Two fundamental limitations in the field video segmentation: the lack of a common dataset with sufficient annotation and the lack of an evaluation metric that is general enough to be employed on a large set of video segmentation subtasks (Goyette et al., 2014).

BACKGROUND

In previous section, a brief introduction of digital images and video has been discussed along with the need and importance of dataset, ground truth and quality measure necessary to compare image and video segmentation methods, A small survey table presented in previous section discuss the unawareness of using quality metrics and dataset ; this section explains the various methods and views regarding image segmentation and subsequently, the next sub-section describes various methods related to video segmentation.

Image Segmentation

Image segmentation has been, and still is, a relevant research area, and hundreds of segmentation algorithms have been proposed in more than last four decades. However, it is well known that elemental segmentation techniques based on boundary or region information often fail to produce accurate segmentation results. Hence, in the last few years, there has been a tendency towards algorithms which take advantage of the complementary nature of such information. (Haralick & Shaprio, 1965) categorize image segmentation as a technique that can be classified as: measurement space guided spatial clustering; single linkage growing schemes, hybrid linkage region growing schemes, centroid linkage growing schemes, spatial clustering schemes, and split and merge schemes, by looking at these techniques of segmentation, it can be observed from this brief typology, image segmentation was considered as a clustering process. The significant difference between image segmentation and clustering is that, in clustering, the grouping is done in measurement space. While, in image segmentation, the grouping is done on the spatial domain of the image (Haralick & Shaprio, 1965)

In allied field of Image and Video processing called Computer Vision, Image and Video Segmentation is one of the most important operations. The aim of image segmentation is the domain-independent partition of the image into a set of regions which are visually distinct and uniform with respect to some property, such as grey level, texture or colour. The problem of image segmentation has been significantly discussed in (Papari & Petkov, 2011). Many segmentation methods are based on two basic properties of the pixels in relation to their local neighbourhood discontinuity and similarity. Methods based on some discontinuity property of the pixels are called boundary-based methods, whereas methods based on some similarity property are called region-based methods. Unfortunately, both techniques, boundary-based and region-based, often fail to produce accurate segmentation results (Papari & Petkov, 2011). With the aim of improving the segmentation process, a large number of new algorithms which integrate region and boundary information have been proposed over the last few years. Among other features, one of the main characteristics of these approaches is the time of fusion: embedded in the region detection or after both processes (Papari & Petkov, 2011). Integration through the definition of new parameters or new decision criteria for segmentation can be referred as Embedded Integration. In the most regular approach, firstly, the edge information is extracted and then been used as a seed point within the segmentation algorithm which is mainly based on regions (Papari & Petkov, 2011). On image segmentation (Jain, 1989) explains that segmentation refers to the decomposition of a scene into its components. It is a key step in image analysis. For example, a document reader would first segment the various characters before proceeding to identify them. Following are different Image Segmentation Techniques discussed in (Jain, 1989).

Amplitude Thresholding or Window Slicing

When the amplitude features sufficiently characterize the object, amplitude thresholding is useful. The appropriate amplitude feature values are adjusted so that a given amplitude interval represents a unique object characteristic. Threshold selection is an important step in this method. Some commonly used approaches for threshold selection are as follows (Jain, 1989)

1. The histogram of the image is examined for locating peaks and valleys. If it is multimodal, then the valleys can be used for selecting thresholds.
2. By selecting the threshold called T so that a predetermined fraction of the total number of samples is below T.
3. Selecting threshold adaptively by examining local neighborhood histograms.
4. Selectively threshold by examining histograms only of those points that satisfy a chosen criterion. For example, in low-contrast images, the histogram of those pixels whose Laplacian magnitude is above a prescribed value will exhibit clearer bimodal features than that of the original image.
5. If a probabilistic model of the different segmentation classes is known, determine the threshold to minimize the probability of error or some other quantity, for instance, Bayes' risk

Component Labeling

The Second method suggested in (Jain, 1989) is simple and effective method of segmentation of binary images. Segmentation is done by finding the connectivity of pixels with their neighbors and then labeling the connected group of pixels or sets. Pixel labeling and Run-length connectivity analysis are two practical algorithms

Boundary-Based Approaches

One of the mostly studied approaches for segmentation are boundary based approaches and clustering out of which, Boundary extraction techniques segment objects on the basis of their profiles. Thus, contour following, connectivity, edge linking, graph searching curve fitting, Hough transform and other techniques applicable to image segmentation (Jain, 1989). Difficulties arise in boundary-based methods when objects are overlapping or touching each other in image or if a break occurs in the boundary detected due to various noises during image acquisition process.

Region-Based Approaches and Clustering

Another important technique used for segmentation is region based growing of pixels. The main idea in region-based segmentation techniques is to identify various regions in an image that have similar features (Jain, 1989). Concept of Clustering techniques taken from pattern-recognition is to cluster similar objects in image and can be applied for image segmentation.

Template Matching

Segmentation of an image using template matching segmentation is a one direct method in which an image to be segmented is to match it against templates from a given list. The detected objects can then be segmented out and the remaining image can be analyzed by other techniques. To segment busy images, such as journal pages containing text and graphics, this method is used. The text can be segmented by template-matching techniques and graphics can be analyzed by boundary following algorithms (Jain, 1989).

Texture Segmentation

When objects in a scene have a textured background texture segmentation is important for image segmentation. Texture often contains a high density of edges; therefore, segmenting objects using boundary-based approaches may become difficult and ineffective unless texture is filtered out. A way out from this problem is to apply, Clustering and region-based approaches to textured features and can be used to segment textured regions. In general, texture classification and segmentation is quite a difficult problem (Jain, 1989).

Video Segmentation

Video Segmentation is a process which refers to the identification of regions in a frame of video that are homogeneous in some sense (Bovik, 2000). There are number of multimedia applications of video segmentations like MPEG video coding standards, content-based image retrieval and representation, virtual reality and video surveillance and tracking which makes a rising attention in object-based video segmentation (Koprinska & Carrato, 2001; Li et al., 2003). Meaningful segmentation generally called as "Semantic Segmentation," is the partitions of images and video in meaningful entities such as regions, objects (e.g., balls, flowers, persons or cars) or object parts so that each segmented region or part in image have a semantic meaning (Zhang, 2006). While simple objects are often segmented automatically by grouping pixels with similar or homogeneous low-level image features, for segmentation of complex objects consisting of multiple and overlapping regions often requires some kind of supervised segmentation, as there is a gap between the low-level image features and high-level semantics and human perception (Zhang, 2006). To make semantic inferences from low level features additional information is required which is either provided by user or by high level modules. Video Segmentation has additional features related to object motion like homogeneity in speed acceleration and direction when compared with 2D image segmentation. Exploiting these features in the pixel region, or object level, often makes the video segmentation and tracking less challenging (Zhang, 2006). Categories of Segmentation for both image and video vary with intent to which the application requires the segmentation process. A common way to categorize segmentation methods is according to the source of attributes which the segmentation is based upon, e.g., pixel-based segmentation, region-based segmentation, content-based segmentation, edge-based segmentation, object-based segmentation, semantic-based segmentation and many more. Pixel-based segmentation methods use low-level image features such as color, intensity, texture, motion, optical flow, depth and disparity from each individual pixel. Region-based segmentation methods employ multi-scale shape, edges, polygon boundaries and texture statistics for extracting regions of interest (Pauwels, 1999). Object-based methods employ features associated with individual objects; these features can be extracted from different levels (Li, 2003; Pan, 2000; Meier, 1998). Hybrid methods can also be created by the combination of some or all these methods. Segmentation methods can also be classified based on whether the method is fully automatic or semiautomatic. Typical examples of semiautomatic methods include a user interactive process, e.g., active contours and image retrieval. In active contour-based segmentation, a user is usually required to select some initial points close to the contour of a region (Zhang, 2006). In hybrid multi-layer image retrievals, a user may add some high-level semantics and make choices from the low-level segmentation results (Wang, 2004; Sun, 2003). A user-interactive process usually can achieve better segmentation with the input of high-level semantics and intelligent choices from the human perception.

Color segmentation, texture segmentation, change detection and motion segmentation of video usually result in different segmentation maps of same video and for different information and applications. Different features and homogeneity criteria generally lead to different segmentations of the same data; furthermore, there is no guarantee that any of the resulting segmentations will be semantically meaningful. A semantically meaningful region may have multiple colors, multiple textures, or multiple motions that leads to notion of semantic segmentation of video and image. In recent years, finally research in the area of semantic segmentation is growing through mutual collaborations in research communities with an aim to segment objects semantically from video and images. The principal concerned in video segmentation is to label image regions which moves independently of another object in the video (motion segmentation) or semantically meaningful image regions (video object plane segmentation) (Bovik, 2000). Motion segmentation which is known with another name called "Optical Flow Segmentation" label pixels or optical flow vectors at each frame that are associated with independently moving part of a scene. For the object to be semantically meaningful in video, object segmentation generally requires user to define the object of interest in at least some key frames. Two other closely related problems of Motion segmentation are motion detection sometimes referred as "Change Detection" in Video and Motion Estimation of Objects. Change detection is a special case of motion segmentation, in the case of where camera is static; the change detection has only two regions, generally called as changed and unchanged regions. In another case where camera is moving, the regions are named as global and local motion regions (Bovik, 2000). If the scene is recorded with a static camera then it makes simpler to distinct between change detection and motion segmentation, Change Detection can be achieved without motion estimation of the object in the case of static camera. Change detection in the case of a moving camera and general motion segmentation, in contrast, require some sort of global or local motion estimation, either explicitly or implicitly (Bovik, 2000). Fundamental limitations of Motion Estimation are occlusion and aperture, Motion detection and segmentation are also weighed down with the same two fundamental limitations (Bovik, 2000). For example, due to aperture effect, pixels in a flat image region may appear stationary even if they are moving or due to occlusion problem, erroneous labels may be assigned to pixels in covered or uncovered image regions as a result of it. Motion Segmentation and Object Segmentation is an integral part of many video analysis problems as discussed by (Bovik, 2000), that includes:

1. Improved Motion Estimation sometimes also called as Optical Flow Estimation.
2. Three-Dimensional (3-D) Motion and structure estimation in the presence of multiple moving objects.
3. Description of the temporal variations or the content of video.

In any segmentation problem image or video, Feature Selection is an important aspect; the use proper feature selection makes motion segmentation or change detection possible effectively. In general, the application of standard image segmentation methods is not used directly to estimated optical flow vectors and may not yield meaningful results, since an object moving in 3-D usually generates a spatially varying optical flow field. Following are different types of video segmentation methods discussed by (Bovik, 2000) in greater detail, these methods are Change Detection in which change is detected, Parametric

Background Modeling and Subtraction, Multiple Motion Segmentation, Simultaneous Estimation and Segmentation, Semantic Video Object (Bovik, 2000).

Change Detection

One of the most frequently used Low-level tasks in Video Processing and its application to the allied field Computer Vision is "Change Detection." Numbers of algorithms have been developed till today for change Detection, yet no widely accepted, realistic, large-scale video data set exists for benchmarking different methods (Goyette et al., 2014). Widely Known dataset available today for Change Detection are discussed in Section "Datasets for Video Segmentation" later in this chapter. In this section a brief overview of Change Detection is discussed. Higher Level Tasks like detection, localization, tracking, and classification of moving objects are closely coupled with change detection and is often considered to be preprocessing step. Importance of change detection can be judged by the large number of algorithms developed till date and numbers are growing year by year.

When the video is taken from the static camera i.e., capturing the video when the camera is still and objects are moving during acquisition of video. For static camera, Change detection methods segment each frame into two regions, namely changed and unchanged regions, while in the case of moving camera, where the camera is moving along with the object or picture to be captured, Change Detection refers to global and local motion regions. In still camera, change detection classifies foreground and background of image as regions which are unchanged regions correspond to the background and regions which are changed corresponds to the foreground objects or areas. In the case of moving camera situation is identical to the former with some advancements (Bovik, 2000).

Dominant Motion Segmentation

Segmentation by dominant motion analysis refers to extracting one object (with the dominant motion) from the scene at a time. Dominant motion segmentation can be considered as a hierarchically structured top-down approach, which starts by fitting a single parametric motion model to the entire frame, and then partitions the frame into two regions, those pixels that are well represented by this dominant motion model and those that are not. The dominant motion may correspond to the camera (background) motion or a foreground object motion, whichever occupies a larger area in the frame. The dominant motion approach may also handle separation of individually moving objects. Once the first dominant object is segmented, it is excluded from the region of analysis, and the entire process is repeated to define the next dominant object. This is unlike the multiple motion segmentation approaches that are discussed in the next section, which start with an initial segmentation mask (usually with many small regions) and refine them according to some criterion function to form the final mask. (Bovik, 2000). It is worth noting that the dominant motion approach is a direct method that is based on spatiotemporal image intensity gradient information.

Multiple Motion Segmentation

Multiple motion segmentation methods let multiple motion models compete against each other at each decision site. They consist of three basic steps, which are strongly interrelated: estimation of the number K of independent motions, estimation of model parameters for each motion, and determination of support of each model (segmentation labels). If we assume that we know the number K of motions and the K sets of motion parameters, then we can determine the support of each model (Bovik, 2000).

Simultaneous Estimation and Segmentation

Until now, we discussed methods to compute the segmentation labels from either precomputed optical flow or directly from intensity values, but we did not address how to compute an improved dense motion field along with the segmentation map. It is clear that the success of optical flow segmentation is closely related to the accuracy of the estimated optical flow field (in the case of using precomputed flow values), and vice versa. (Bovik, 2000). It follows that optical flow estimation and segmentation have to be addressed simultaneously for best results.

Semantic Video Object Segmentation

So far we discussed methods for automatic motion segmentation. However, it is difficult to achieve semantically meaningful object segmentation by using fully automatic methods based on low-level features such as motion, color, and texture. This is because a semantic object may contain multiple motions, colors, textures, and so on, and the definition of semantic objects may depend on the context, which may not be possible to capture by using low-level features. Thus, in this section, we present two approaches that can extract semantically meaningful objects by using capture-specific information or user interaction (Bovik, 2000).

MAIN FOCUS OF THE CHAPTER

The importance as well as the necessity of largely available dataset and qualitative measures for image and video segmentation is discussed in previous sections of this chapter. This section of this chapter mainly focuses on the datasets and qualitative metrics available for evaluation of image and video segmentation methods. Well known datasets like Berkeley Segmentation Dataset for images and objective evaluation measures like, precision-recall and performance ratio are discussed in this section.

Datasets for Image Segmentation

For still image segmentation, many algorithms have been proposed from early 70's till today, due to lack of segmentation dataset and respective ground truth, it was difficult for researchers in the field of computer vision, image processing, to evaluate their performance in the absence of ground-truth images. Ground-truth for a still image defines the support of the objects, which corresponds to the location, size, and the shape of the object. This section discusses renowned and widely used publicly available segmentation datasets called Berkeley Segmentation Dataset (BSD).

Berkeley Segmentation Dataset

In 2001 (Martin and et al., 2001) introduced the Berkeley Segmentation Dataset (BSD) as a repository of natural images with corresponding human segmentations. Since then, BSD has been mostly used by researchers to develop and evaluate boundary extraction and segmentation algorithms based on image statistics. Concurrently with the development of the BSD several image segmentation algorithms have been made available by their authors. This has allowed different research groups to extend and improve known segmentation methods, and to visually compare the resulting segmentations. An example of BSD images and Ground truth is given in Figure 1 and Figure 2 respectively, the output of well-known canny method for edge detection is given in Figure 3. Canny method is also used for boundary detection in many applications. The output of canny method is compared with ground image pixel by pixel, using metrics F- Measure and Performance Ratio (PR), values obtained for both these metrics after calculation are 0.0036 and 7.95 respectively. Equations and formulas for calculation of F-measure and Performance Ratio (PR) is given in next section. This example gives a brief idea of using ground truth image and calculation of metrics with other method. This is how researchers can compare their methods or technique of boundary detection or feature extraction using ground truth and evaluation metrics.

Figure 1. Sample image from Berkeley segmentation dataset
Source: Martin et al., 2001.

Figure 2. Corresponding ground truth of Figure 1 for segmentation
Source: Martin et al., 2001.

Figure 3. Output of Canny method on sample image with PR = 7.95, F-measure = 0.0036

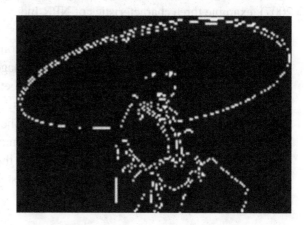

Segmentation is a frequent pre-processing step in many image understanding algorithms and practical vision systems. In an effort to compare the performance of current segmentation algorithms to human perceptual grouping as well as understand the cognitive processes that govern grouping of visual elements in images, much work has gone into gathering hand-labeled segmentations of natural images (Martin et al., 2001). Quantifying the performance of a segmentation algorithm, however, remains a challenging task. This is largely due to image segmentation being an ill-defined problem – there is no single ground truth segmentation against which the output of an algorithm may be compared. Rather the comparison is to be made against the set of all possible perceptually consistent interpretations of the image, of which only a minuscule fraction is usually available. Measures of similarity that quantify the extent to which two segmentations agree may also depend on the type and cardinality of the labels. For example, supervised segmentations of objects into semantic categories (e.g. 'Sky,' 'road,' 'grass,' etc.) must be treated differently from unsupervised clustering of pixels into groups with unordered and permutable labels (Martin et al., 2001).

Measures of Image Segmentation

In the previous sections we discussed the requirement of image dataset and a well-known Berkeley Segmentation Dataset (BSD). This section discusses distinguished measures and evaluation parameters of image segmentation, like F- Measure, Performance Ratio and Non Probabilistic Rand Index

Non Probabilistic Rand Index

Over the years unsupervised image segmentation algorithms have reached to the level, where they generate reasonable segmentations, and thus can begin to be incorporated into larger systems (Unnikrishnan et al., 2007). Numbers of algorithms for unsupervised segmentation are available and the system designer can select any of them, however, very few objective numerical evaluations exist of these segmentation algorithms. As a step towards filling the gap of algorithms and their objective evaluation (Unnikrishnan et al., 2007), presents an evaluation method which is quantitative and called as Normalized Probabilistic Rand (NPR) index which allows a principled comparison between segmentations created by different algorithms, as well as segmentations on different images. NPR Index produces correct segmentation results with a wide array of parameters on any one image, as well as correct segmentation results on multiple images with the same parameters. During the evaluation of algorithms, (Unnikrishnan et al., 2007), examine three characteristics of NPR Index:

1. **Correctness:** Ability to produce segmentations which agree with human intuition, segmentations which correctly identify structures in the image at neither too fine nor too coarse a level of detail
2. **Stability with Respect to Parameter Choice:** The ability to produce segmentations of consistent correctness for a range of parameter choices
3. **Stability with Respect to Image Choice:** The ability to produce segmentations of consistent correctness using the same parameter choice on a wide range of different images. If a segmentation scheme satisfies these three characteristics, then it will give useful and predictable results which can be reliably incorporated into a larger system (Unnikrishnan et al., 2007)

F-Measure

In statistical analysis of binary classification, the F-Measure (also F1-score or F-measure) is a measure of a test's accuracy. It considers both the precision (P) and the recall (R) of the test to compute the score, P is the number of correct positive results divided by the number of all positive results, and R is the number of correct positive results divided by the number of positive results that should have been returned.(Wikipedia, 2016). The F-Measure can be interpreted as a weighted average of the precision and recall, where an F-Measure score reaches its best value at 1 and worst at 0. The F-measure can be calculated using the Equation 1.

The traditional F-measure is the harmonic mean of precision and recall

$$F = 2 \cdot \frac{Precision \cdot Recall}{Precision + Recall} \tag{1}$$

where the precision is P is the proportion of the predicted positive cases that were correct, as calculated using the Equation 2.

$$Precision = \frac{True\ Positive}{True\ Positive + False\ Positive} \tag{2}$$

and recall is Recall or Sensitivity or True Positive Rate (TPR): It is the proportion of positive cases that were correctly identified; recall can be calculated using Equation 3.

$$Recall = \frac{True\ Positive}{True\ Positive + False\ Negative} \tag{3}$$

Performance Ratio (PR)

To objectively evaluate edge detection or boundary segmentation algorithms, (Khaire & Thakur, 2012) proposed an evaluation metric which is used to evaluate edges and boundaries, obtained by edge detector or boundary detector methods, which is generally used as a preprocessing step for image segmentation. For evaluation, the reference image, which is the ground truth of the image and obtained image using any edge or boundary algorithms are compared. Ground truth image is the reference or ideal image and pixel comparisons are being measured between ideal image and image obtained by any edge or boundary based algorithm.

For example, Let *I*, be the original image to be given to the edge or boundary extractor algorithm. Image *I*, may be a gray image or Color Image. Let *G*, be the corresponding ideal ground truth of edges or boundaries of that image. Let *X*, be the image obtained by the boundary or edge extractor. To calculate PR, comparison is done between pixels of image *G* and pixels of the image *X*. Let us consider following notions to calculate Performance Ratio

1. *E (e)* be (Correct) true edges identified by the given boundary or edge based method of image *X* with reference to ground truth *G*.
2. *N (e)* be the non edge pixels identified as edges in *X* with reference to ground truth *G*.
3. *E (n)* be the edge pixels identified as non edge pixels in *X* with reference to ground truth *G*.

Using following Notations, E*(e), N(e)* and *E(n)* to image and their pixels the Performance Ratio can be calculated as given in Equation 4.

$$PR = \frac{E\left(e\right)}{N\left(e\right) + E\left(n\right)} \times 100 \qquad (4)$$

Dataset for Video Segmentation

Video segmentation is a fundamental problem with many applications such as object recognition, action recognition, 3D reconstruction, classification, or video indexing. A good video segmentation dataset should consist of a large number of diverse sequences with the diversity spanning across different aspects. While there are standard benchmarks datasets for still image segmentation, such as the Berkeley segmentation dataset BSDS, a similar standard is missing for video segmentation. Recent influential works have introduced video datasets that specialize on sub-problems in video segmentation, such as motion segmentation, occlusion boundaries, or video super-pixels. Current video segmentation datasets are limited in some aspects as stated by (Fragkiadaki & Shi, 2012) figment, it only includes equally sized basketball players; CamVid fulfills appearance heterogeneity but only includes 4 sequences, all of them recorded from a driving car. Similarly, only include few sequences and even lacks annotation. On motion segmentation, the Hopkins155 dataset provides many sequences, but most of them show artificial checkerboard patterns and ground truth is only available for a very sparse set of points. The dataset in offers dense ground truth for 26 sequences, but the objects are mainly limited to people and cars. Moreover, the motion is notably translational. The field of motion segmentation lacks a sufficiently large and realistic benchmark dataset. There is the Hopkins 155 benchmark (Tron & Vidal, 2007), but it focuses on short sequences with little occlusion and allows evaluation only of sparse, complete trajectories. This Section describes well- known datasets used for Video Segmentation Following are some of the mostly used dataset for video segmentation some of them are discussed in detail including famous Cdnet2014 for Change Detection.

Freiburg-Berkeley Motion Segmentation Dataset (Galasso et al., 2014)

It provides an analysis based on annotations of a large video dataset, where each video is manually segmented by multiple persons. They introduce a new volume-based metric that includes the important aspect of temporal consistency, that can deal with segmentation hierarchies, and that reflects the trade-off between over-segmentation and segmentation accuracy (Galasso et al., 2014). Snapshots of motion segmentation and corresponding ground truth are shown in Figure 3 and Figure 4 respectively.

Figure 4. Snapshots of motion segmentation   *Figure 5. Corresponding ground truth of Figure 4*

CDnet 2012 Dataset

The 2012 dataset created by (Wang et al., 2014) was composed of nearly 90,000 frames in 31 video sequences representing various challenges divided into 6 categories

1. Baseline contains 4 videos with a mixture of mild challenges of the next 4 categories. These videos are fairly easy and are provided mainly as reference.
2. Dynamic Background contains 6 videos depicting outdoor scenes with strong background motion.
3. Camera Jitter represents 4 videos captured with unstable cameras.
4. Shadow is composed of 6 videos with both strong and soft moving and cast shadows.
5. Intermittent Object Motion contains 6 videos with scenarios known for causing ghosting artifacts (e.g. contains still objects that suddenly start moving).
6. Thermal is composed of five videos captured by far infrared cameras.

CDnet 2014 Dataset

CDnet 2014 Dataset is the extended dataset of CDnet 2012 for change detection was presented by (Wang et al., 2014) that consist of nearly 90 000 frames in 31 video sequences representing six categories selected to cover a wide range of challenges in two modalities (color and thermal infrared). One of the distinguishing characteristic of this benchmark video data set is that each frame is meticulously annotated by hand for ground-truth foreground, background, and shadow area boundaries an effort that goes much beyond a simple binary label denoting the presence of change. This enables objective and precise quantitative comparison and ranking of video-based change detection algorithms (Wang et al., 2014). Due to the tremendous effort required to build an inclusive benchmark dataset that supplies pixel-precision ground truth labels and provides a balanced coverage of the representative challenges, prior attempts to attain an objective evaluation of change detection methods have been confined to limited partial assessments.

This dataset contains 11 video categories with 4 to 6 videos sequences in each category (Wang et al., 2014) each individual video file (.zip or .7z. Alternatively, all videos files within one category can be downloaded as a single .zip or .7z file Each video file when uncompressed becomes a directory which contains a sub-directory named "input" containing a separate JPEG file for each frame of the input video, sub-directory named "ground truth" containing a separate BMP file for each frame of the ground truth, files named "ROI.bmp" and "ROI.jpg" showing the spatial region of interest, a file named "temporal-ROI.txt" containing two frame numbers. First version of the changedetection.net (CDnet) dataset was released in 2012. The goal was to provide a balanced dataset depicting various scenarios that arc common in change detection. For this, they prepared a dataset of 31 videos (~90, 000 manually annotated frames) categorized into 6 challenges. Since each category is associated with a specific change detection problem, e.g., dynamic background, shadows, etc., CDnet enabled an objective identification and ranking of methods that are most suitable for a specific problem as well as competent overall. In addition, online evaluation facility to help researchers compare their algorithms with the state-of-the art including evaluation tools, pixel-accurate ground-truth frames and online ranking tables are provided. (Wang et al., 2014). A second version of the dataset was created i.e. the 2014 CDnet which includes 22 additional videos in 5 new categories. In total, more than 70; 000 frames have been captured, and then manually segmented and annotated by a team of 13 researchers from 7 universities. The 2014 CDnet dataset provides realistic, camera captured diverse set of indoor and outdoor videos like the 2012 CDnet. These videos have been recorded using cameras ranging from low-resolution IP cameras, higher resolution consumer grade camcorders, and commercial PTZ cameras to near-infrared cameras. As a consequence, spatial resolutions of the videos in the 2014 CDnet vary from 320240 to 720486. Due to the diverse lighting conditions present and compression parameters used, the level of noise and compression artifacts significantly varies from one video to another. (Yi Wang et al., 2014) Duration of the videos are from 900 to 7,000 frames. Videos acquired by low-resolution IP cameras suffer from noticeable radial distortion. Different cameras have different hue bias due to different white balancing algorithms employed. Some cameras apply automatic exposure adjustment resulting in global brightness fluctuations in time. Frame rate also varies from one video to another, often as a result of limited bandwidth. Since these videos have been captured under a wide range of settings, the extended 2014 CDnet dataset does not favour a certain family of change detection methods over others (Wang et al., 2014). There are 5 new categories as, similarly to the 2012 dataset, the change detection challenge in a category is unique to that category. Such a grouping is essential for an unbiased and clear identification of the strengths and weaknesses of different methods. (Wang et al., 2014) These categories are:

1. **Challenging Weather:** This category contains 4 outdoor videos showing low-visibility winter storm conditions. This includes two traffic scenes in a blizzard, cars and pedestrians at the corner of a street and people skating in the snow. These videos present a double challenge: in addition to snow accumulation, the dark tire tracks left in the snow have potential to cause false positives (Wang et al., 2014).

2. **Low Frame-Rate:** In this category 4 videos, all recorded with IP cameras, are included. The frame rate varies from 0.17 fps to 1 fps due to limited transmission bandwidth. By nature, these videos show "erratic motion patterns" of moving objects that are hard (if not impossible) to correlate. Optical flow might be ineffective for these videos. One sequence is particularly challenging (port 0 17fps), which shows boats and people coming in and out of a harbour, as the low frame rate accentuates the wavy motion of moored boats causing false detections (Wang et al., 2014).

3. **Night:** This category has 6 motor traffic videos. The main challenge is to cope with low-visibility of vehicles yet their very strong headlights that cause over saturation. Headlights cause halos and reflections on the street (Wang et al., 2014).
4. **PTZ:** We included 4 videos in this category: one video with a slow continuous camera pan, one video with an intermittent pan, one video with a 2-position patrol-mode PTZ, and one video with zoom-in/zoom-out. The PTZ category by itself requires different type of change detection techniques in comparison to static camera videos (Wang et al., 2014).
5. **Air Turbulence:** This category contains 4 videos showing moving objects depicted by a near-infrared camera at noon during a hot summer day. Since the scene is filmed at a distance (5 to 15 km) with a telephoto lens, the heat causes constant air turbulence and distortion in frames. This results in false positives. The size of the moving objects also varies significantly from one video to another. The air turbulence category presents very similar challenges to those arising in long-distance remote surveillance applications (Wang et al., 2014).

Some of the Datasets fairly discussed by (Goyette et al., 2014) includes Wallflower: A very well-known dataset for video segmentation that is used today is Wallflower Dataset. It contains 7 short video clips, each representing a specific challenge such as illumination change, background motion, etc. The limitations of these dataset is that only one frame per video has been labeled. PETS: The Performance Evaluation of Tracking and Surveillance (PETS) program was launched with the goal of evaluating visual tracking and surveillance algorithms. The program has been collecting videos for the scientific community since the year 2000 and now contains several dozen videos. Many of these videos have been manually labeled by bounding boxes with the goal of evaluating the performance of tracking algorithms.

CAVIAR

CAVIAR dataset has videos representing more than all behavior of human like browsing, walking, shopping etc. It contains more than 80 staged indoor videos of them associated with a bounding box (Goyette et al., 2014) i-LIDS: This dataset contains 4 scenarios (parked vehicle, abandoned object, people walking in a restricted area, doorway). Due to the size of the videos (more than 24 hours of footage) the videos are not fully labeled (Goyette et al., 2014)

BEHAVE 2007

It contains videos with human interactions such as walking in a group, meeting with each other etc. Ground truth also consists of bounding boxes which surrounds moving object. This dataset contain 7 real videos shot by the same camera. VSSN 2006, videos in this dataset contain animated background, illumination changes and shadows, this dataset contains 9 semi-synthetic videos composed of a real background and artificially moving objects. (Goyette et al., 2014).

IBM

This dataset videos in which frame is labeled with a bounding box around each foreground moving object. It contains 15 indoor and outdoor videos taken from PETS 2001 plus additional videos.

Karlsruhe

This dataset contains 4 grayscale videos from the Institut für Algorithmen und Kognitive Systeme. These videos show traffic scenes under various weather conditions. The author's ground-truth labeled 10 frames for each video. (Goyette et al., 2014).

Measures of Video Segmentation

The evaluation metric like ROC Curve, NPRI and Precision- Recall is most popular in the BSDS benchmark for image segmentation. It casts the boundary detection problem as one of classifying boundary from non-boundary pixels and measures the quality of a segmentation boundary in the precision-recall framework. In this section evaluation metrics for video segmentation, using ground truth and without ground truth are discussed including metrics like F- measure for Change Detection, Boundary Precision-Recall and Volume Precision-Recall

Performance Measures with Ground Truth

In previous section we discuss necessity of objective evaluation parameters to evaluate video segmentation, in this section we talk about well-known objective evaluation parameters which uses ground truth for evaluation of video segmentation methods

Boundary Precision: Recall

The metric is of limited use in a video segmentation benchmark, as it evaluates every frame independently, i.e., temporal consistency of the segmentation does not play a role. Moreover, good boundaries are only half the way to a good segmentation, as it is still hard to obtain closed object regions from a boundary map (Galasso et al., 2014).

Volume Precision: Recall

VPR optimally assigns spatial-temporal volumes between the computer generated segmentation and the human annotated segmentations (Galasso et al., 2014).

   Properties of Boundary Precision Recall and Volume Precision Recall: Boundary Precision Recall (BPR) and Volume Precision Recall (VPR) satisfy some important requirements of evaluation metrics for video segmentation (Galasso et al., 2014).

1.   **Non-Degeneracy:** Measures are low for degenerate segmentations.
2.   **No Assumption about Data Generation:** The metrics do not assume a certain number of labels and apply therefore to cases where the computed number of labels is different from the ground truth.
3.   **Multiple Human Annotations:** Inconsistency among humans to decide on the number of labels is integrated into the metrics, which provides a sample of the acceptable variability.

4.   **Adaptive Accommodation of Refinement:** Segmentation outputs addressing different coarse-to-fine granularity are not penalized, especially if the refinement is reflected in the human annotations, but granularity levels closer to human annotations score higher than the respective over- and under-segmentations; this property draws directly from the humans, who psychologically perceive the same scenes to a different level of detail.

Performance Measures without Ground Truth

Following are the metrics used for evaluation discussed by (Murat Tekalp et al., 2004)

1.   **Combining Color and Motion Metrics:** A single numerical measure can be obtained to evaluate the performance of spatio-temporal segmentation of a video object by combining the color and motion metrics. (Murat Tekalp et al., 2004)
2.   **Temporal Localization:** The temporal performance localization can be achieved by checking, per frame, color, and motion measures, as a function of time.
3.   **Spatial Localization:** We can further identify incorrectly tracked boundary portions within any frame whose score is above the threshold, using only the color and motion measures. (Murat Tekalp et al., 2004)

For Change Detection, finding the right metric to accurately measure the ability of a method to detect motion or change without producing excessive false positives and false negatives is not trivial. For instance, recall favors methods with a low False Negative Rate. On the contrary, specificity favors methods with a low False Positive Rate (Goyette et al., 2014). Having the entire precision-recall tradeoff curve or the ROC curve would be ideal, but not all methods have the flexibility to sweep through the complete gamut of tradeoffs. In addition, one cannot, in general, rank-order methods based on a curve (Goyette et al., 2014).

Solutions and Recommendations

The main purpose of presenting this work is to help researchers in this field to obtain implementations of other researchers by using available datasets, ground truth and quality metrics for better understanding the future and the limitations of research area. While a considerable number of segmentation schemes have been devised towards this end no comprehensive attempt has been made to compare the various schemes available. Such a comparison is necessary in order to establish the performance limitations of each scheme under the various conditions encountered. Results are often displayed with very few images and without objective evaluation. It is very difficult for a researcher to evaluate and compare his algorithm or method, given the results with subjective evaluation. The aim is to encourage researchers to compare and evaluate their image and video segmentation techniques on renowned dataset using quality metrics for advancement of research in Image and Video Segmentation

FUTURE RESEARCH DIRECTIONS

In recent days the evolving of dataset and quality metrics for Image and Video segmentation makes a quality impact in the advancement of this topic. In future the datasets and measures will be required for sub-methods of image and videos segmentation Like Edge Detection and Change Detection. A more work is left in the field of Semantic segmentation, an appropriate dataset and evaluation measures will rapidly advance the research in the field of Semantic Segmentation

CONCLUSION

The main purpose of presenting this work, so that, researchers in this field, should make an effort to obtain implementations of other researchers and use available datasets, ground truth and quality measures, results in better understanding of scope and limitations of image and video segmentation. Idea of this chapter is to present the most well-known image and video segmentation dataset and measures for the readers and to encourage researchers to compare and evaluate their image and video segmentation techniques on renowned dataset using quality metrics for advancement of research in Image and Video Segmentation

REFERENCES

Alshennawy, A. A., & Aly, A. A. (2009). *Edge Detection in Digital Images Using Fuzzy Logic Technique.* World Academy of Science, Engineering and Technology.

Bashir, F., & Porikli, F. (2006). Performance evaluation of object detection and tracking systems.*Proc. IEEE Int. Workshop on Performance Evaluation of Tracking Systems.*

Bhardwaj, S., & Mittal, A. (2012). A Survey on Various Edge Detector Techniques. *Procedia Technology, 4,* 220–226. doi:10.1016/j.protcy.2012.05.033

Biswas, R., & Sil, J. (2012). An Improved Canny Edge Detection Algorithm Based on Type-2 Fuzzy Sets. *Procedia Technology, 4,* 820–824. doi:10.1016/j.protcy.2012.05.134

Bouwmans, T. (2011). Recent advanced statistical background modeling for foreground detection: A systematic survey. *Recent Patents on Computer Science, 4*(3).

Bouwmans, T., Porikli, F., Hferlin, B., & Vacavant, A. (2014). *Background Modeling and Foreground Detection for Video Surveillance.* Chapman and Hall/CRC. doi:10.1201/b17223

Bovik, A. C. (2000). *Handbook of Image and Video Processing.* Academic Press.

Bowyer, K., Kranenburg, C., & Dougherty, S. (2001). Edge detector evaluation using empirical ROC curves. *Journal on Computer Vision and Understanding, 84*(1), 77–103. doi:10.1006/cviu.2001.0931

Brutzer, S., Hoferlin, B., & Heideman, G. (2011). Evaluation of background subtraction techniques for video surveillance.*Proc. IEEE Conf. Computer Vision Pat. Recog.* (pp. 1937–1944). doi:10.1109/CVPR.2011.5995508

Canny, J. (1986). A computational approach to edge detection. *IEEE Transactions on Pattern Analysis and Machine Intelligence, 6*(6), 679–698. doi:10.1109/TPAMI.1986.4767851 PMID:21869365

Erdem, C. E., Sankur, B., & Tekalp, A. M. (2004). Performance Measures for Video Object Segmentation and Tracking. *IEEE Transactions on Image Processing, 13*(7), 937–951. doi:10.1109/TIP.2004.828427 PMID:15648860

Fawcett, T. (2006). An Introduction to ROC Analysis. *Pattern Recognition Letters, 27*(8), 861–874. doi:10.1016/j.patrec.2005.10.010

Fragkiadaki, K., & Shi, J. (2012). Video segmentation by tracing discontinuities in a trajectory embedding.*Proceedings of the 2012 IEEE Conference on Computer Vision and Pattern Recognition* (pp. 1946-1853). doi:10.1109/CVPR.2012.6247883

Fram, J. R., & Deutsch, E. S. (1975). On the Quantitative Evaluation of Edge Detection Schemes and Their Comparison with Human Performance. *IEEE Transactions on Computers, 100*(6), 616–628. doi:10.1109/T-C.1975.224274

Galasso, F., Shankar Nagaraja, N., Jimenez Cardenas, T., Brox, T., & Schiele, B. (2013). A unified video segmentation benchmark: Annotation, metrics and analysis.*Proceedings of the IEEE International Conference on Computer Vision* (pp. 3527-3534). doi:10.1109/ICCV.2013.438

Gao, W., Zhang, X., Yang, L., & Liu, H. (2010) An Improved Sobel Edge Detection.*Proceedings of the 2010 3rd IEEE International Conference on Computer Science and Information Technology (ICCSIT)* (Vol. 5, pp. 67-71).

Gonzalez, R. C., & Woods, R. E. (2002). *Digital image processing* (2nd ed.). Prentice Hall.

Goyette, N., Jodoin, P.-M., Porikli, F., Konrad, J., & Ishwar, P. (2012). Changedetection.net: A new change detection benchmark dataset.*Proceedings of the IEEE CVPR change detection workshop*. doi:10.1109/CVPRW.2012.6238919

Goyette, N., Jodoin, P.-M., Porikli, F., Konrad, J., & Ishwar, P. (2014). A Novel Video Dataset for Change Detection Benchmarking. *IEEE Transactions on Image Processing, 23*(11), 4663–4679. doi:10.1109/TIP.2014.2346013 PMID:25122568

Haralick, R. M., & Shapiro, L. G. (1985). Image segmentation techniques. *CVGIP, 29*, 100–132.

Heath, M., Sarkar, S., Sanocki, T., & Bowyer, K. (1998). Comparison of Edge Detectors. *Computer Vision and Image Understanding, 69*(1), 38–54. doi:10.1006/cviu.1997.0587

Hoover, A., Jean-Baptiste, G., Jiang, X., Flynn, P. J., Bunke, H., Goldgof, D. B., & Fisher, R. B. et al. (1996). An experimental comparison of range image segmentation algorithms. *IEEE Transactions on Pattern Analysis and Machine Intelligence, 18*(7), 673–689. doi:10.1109/34.506791

Jain, A. K. (1989). *Fundamentals of Digital Image Processing*. Prentice-Hall, Inc.

Jain, R., & Binford, T. (1991). Ignorance, myopia, and naivete in computer vision systems. *CVGIP. Image Understanding, 53*(1), 112–117. doi:10.1016/1049-9660(91)90009-E

Khaire, P. A., & Thakur, N. V. (2012). A Fuzzy Set Approach for Edge Detection. *International Journal of Image Processing*, *6*(6), 403–412.

Khaire, P. A., & Thakur, N. V. (2012). An Overview of Image Segmentation Algorithms. *International Journal of Image Processing and Vision Sciences*, *1*(2), 62–68.

Koprinska, I., & Carrato, S. (2001). Temporal video segmentation: A survey. *Signal Processing Image Communication*, *16*(5), 477–500. doi:10.1016/S0923-5965(00)00011-4

Li, H., & Ngan, K. N. (2007). Automatic video segmentation and tracking for content-based applications. *IEEE Communications Magazine*, *45*(1), 27–33. doi:10.1109/MCOM.2007.284535

Li, L., Huang, W., Gu, I. Y. H., Leman, K., & Tian, Q. (2003, October 5-8). Principal color representation for tracking persons.*Proceedings of the IEEE International Conference on Systems, Man and Cybernetics*, (pp. 1007-1012).

Martin, D., Fowlkes, C., Tal, D., & Malik, J. (2001). A database of human segmented natural images and its application to evaluating segmentation algorithms and measuring ecological statistics.*Proceedings of the 8th International Conference Computer Vision* (pp. 416-423). doi:10.1109/ICCV.2001.937655

Meier, T., & Ngan, K. N. (1998). Automatic segmentation of moving objects for video object plane generation. *IEEE Transactions on Circuits and Systems for Video Technology*, *8*(5), 525–538. doi:10.1109/76.718500

Mendhurwar, K., Patil, S., Sundani, H., Aggarwal, P., & Devabhaktuni, V. (2011). Edge-Detection in Noisy Images Using Independent Component Analysis. *ISRN Signal Processing*.

Nascimento, J., & Marques, J. (2006). Performance evaluation of object detection algorithms for video surveillance. *IEEE Transactions on Multimedia*, *8*(8), 761–774. doi:10.1109/TMM.2006.876287

Ou, Y., & GuangZhi, D. (2001). Color Edge Detection Based on Data Fusion Technology in Presence of Gaussian Noise. *Procedia Engineering*, *15*, 2439–2443. doi:10.1016/j.proeng.2011.08.458

Pal, N. R., & Pal, S. K. (1993). A review on image segmentation techniques. *Pattern Recognition*, *26*(9), 1277–1294. doi:10.1016/0031-3203(93)90135-J

Pan, J., Li, S., & Zhang, Y. (2000). Automatic extraction of moving objects using multiple features and multiple frames.*Proceedings of IEEE Symposium Circuits and Systems*, (Vol. 1, pp. 36-39).

Papari, G., & Petkov, N. (2011). Edge and line oriented contour detection: State of the art. In Image and Vision Computing (pp. 79–103).

Patel, J., Patwardhan, J., Sankhe, K., & Kumbhare, R. (2011). Fuzzy Inference based Edge Detection System using Sobel and Laplacian of Gaussian Operators.*Proceedings of the International Conference & Workshop on Emerging Trends in Technology ICWET'11* (pp. 694-697). doi:10.1145/1980022.1980171

Pauwels, E. J., & Frederix, G. (1999). Finding salient regions in images-non-parametric clustering for image segmentation and grouping. *Computer Vision and Image Understanding*, *75*(1-2), 73–85. doi:10.1006/cviu.1999.0763

Piccardi, M. (2004). *Background subtraction techniques: a review*. Academic Press.

Pratt, W. K. (2007). *Digital Image Processing* (4th ed.). Wiley & Sons, Inc. doi:10.1002/0470097434

Price, K. (1986). Anything You Can Do, I Can Do Better (No You Can't). *Computer Vision Graphics and Image Processing, 36*(2-3), 387–391. doi:10.1016/0734-189X(86)90083-6

Radke, R., Andra, S., Al-Kofahi, O., & Roysam, B. (2005). Image change detection algorithms: A systematic survey. *IEEE Transactions on Image Processing, 14*(3), 294–307. doi:10.1109/TIP.2004.838698 PMID:15762326

Sekhar Panda, C., & Patnaik, S. (2010). Filtering Corrupted Image and Edge Detection in Restored Grayscale Image Using Derivative Filters. *International Journal of Image Processing, 3*(3), 105–119.

Setayesh, M., Zhang, M., & Johnston, M. (2011). Detection of Continuous, Smooth and Thin Edges in Noisy Images Using Constrained Particle Swarm Optimization.*Proceedings of the 13th annual conference on Genetic and evolutionary computation* (pp. 45-52).

Shi, J., & Malik, J. (2000). Normalized cuts and image segmentation. *IEEE Transactions on Pattern Analysis and Machine Intelligence, 22*(8), 888–905. doi:10.1109/34.868688

Sun, S., Haynor, D. R., & Kim, Y. (2003). Semiautomatic video object segmentation using Vsnakes. *IEEE Transactions on Circuits and Systems for Video Technology, 13*(1), 75–82. doi:10.1109/TC-SVT.2002.808089

Tekalp, A. M. (1995). *Digital Video Processing*. Prentice-Hall, Inc.

Thakkar, M., & Shah, H. (2010). *Automatic thresholding in edge detection using fuzzy approach*. IEEE. doi:10.1109/ICCIC.2010.5705868

The Berkeley Segmentation Dataset and Benchmark. (n.d.). Retrieved from https://www.eecs.berkeley.edu/Research/Projects/CS/vision/bsds/

Toklu, C., Tekalp, M., & Erdem, A. T. (2000). Semi-automatic video object segmentation in the presence of occlusion. *IEEE Transactions on Circuits and Systems for Video Technology, 10*(4), 624–629. doi:10.1109/76.845008

Toyama, K., Krumm, J., Brumitt, B., & Wallflower, B. M. (1999). Principles and practice of background maintenance.*Proc. IEEE Int. Conf. Computer Vision* (Vol. 1, pp. 255–261).

Tron, R., & Vidal, R. (2007). A benchmark for the comparison of 3-D motion segmentation algorithms. *Proceedings of the 2007 IEEE Conference on Computer Vision and Pattern Recognition*. doi:10.1109/CVPR.2007.382974

Unnikrishnan, R., Pantofaru, C., & Hebert, M. (2007). Toward Objective Evaluation of Image Segmentation Algorithms. *IEEE Transactions on Pattern Analysis and Machine Intelligence, 29*(6), 929–944. doi:10.1109/TPAMI.2007.1046 PMID:17431294

Vezzani, R., & Cucchiara, R. (2010). Video surveillance online repository (visor): An integrated framework. *Multimedia Tools and Applications, 50*(2), 359–380. doi:10.1007/s11042-009-0402-9

Wang, J., Thiesson, B., Xu, Y., & Cohen, M. (2004). Image and video segmentation by anisotropic kernel mean shift.*Proceedings of the European Conference of Computer Vision (ECCV '04)*, (LNCS), (Vol. 3022, pp. 238-249). doi:10.1007/978-3-540-24671-8_19

Wang, Y., Jodoin, P. M., Porikli, F., Konrad, J., Benezeth, Y., & Ishwar, P. (2014). CDnet 2014: An Expanded Change Detection Benchmark Dataset.*Proceedings of the IEEE Conference on Computer Vision and Pattern Recognition Workshops* (pp. 387-394). doi:10.1109/CVPRW.2014.126

Wikipedia. (n.d.). *F1 Score*. Retrieved from https://en.wikipedia.org/wiki/F1_score

Yang, G., & Xu, F. (2011). Research and analysis of Image edge detection algorithm Based on the MATLAB. *Procedia Engineering*, *15*, 1313–1318. doi:10.1016/j.proeng.2011.08.243

Young, D., & Ferryman, J. (2005). PETS metrics: Online performance evaluation service.*Proc. IEEE Int. Workshop on Performance Evaluation of Tracking Systems* (pp. 317–324).

Zhang, X., Zhang, Y., & Zheng, R. (2011). Image edge detection method of combining wavelet lift with Canny operator. *Procedia Engineering*, *15*, 1335–1339. doi:10.1016/j.proeng.2011.08.247

Zhang, Y.-J. (2006).*Advances in Image and Video Segmentation*. Hershey, PA: IGI Global. doi:10.4018/978-1-59140-753-9

ADDITIONAL READING

Akutsu, A., & Tonomura, Y. (1994, October 15-20). Video tomography: an efficient method for camerawork extraction and motion analysis. *Presented at ACM Multimedia '94*, San Francisco, CA. doi:10.1145/192593.192697

Arbelaez, P., Maire, M., Fowlkes, C., & Malik, J. (2011). Contour detection and hierarchical image segmentation.*IEEE Transactions on Pattern Analysis and Machine Intelligence*, *33*(5), 898–916. doi:10.1109/TPAMI.2010.161 PMID:20733228

Ren, X., Fowlkes, C. C., & Malik, J. (2005, October). Scale-invariant contour completion using conditional random fields.*Proceedings of the Tenth IEEE International Conference on Computer Vision ICCV'05* (Vol. 2, pp. 1214-1221). IEEE.

Smith, M. A., & Kanade, T. (1997). Video skimming and characterization through the combination of image and language understanding techniques. *Presented at the Computer Vision and Pattern Recognition,* San Juan, Puerto Rico. doi:10.1109/CVPR.1997.609414

KEY TERMS AND DEFINITIONS

Edge Detection: The process of finding the boundaries and edges in an image.

Ground Truth: Image or video with ideal segmentation characteristics (like edges, regions, colors, etc.) provided in correspondence with the image or video to be segmented.

Image Dataset: Collection of different images, for evaluating methods of segmentation, etc.

Image Segmentation: Process of segmenting image into parts or regions.

Performance Measures: The metrics and parameters required for evaluation of segmentation methods.

Performance Ratio: A ratio of true to false edges detected to evaluate image segmentation using ground truth.

Video Dataset: Collection of different videos, for evaluating methods of video segmentation, etc.

Video Segmentation: Segmenting video frames into corresponding objects.

Chapter 3
Automated System for Crops Recognition and Classification

Alaa M. AlShahrani
Taif University, Saudi Arabia

Areej S. Al-Malki
Taif University, Saudi Arabia

Manal A. Al-Abadi
Taif University, Saudi Arabia

Amira S. Ashour
Taif University, Saudi Arabia & Tanta University, Egypt

Nilanjan Dey
Techno India College of Technology, India

ABSTRACT

Marketing profit optimization and preventing the crops' infections are a critical issue. This requires crops recognition and classification based on their characteristics and different features. The current work proposed a recognition/classification system that applied to differentiate between fresh (healthy) from rotten crops as well as to identify each crop from the other based on their common feature vectors. Consequently, image processing is employed to perform the statistical measurements of each crop. ImageJ software was employed to analyze the desired crops to extract their features. These extracted features are used for further crops recognition and classification using the Least Mean Square Error (LMSE) algorithm in Matlab. Another classification method based on Bag of Features (BoF) technique is employed to classify crops into classes, namely healthy and rotten. The experimental results are applied of databases for orange, mango, tomato and potatoes. The achieved recognition (classification) rate by using the LMSE for all datasets (healthy and rotten) has 100%. However, after adding 10%, 20%, and 30% Gaussian noise, the obtained the average recognition rates were 85%, 70%, and 25%; respectively. Moreover, the classification (healthy and rotten) using BoF achieved accuracies of 100%, 88%, 94%, and 75% for potatoes, mango, orange, and tomato; respectively. Furthermore, the classification for all the healthy datasets achieved accuracy of 88%.

DOI: 10.4018/978-1-5225-1022-2.ch003

INTRODUCTION

Agricultural crops such as cotton, fruits and vegetables are important raw materials for many industries. However, these crops can be damaged by insects, soil pollution, weather and other many factors. Insects and diseases cause vegetables and fruit devastating problem, which results in economic losses and affect the production in agricultural industry worldwide. Now days, with growing the population, it is required to categorize and to distinguish good and bad crops. Moreover, the agriculture sector is interested with several aspects, such as product's quality and the crops status measurements based on the farmers' point of view. However, the availability of experts and their services may consume time and efforts. Previously, the use of traditional methods led to waste time and to inaccurate results. Thus, to overcome these drawbacks automated image processing techniques are conducted to analyze the crops and to classify them into healthy and rotten classes in a shorter time, with less effort and more accurately. In addition, image processing along with communication network availability facilities the communication with experts within less time consuming and at affordable cost compared to the traditional methods.

Currently, computer vision systems are used for automatic inspection of vegetables/fruits based on various methods that are equipped to detect defects on the crops' surface. Recognition system is a 'grand challenge' for the computer vision to achieve near human levels of recognition. The fruits and vegetables classification is useful in the super markets to classify them, and then assign prices automatically. Fruits and vegetables classification can also be used in computer vision for the automatic sorting of fruits from a set, consisting of different kinds of fruit (Song *et al.*, 2014).

Consequently, the current work proposed an automated approach to analyze, recognize, and classify crops including fruits and vegetables by using image processing techniques. The proposed approach is used to measure the characteristics of the crop under concern as well as to identify the rotten parts based on the crop's image using digital camera. The extracted features such as the area, standard deviation, and the perimeter are then used in the identification and classification of the crops. The datasets used in the proposed work were manually obtained by capture images for different crops (fruits/vegetables), including tomato, potatoes, orange, and mango. The recognition/classification results using the LMSE approach is compared to the classification using the BoF model.

RELATED WORK

Researchers are interested with applying various image processing techniques in the agriculture applications due to its accuracy. Various studies were conducted that use image processing to analyze one type of crops by extracting certain feature such as the color, size or texture. Jiménez *et al.* (1999) introduced a review of various vision systems to recognize fruits for automated harvesting. The survey concluded that under difficult conditions there is a feasibility of practical implementations of computer vision systems for the analysis of agricultural scenes to locate natural objects. The authors discussed the basic considerations about the distributions and characteristics of the fruits in natural orange crops.

Zhao *et al.* (2005) introduced a vision based algorithm to locate apples in a single image as a prelude to the use of stereo vision to correctly locate apples in an orchard. The authors considered on-tree situations of contrasting red and green apples as well as green apples in the orchard with poor contrast. The results depicted that redness in both cases of red and green apples can be used to distinguish apples from the rest of the orchard. In addition, texture based edge detection measurements were conducted as well as area thresholding followed by circle fitting in order to determine the apples' location in the image plane. The Laplacian filters were used for further cluttering the foliage arrays by edge enhancement. Thus, texture differences between the foliage and the apples increased thereby facilitating the separation of apples from the foliage. The experimental results established that the recognition of red and green apples in different situations as well as apple that are clustered together and/or occluded. Since, advances in computer technology give the facility to create databases of large number of images. Thus, achieving Content-based image retrieval (CBIR) become an interesting research field, which is known as the use of low-level visual features of the image data to segment, index and retrieve relevant images from the image database. Zhongzhi *et al.* (2012) presented multiple variable analysis approach for skin defects in citrus fruits. Song *et al.* (2014) proposed a method for fruits recognition and counting from images in cluttered greenhouses. The authors tested fruits of complex shapes and varying colors similar to the plant canopy. A new two-step method to locate and count pepper fruits by finding fruits in a single image using a bag-of-words model, and the then aggregating estimates from multiple images using a novel statistical approach to cluster repeated incomplete observations. Ninawe, and Pandey (2014) suggested new fruits recognition approach that combined four features using analysis methods. The extracted features were shape, size, color and texture to increase the recognition accuracy. The proposed approach used the nearest neighbor classification algorithm to classify and recognize the fruit images from the nearest training fruits. Agrawal *et al.* (2014) introduced image features that can be integrated with the fruit disease recognition.

Classification and sorting of the vegetables'/fruits' quality are commonly related to size, mass, shape, firmness, bruises and color. However, high cost of the equipment, software as well as operational costs, obstacle the implementation producers used to assess the required quality. Lino et al. (2008) evaluated new open software which classifies and recognizes the fruit shape, border detection, color, volume and possibly bruises at a unique glance using ImageJ. Blasco *et al.* (2009) developed a system to recognize and classify external defects in citrus. Images of the defects were obtained in five spectral areas to discriminate between 11 types of defects such as ultraviolet induced fluorescence and the near infrared reflectance. Morphological estimations for the defects are measured to classify the fruits in categories. This proposed fruit-sorting algorithm was tested to identify defects in about 2000 citrus fruits, including oranges and mandarins. The results established overall classification success rate of 86%.

Color, texture, flavor and the nutritional value of fresh-cut fruits/vegetables are critical factors from the consumers' point of view. Thus, Barrett *et al.* (2010) introduced the attributes of the desirable and undesirable quality of fresh-cut fruit and vegetable products. The authors described the advantages and disadvantages of the instrumental quality measurements. A review of typical unit operations occupied in the fresh-cut products production was presented. The effects of fresh-cut processing techniques including the texture, appearance and flavor of vegetables, and fruits were detailed. Generally, several fruit images' recognition is considered a major challenge for researchers. Typically, fruit recognition techniques combine different analysis methods, such as color-, shape-, size- and texture- based. Different fruit images' color/shape values are same, thus they are not robust and effective to recognize and identify the images.

Mente *et al.* (2014) used the K-means algorithm for segmentation of color images, afterwards shape descriptor eccentricity and color features for achieving effective and efficient retrieval performance. This approach was applied to an image database containing 2600 fruit images.

Diseases in the different fruit kinds lead to economic and production losses in the agricultural industry worldwide. Dubey,& Jalal (2014) proposed an approach for fruit diseases classification using image processing based proposed approach, which composed of three main steps. The first step, involved the

K-Means clustering technique for the defect segmentation. Then, in the second step from the segmented image, color and textural cues were extracted and fused. Finally, images were classified into one of the classes by using a Multi-class Support Vector Machine (SVM). The authors considered apples' diseases as a test case and evaluated the proposed approach for three types of apple diseases, namely apple blotch, apple scab, apple rot and normal apples without diseases. The results established that the proposed fusion scheme can significantly support accurate detection and automatic classification of fruit diseases.

Anthimopoulos *et al.* (2014) suggested a technique for automatic food recognition using the bag-of-features (BoF) model. A widespread investigation was done for identifying and optimizing the best performing components in the BoF architecture. Moreover, an estimation of the corresponding parameters was performed. A visual dataset of about 5000 food images was created and categorized into 11 classes. The scale-invariant feature transform on the HSV color space was computed to build a visual dictionary of 10000 visual words based on the k-means clustering. Finally, the linear support vector machine classifier was employed to classify the food images. The experiential results proved 78% classification accuracy.

From the preceding survey, it is concluded that image processing has several techniques that attracted the focus of various researchers to develop for agriculture applications. In addition, several studies were conducted in the agriculture domain on only some kind of crops, such as tomato and lemon in order to recognize them, while other related work were extracted some features such as shape, color, size and texture in order to recognize the fruit and/or vegetable. Moreover, other applications were focused on demonstrating the infected regions by using various methods, such as K-Means clustering, fuzzy C-Means, artificial neural network and other techniques. Consequently, the proposed work is concerned with building an integrated system to recognize/classify different types of fruits and vegetables by their features such as area, perimeter, standard deviation, mean and shape descriptors in order to determine whether they are healthy or rotten. Moreover, classification process is conducted based on the BoF model.

METHODOLOGY

The datasets in this work were captured by a digital camera. It is consisting of four different crops (mango, tomato, orange and potato) each crop has its healthy and rotten dataset which consists of 15 images per each. The selection process of the previous types is based on their similar sizes and shapes to be approximately rounded. Therefore, the recognition process is going to be more difficult and such a challenge for our integrated system to classify and recognize each type from the other. Some samples from each category in the used dataset are shown in Figure 1.

This datasets were used with the proposed system. The proposed system in the current work included two methods, namely the recognition/classification method using the least Mean Square Error (LMSE), and another classification method using Bag of Feature (BoF) model. These two methods are implemented using the Matlab software. Prior to the any of the classification methods, the captured images are analyzed capture using ImageJ software. A detailed description of each phase is represents as follows.

First Method: Recognition/Classification Using Image Analysis

Image analysis involves extract the fundamental components that represent the image in order to extract statistical data. It includes several tasks, such as detecting edges, finding shapes, removing noise, counting objects, and measuring region and image properties of an object. Generally, image analysis is

Figure 1. Used dataset: (a), (c), (e) and (g) are samples from healthy fruit/vegetables and (b), (d), (f) and (h) are samples from rotten fruit/vegetables.

an extensive term that covers a range of techniques to perform several tasks including region analysis to extract statistical data. In the present work, ImageJ is a powerful image analysis software program which used to facilitate the required analysis to recognize fruits/vegetables. The block diagram of the proposed steps for crops' images analysis using ImageJ is demonstrated in Figure 2.

The steps used for the image analysis in the proposed system are explained as follows:

Median Filter

Typically, there is several numbers of filters that used in the pre-processing for any image enhancement. Such filters are the Gaussian filter, maximum filter, and the median filter. Median filter is a way to remove noise that results during the capturing image by light and shade. Thus, the median filter assists

Figure 2. Block diagram of the proposed steps for crops' images analysis using ImageJ

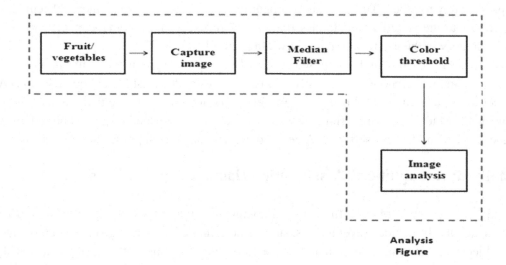

the correct edges determination. Generally, the median filter is widely used in digital image processing as it preserves image edges during the noise removal. Also, it is considered a non-linear local filter. The median filter filters each pixel in the image and used its nearby neighbors are to decide if this pixel is representative of its surroundings of not (Motwani *et al.*, 2004). Compared to other smoothing filters, the median filter offers the following advantages:

1.   Does not reduce the contrast across steps,
2.   Does not shift boundaries, and
3.   Is less sensitive to the extreme values. Therefore, in the proposed work, the median is employed to extract edges as it concerns with color intensity, thus it is appropriated in the present study.

Color Threshold

In the present work, in order to remove parts of the image that fall within a specific color range, the Color Threshold module is used. It is used to detect objects of reliable color values. The histogram display is based on the selected color threshold as it refers to the pixel intensity values (Sural et al., 2002). This histogram is a graphical representation that show the pixels number in an image at each different intensity value. Therefore, in the proposed work, the color threshold is employed to determine regions of interest within the image.

Feature Extraction

Feature extraction (Hong, 1991) is efficiently representing the significant parts of an image for dimensionality reduction by generating feature vector. The extracted features using the ImageJ for the image under analysis included the following extracted features: area, perimeter, mean gray value, standard deviation, IntDen (product of area and mean gray value), and the median.

Recognition/Classification Using Least Mean Square Error (LMSE)

Image recognition is the process of identifying and detecting an object in an image. Typical, image recognition algorithms, include optical character recognition, face recognition and pattern and gradient matching. The LMSE is used for the recognition and thus classification of the corps based on the steps shown in Figure 3.

Figure 3. Block diagram of the proposed steps for crops' images recognition/classification using Matlab

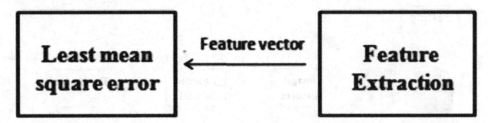

After, the selection of the statistical features, which extracted in the image analysis step, the LMSE is applied to recognize the fruits/vegetables and their status (healthy/rotten). The LMSE algorithm (Liu *et al.*, 2008) is very widely used because of its low computational complexity. It is an estimation method that minimizes the mean square error (MSE). Generally, the LMSE filter for recognition problems is designed to locate the target by producing a delta function output at the target position. The filter minimizes the MSE of the difference between the output and the input to the filter.

Second Method: Classification Using Bag of Features (BoF)

Image classification analyzes the numerical properties of various image features to classify the data into categories. Classification algorithms have two phases: *training* and *testing*. In supervised classification, *statistical* processes or *distribution-free* processes can be used to extract class descriptors. Unsupervised classification relies on *clustering* algorithms to automatically segment the training data into prototype classes. Some of supervised methods used are Naïve Bayes classifier, J48 Decision Trees and Support Vector Machines, Bag of Feature, whereas the unsupervised method is an adaptation of the K-means clustering method (Sonka *et al.*, 2014). In this phase, the fruit/vegetables classification is performed to classify rotten fruit/vegetable from healthy using BoF as shown in Figure 4.

The BoF is considered a simple and accurate classifier. For features detection, good descriptors that handle intensity, rotation, scale and affine variations are used. Then, the k-means clustering is used to reduce the number of features that used in the classification process. The feature vector of an image

Figure 4. The block diagram of the classification using Bag of Feature

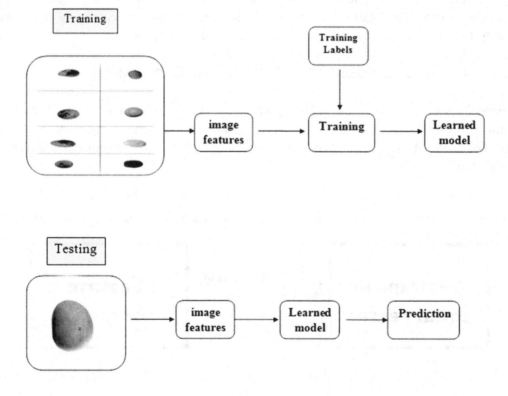

represents the histogram of visual word occurrences contained in the image. This histogram considered a basis for training the classifier, as it encoded the image into a feature vector. For classification purpose each image set is divided into 20% of the images (2 images) are used for feature extraction and training, 80% for validation and testing (8 images) in each image set.

RESULTS AND DISCUSSION

The results of the previous two methods: the recognition/classification using the LMSE and the classification using BoF were implemented by using Matlab. In order to prepare the inputs of the dataset, ImageJ software which is an effective tool in image processing domain was used to perform the analysis part and the obtained results (feature vectors) from the analysis part was used for the recognition/classification using the LMSE. The resultant feature vector of each type was selected in this system based on the average of the whole measurements of a particular crop including its healthy and rotten measurements.

First Method: Recognition/Classification Using Least Mean Square Error (LMSE)

The steps used to perform the analysis using imageJ are shown with both types (healthy and rotten) for a mango as a sample from the dataset as follows.

Analyzing a Sample Healthy Mango (Without Noise)

The image analysis steps for healthy mango are illustrated in Figure 5, where (A) Healthy mango sample image and applying median filter (B) After changing the brightness and saturation (C) After selecting the whole area.

In Figure 5 (A), the healthy mango image is captured and is enhanced by reducing the noise through the Median Filter. This filter preserves the edges while removing the noise and provides better de-noising compared to other filters that have been tried. In Figure 5 (B), the threshold of the colored image is adjusted to select the region of the whole fruit by changing the brightness and Saturation level. In Figure 5 (C), the border of the whole fruit is selected/ is bounded. Finally, analysis measurements of the whole fruit are calculated.

Figure 5. Image analysis for healthy mango without noise

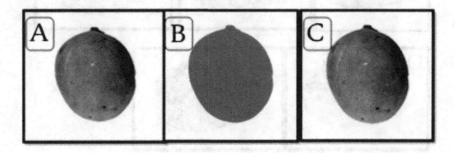

Afterward, for recognition/classification, choose only the feature vector of the desired crop. For example, the mango's feature vector was selected as shown in Table 1 to be [Area, Min, Max, Perimeter, BX, BY, Angle, Circularity, Integrated Density, Skewness, Feret x, Feret y, Feret Angle] to recognize healthy and rotten mangoes.

Analyzing a Sample Rotten Mango (without Noise)

Figure 6 illustrates the steps of analyzing a sample rotten mango with noise. Figure 6 (A) is the captured rotten mango image and applying median filter, (B) changing the brightness and saturation levels to select the whole area, (C) after selecting the whole area, (D) changing the brightness and saturation level to fill only on the healthy area of the rotten mango, and finally (E) selecting only healthy areas.

In Figure 6, the captured defected (rotten) mango is enhanced by using the median filter. Threshold the colored image to select the region of the whole fruit by changing the brightness and saturation levels is performed. After select the border, the whole fruit is bounded. Thus, the whole mango including the rotten regions is measured. In order to measure only the healthy region, so repeat the previous step but changing only the saturation level in order to fill only the rotten regions. Finally, get measurements of

Table 1. The selected feature vectors of each fruit/vegetable set

Selected Feature Vector	Crops Recognition/Classification Cases
[Area, Min, Max, Perimeter, Angle, Circularity, Feret, Integrated Density, Skew, Kurt, FeretX, FeretAngle]	Healthy and Rotten Mango
[Area, Standard Deviation, Min, Max, Perimeter, Angle, Circularity, Integrated Density, FeretX, FeretY, FeretAngle]	Healthy and Rotten Tomato
[Min, Max, Perimeter, Angle, Circ., Feret, Integrated Density, Skew, Kurt, RawIntDen, FeretX, FeretY, FeretAngle]	Healthy and Rotten Orange
[Area, mean, Standard Deviation, mode, Perimeter, Width, Angle, Circularity, Integrated Density, Median, Kurt, RawIntDen, Feretx, Feret y, Feret Angle].	Healthy and Rotten Potato

Figure 6. Block diagram of measuring a rotten mango without noise

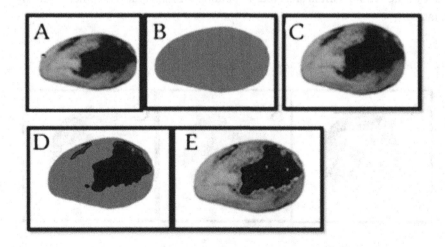

the healthy regions and select only the feature vector of mango [Area, Min, Max, Perimeter, BX, BY, Angle, Circularity, Integrated Density, Skewness, Feret x, Feret y, Feret Angle).

After applying the previous steps on the 15 healthy mangos and 15 rotten mangos, and then testing only 5 from each type, the LMSE algorithm is used to recognize/classify each mango whether it is healthy or rotten. Table 2 illustrates the recognition/classification results after applying the LMSE on the mango datasets (healthy and rotten).

The result of the recognition and matching is 100% as shown in Table 2. The same steps were implemented with the other three kinds (tomato, potatoes, and orange). The same recognition/classification of 100% is obtained with the other cases. Moreover, the LMSE algorithm is used to distinguish each kind of fruit and/or vegetable from the other. The datasets are composed of 10 healthy mangos and 10 rotten mangos, 10 healthy tomatos and 10 rotten tomatos, 10 healthy potatoes and 10 rotten potatoes, 10 healthy oranges and 10 rotten oranges. The LMSE algorithm was able to identify and recognize each crop correctly with its correct status (healthy/ rotten). Thus, recognition rate of 100% is achieved with all cases. Thus, to study the performance of the proposed LMSE, different noise levels are added and the recognition is tested with these noisy images as follows.

Analyzing a Sample Healthy Mango (with Noise)

Figure 7 illustrates the steps for analyzing healthy mango after adding Gaussian noise, where (A) the healthy mango sample image after applying Gaussian noise, (B) noised image after changing the brightness and saturation, and (C) selecting the whole area.

Afterward, the proposed recognition/classification system is tested with noised healthy datasets at five levels of Gaussian noise, namely 10%, 20%, 30%, 40% and 50%. These noise levels are added to the four original sets, namely mango, tomato, potatoes, and orange with only healthy type by using Gaussian noise as shown in Figure 8.

The effect of the noise level on the recognition rate is shown in Table 3 for the different levels of noise. In addition, the average effect of the noise level on the recognition rate using the LMSE algorithm is shown in Figure 9.

Table 2. Recognition/classification result of LMSE mango (healthy and rotten)

Tested Actual Status	Tested Mango's Actual Image No.	Recognized Mango's Image No.	Recognized Status
Healthy Mango	1	1	Healthy
	2	2	Healthy
	3	3	Healthy
	4	4	Healthy
	5	5	Healthy
Rotten Mango	1	1	Rotten
	2	2	Rotten
	3	3	Rotten
	4	4	Rotten
	5	5	Rotten

Figure 7. Block diagram of measuring a healthy mango with noise

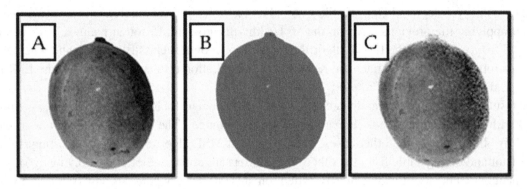

Figure 8. Sample images of mango datasets with several levels of Gaussian noise

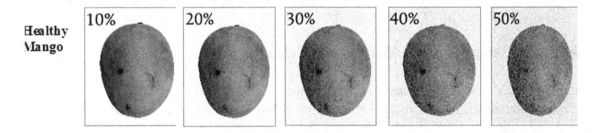

Table 3. Average recognition/classification of the LMSE with four noised datasets (healthy)

Noised Dataset (Healthy)	10% Gaussian Noise	20% Gaussian Noise	30% Gaussian Noise	40% Gaussian Noise	50% Gaussian Noise
Mango	4 of 5	4 of 5	1 of 5	0 of 5	-
Tomato	3 of 5	0 of 5	-	-	-
Orange	5 of 5	5 of 5	2 of 5	2 of 5	0 of 5
Potatoes	5 of 5	5 of 5	2 of 5	0 of 5	-

Second Method: Results of Classification Using Bag of Features (BoF)

The BoF model is used, which extract SURF features from the selected feature point locations. About 80% of the strongest features from each image set are employed in the classification process. The K-Means clustering is used to create 500 word visual vocabularies. In the classification phase, 20% of the images (2 images) are used for feature extraction and training, 80% for validation and testing (8 images) in each image set.

Mango Binary Classification Using BoF (Healthy From Rotten)

The clustering process within BoF is completed and converged after 46 iterations (~7.16 seconds/iteration). The confusion matrix for classifying mango (healthy/rotten) is demonstrated in Table 4 in the test set.

Figure 9. Average recognition rate of LMSE with four noised datasets (healthy)

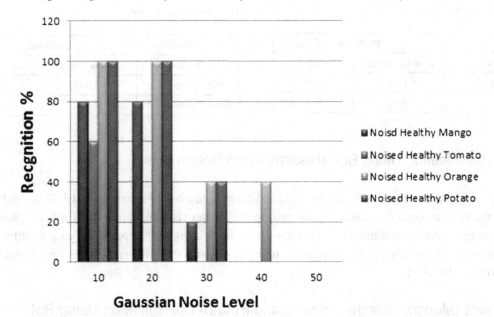

The average accuracy obtained by using the BoF classifier is 88% for classify healthy mango and rotten one.

Orange Binary Classification Using BoF (Healthy from Rotten)

The clustering process within BoF is completed and converged after 14 iterations (~16.96 seconds/iteration). The confusion matrix for classifying orange (healthy/rotten) is demonstrated in Table 5 in the test set.

The average accuracy obtained by using the BoF classifier is 94% for classifying healthy and rotten oranges.

Tomato Classification Using BoF (Healthy From Rotten)

The clustering process within BoF is completed and converged after 24 iterations (~7.78 seconds/iteration). The confusion matrix for classifying tomato (healthy/rotten) is demonstrated in Table 6 in the test set.

The average accuracy obtained by using the BoF classifier is 75% for classifying healthy tomato from rotten one.

Table 4. Confusion matrix when classifying healthy and rotten mangoes

Known	Predicted	
Mango	Healthy Mango	Rotten
Healthy Mango	1.00	0.00
Rotten Mango	0.25	0.75

Table 5. Confusion matrix when classifying healthy and rotten oranges

Known	Predicted	
Orange	Healthy Orange	Rotten
Healthy Orange	1.00	0.00
Rotten Orange	0.11	0.89

Table 6. Confusion matrix when classifying healthy and rotten tomatoes

Known	Predicted	
Tomato	Healthy Tomato	Rotten
Healthy Tomato	1.00	0.00
Rotten Tomato	0.50	0.50

Table 7. Confusion matrix when classifying healthy and rotten tomatoes

Known	Predicted	
Potato	Healthy Tomato	Rotten
Healthy Potato	1.00	0.00
Rotten Potato	0.00	1.00

Potato Classification Using BoF (Healthy From Rotten)

The clustering process within BoF is completed and converged after 19 iterations (~17.52 seconds/iteration). The confusion matrix for classifying tomato (healthy/rotten) is demonstrated in Table 7 in the test set.

The average accuracy obtained by using the BoF classifier is 100% for classifying healthy potato from rotten one. Consequently, the average accuracy to classify the different at each crop (healthy/rotten) is shown in Figure 10.

All Datasets (Mango, Orange, Tomato, and Potato) Classification Using BoF

The datasets for the healthy crops are classified into one of the four classes (mango, orange, tomato, or potatoes). The clustering process within BoF is completed and converged after 26 iterations (~13.61 seconds/iteration). The confusion matrix for classifying the four healthy crops is demonstrated in Table 8 in the test set.

The average classification accuracy of the proposed system for all the healthy datasets including mango, potato, tomato and orange obtained by using the BoF is 88%.

Figure 10. Average accuracy of each image set using BoF

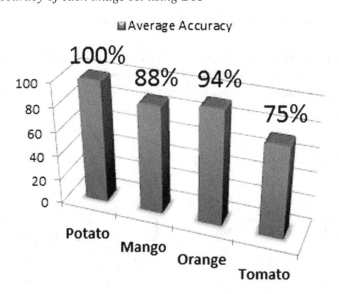

Table 8. All healthy crops confusion matrix

Known	Predicted			
	Mango	**Potato**	**Orange**	**Tomato**
Mango	0.88	0.00	0.13	0.00
Potato	0.00	1.00	0.00	0.00
Orange	0.00	0.38	0.63	0.00
Tomato	0.00	0.00	0.00	1.00

The preceding results established the effectiveness of using ImageJ for features selection. The successfully extracted features achieved 100% recognition/classification using the LMSE algorithm. Moreover, by employing the BoF model on the datasets' images using SURF features, 100%, 88%, 94%, and 75% classification accuracies were achieved during the classification of the healthy/rotten potato, mango, orange and tomato each; respectively. This establishes that using the statistical analysis extracted using ImageJ is superior to the BoF model in terms of the used number of features and the recognition/classification accuracy.

In the future work, other methods for classification such as the neural network and the support vector machine can be used. Also, other features such as the color and texture features can be extracted. The dataset can be extended for superior results. In addition, the proposed current system can be applied on various crops with larger datasets' size.

CONCLUSION

Image processing is an approach used to convert an image into digital form and performing some operations to enhance the image or to extract useful information from it. Recognizing different crops based on their features is a very effective technique that has been proposed in several times in the past. In the proposed system the recognition process was performed to differentiate between the healthy from rotten crop and to identify each crop from the other based on their statistical features. In order to get the measurements of each crop, image processing tools are used. Then, the LMSE algorithm in Matlab was applied for the recognition phase. In addition, the classification phase using BoF technique in Matlab was employed for SURF features extraction and then features selection was done.

In general, the proposed system achieved good recognition rate after adding 10% of Gaussian noise with average 85%. Also, a good recognition rate with 20% of Gaussian noise with average = 70%. It gave a bad recognition rate with 30% of Gaussian noise with average = 25%. It always gave 100% for the original images, which established the superiority of the statistical features extracted using ImageJ.

In the classification phase, the average accuracy obtained by using BoF was 100% for Potato (healthy/rotten), 88% for Mango (healthy/rotten), 94% for Orange (healthy/rotten) and 75% for Tomato (healthy/rotten). The average classification accuracy of the proposed system for all the healthy datasets including by using the BoF was 88%.

REFERENCES

Agrawal, D.D., Dubey, S.R., & Jalal, A.S. (2014). Emotion Recognition from Facial Expressions based on Multi-level Classification. *International Journal of Computational Vision and Robotics*.

Anthimopoulos, M. M., Gianola, L., Scarnato, L., Diem, P., & Mougiakakou, S. G. (2014). A food recognition system for diabetic patients based on an optimized bag-of-features model. *Biomedical and Health Informatics. IEEE Journal of, 18*(4), 1261–1271.

Barrett, D. M., Beaulieu, J. C., & Shewfelt, R. (2010). Color, flavor, texture, and nutritional quality of fresh-cut fruits and vegetables: Desirable levels, instrumental and sensory measurement, and the effects of processing. *Critical Reviews in Food Science and Nutrition, 50*(5), 369–389. doi:10.1080/10408391003626322 PMID:20373184

Blasco, J., Aleixos, N., Gómez-Sanchis, J., & Moltó, E. (2009). Recognition and classification of external skin damage in citrus fruits using multispectral data and morphological features. *Biosystems Engineering, 103*(2), 137–145. doi:10.1016/j.biosystemseng.2009.03.009

Dubey, S. R., & Jalal, A. S. (2014). *Fusing Color and Texture Cues to Categorize the Fruit Diseases from Images*. Academic Press.

Hong, Z. Q. (1991). Algebraic feature extraction of image for recognition. *Pattern Recognition, 24*(3), 211–219. doi:10.1016/0031-3203(91)90063-B

Jiménez, A. R., Jain, A. K., Ceres, R., & Pons, J. L. (1999). Automatic fruit recognition: A survey and new results using range/attenuation images. *Pattern Recognition, 32*(10), 1719–1736. doi:10.1016/S0031-3203(98)00170-8

Lino, A. C. L., Sanches, J., & Fabbro, I. M. D. (2008). Image processing techniques for lemons and tomatoes classification. *Bragantia, 67*(3), 785–789. doi:10.1590/S0006-87052008000300029

Liu, W., Pokharel, P. P., & Principe, J. C. (2008). The kernel least-mean-square algorithm. *Signal Processing. IEEE Transactions on, 56*(2), 543–554. doi:10.1109/TSP.2007.907881

Mente, R., Dhandra, B. V., & Mukarambi, G. (2014). Color Image Segmentation and Recognition based on Shape and Color Features. *International Journal on Computer Science and Engineering, 3*(01).

Motwani, M. C., Gadiya, M. C., Motwani, R. C., & Harris, F. C. (2004, September). Survey of image denoising techniques. In *Proceedings of GSPX* (pp. 27-30).

Ninawe, P., & Pandey, M. S. (2014). A Completion on Fruit Recognition System Using K-Nearest Neighbors Algorithm. *International Journal of Advanced Research in Computer Engineering & Technology, 3*(7).

Song, Y., Glasbey, C. A., Horgan, G. W., Polder, G., Dieleman, J. A., & van der Heijden, G. W. A. M. (2014). Automatic fruit recognition and counting from multiple images. *Biosystems Engineering, 118*, 203–215. doi:10.1016/j.biosystemseng.2013.12.008

Sonka, M., Hlavac, V., & Boyle, R. (2014). *Image processing, analysis, and machine vision*. Cengage Learning.

Sural, S., Qian, G., & Pramanik, S. (2002). Segmentation and histogram generation using the HSV color space for image retrieval. In *Image Processing. 2002. Proceedings. 2002 International Conference on* (Vol. 2, pp. II-589). IEEE. doi:10.1109/ICIP.2002.1040019

Zhao, J., Tow, J., & Katupitiya, J. (2005, August). On-tree fruit recognition using texture properties and color data. In *Intelligent Robots and Systems, 2005.(IROS 2005). 2005 IEEE/RSJ International Conference on* (pp. 263-268). IEEE. doi:10.1109/IROS.2005.1545592

Zhongzhi, H., Jing, L., Yougang, Z., & Yanzhao, L. (2012). Grading System of Pear's Appearance Quality Based on Computer Vision. *International Conference on System and Informatics (ICSAI-2012).*

KEY TERMS AND DEFINITIONS

Bag of Features: Known also as the 'bag-of-words' model that can be applied for image classification. It treats the image features as words. It can be defined as a histogram depiction based on independent features. This model has three steps, namely feature detection, feature description, and codebook generation.

Classification: The process of extracting information classes from an image of multi-classes. It is executed on the base of spectrally defined features, such as texture, color, and density.

Feature Extraction: The key step in pattern recognition, machine learning, and image processing. It starts by measuring data and building derived values (features) that are informative and non-redundant. Thus, it is related to dimensionality reduction. It selects a subset of relevant features (variables) that can be used in model construction.

Gaussian Noise: A statistical noise that has a probability density function equal to that of the normal distribution.

Recognition: The identification of objects in an image. This process starts with image pre-processing techniques for noise removal, followed feature extraction.

Chapter 4
Moving Object Classification in a Video Sequence

S. Vasavi
V. R. Siddhartha Engineering College, India

T. Naga Jyothi
V. R. Siddhartha Engineering College, India

V. Srinivasa Rao
V. R. Siddhartha Engineering College, India

ABSTRACT

Now-a-day's monitoring objects in a video is a major issue in areas such as airports, banks, military installations. Object identification and recognition are the two important tasks in such areas. These require scanning the entire video which is a time consuming process and hence requires a Robust method to detect and classify the objects. Outdoor environments are more challenging because of occlusion and large distance between camera and moving objects. Existing classification methods have proven to have set of limitations under different conditions. In the proposed system, video is divided into frames and Color features using RGB, HSV histograms, Structure features using HoG, DHoG, Harris, Prewitt, LoG operators and Texture features using LBP, Fourier and Wavelet transforms are extracted. Additionally BoV is used for improving the classification performance. Test results proved that SVM classifier works better compared to Bagging, Boosting, J48 classifiers and works well in outdoor environments.

INTRODUCTION

Now-a-day's monitoring the objects (human beings, animals, buildings, vehicles etc.,) in a video is a major issue in the areas such as airports, banks, military installations etc., Object identification and recognition are considered as two important tasks in such areas. To do these two tasks, video is to be taken with a wide landscape of the scene, which results in a small low resolution and occluded images for objects. Basic video analysis operations such as object detection, classification and tracking require scanning the entire video. But this is a time consuming process and hence we require a method to detect and classify

DOI: 10.4018/978-1-5225-1022-2.ch004

the objects that are present in the frames extracted from a real time video. Outdoor environments are more challenging for moving object classification because of incomplete appearance details of moving objects due to occlusions and large distance between the camera and moving objects. So, there is a need to monitor and classify the moving objects by considering the challenges of video in the real time.

Motivation

Visual surveillance is an active area of research topic because of its importance in the areas such as security, law enforcement, and military applications. Surveillance cameras are installed in sensitive areas such as airports, banks, military installations, railway stations, highways, and in public places. Data that is collected from these cameras have to be monitored either manually or using intelligent systems. Human operators monitoring manually for long durations is infeasible due to monotony and fatigue. As such, recorded videos are inspected when any suspicious event is notified. But this method only helps for recovery and does not avoid any unwanted events. "Intelligent" video surveillance systems can be used to identify various events and to notify concerned personal when any unwanted event is identified. As a result, such a system requires algorithms that are fast, robust and reliable during various phases such as detection, tracking, classification etc. This can be done by implementing a fast and efficient technique to classify the objects that are present in the video in the real time.

Problem Statement

Basic video analysis operations such as object detection, classification and tracking require scanning the entire video. But this is a time consuming process and hence we require a method to detect and classify the objects that are present in the frames extracted from a real time video. Object classification is to be done which is an important building block that significantly impacts reliability of its applications. Outdoor environments are more challenging for moving object classification because of incomplete appearance details of moving objects due to occlusions and large distance between the camera and moving objects. So, there is a need to monitor and classify the moving objects by considering all the challenges as mentioned above and object classification is crucial that is done based on the features extracted from the objects in the video, here classification is a tough task when we consider video in the real time.

Background

The field of computer vision requires understanding of key terms such as color spaces, key frame extraction, foreground subtraction, feature extraction, blob detection, Bag of Visual Words (BoV), Occlusions etc. as described in the following sections.

Color Spaces

Color is a phenomenon that relates to the physics of light, chemistry of matter, geometric properties of object and human visual perception. Color is a purely psychological phenomenon. Color model is also called as Color Space or Color System. A color model is a mathematical model and it describes in the way colors can be represented as tuples of numbers, as four or three values of color components. There are different types of color spaces that are RGB, CMYK, HSV, HSI, etc. from which RGB color space and HSV color space are most commonly used.

RGB Color Space

In the RGB model, each color appears in its primary spectral components of red, green, and blue. RGB model is based on a Cartesian coordinate system. RGB image is an image in which each of the red, green, and blue images are an 8-bit image. The term full-color image is used often to denote a 24-bit RGB color image as shown in Figure 1.

HSV Color Space

The HSV color space model represents the Hue, Saturation and Value in 24 bit colors and arranges the colors in a more virtually appropriate modeling. The Hue value is in between 0 and 360 degree rotated along the axis while the Saturation is the distance from the axis, and the Value is a value along the axis where the axis is extended from H: 0, S: 0, V: 0 (Poorani, Prathiba, & Ravindran, 2013). As hue varies from 0 to 1.0, the corresponding colors vary from red through yellow, green, cyan, blue, magenta, and back to red, so that there are actually red values both at 0 and 1.0 as shown in Figure 2. As saturation varies from 0 to 1.0, the corresponding colors (hues) vary from unsaturated (shades of gray) to fully saturated (no white component). As value, or brightness, varies from 0 to 1.0, the corresponding colors become increasingly brighter.

Blob Detection

In computer vision, blob detection methods are based on detecting regions in a digital image that differs in the properties, such as brightness, color compared to surrounding regions.

Figure 1. RGB color space
Source: Poorani, Prathiba, & Ravindran, 2013.

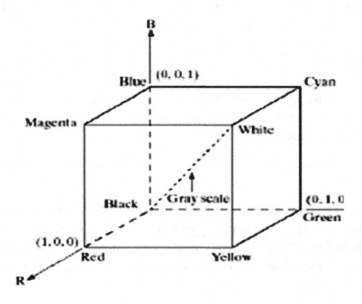

Figure 2. HSV color space
Source: Poorani, Prathiba, & Ravindran, 2013.

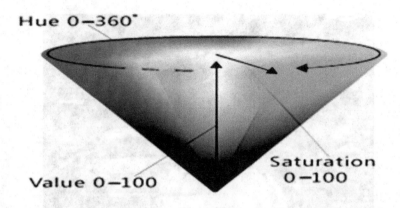

Feature Extraction

For the object classification process, feature extraction is done based on three features such as color, structure and texture.

- Color is a measure of pixel level color intensities.
- Structure provides information about edges and boundaries detected in an image.
- Texture account for information about the spatial arrangement of color or intensities in an image or selected region of an image.

Bag of Visual Words (BoV)

In the field of computer vision, the bag of visual words (BoV) approach can be applied to object classification, by considering image features as words. Bag of words is a sparse vector of occurrence counts of words that is a sparse histogram over the vocabulary. In computer vision, a bag of visual words is a vector of occurrence counts of a vocabulary of local image features.

Occlusions

Occlusion can be explained by taking few examples, if you are developing a system which tracks objects (people, cars), occlusion occurs if an object hides another object. Like two persons walking past each other, or a car that drives under a bridge. The challenge and the problem in the above cases is what we do when an object is hidden or disappears and reappears again. Occlusion is of two kinds.

1. Self-occlusion (one object occludes another), Ex: During tracking, smaller size object (Bicycle) is covered by the large object (Lorry) due to the difference in their speed and velocity as shown in Figure 3.
2. Inter object occlusion (structure in the background occludes the tracked object in the foreground). Ex: During tracking, foreground object (Bicycle) is covered by the background object (Flexi banners or iron gates) as shown in Figure 4.

Figure 3. Self-occlusion
Source: Pepik, Stark, Gehler, & Schiele, 2013.

Figure 4. Inter object occlusion
Source: Pepik, Stark, Gehler, & Schiele, 2013.

Classification of objects in a video involves the process of searching, retrieving and indexing. Object classification is done based on visual features such as color, texture and shape is difficult and has proven to have its own set of limitations under different conditions. Techniques such as edge detection using various filters, edge detection operators, CBIR (Content Based Image Retrieval) and Bag-of visual words (BoV) are used to classify videos into fixed broad classes which would assist searching and indexing using semantic keywords.

The main objective of proposed approach is to classify the objects in a Video taken from the CCTV footage on highways having dynamic background.

1. Extracting 29 fps from the video.
2. Identifying object locations by using blobs (regions) in the frames.
3. Preprocessing the identified blobs by having each blob with constant resolution (256*256).

4. Features are extracted based on color, structure and texture.
5. PCA based dimensionality reduction technique to reduce the feature set.
6. Classification is done by training the SVM, AdaBoost, J48 and bagging classifiers.

LITERATURE REVIEW

The main phases in moving object classification are:

1. **Frame Extraction:** Frames are extracted from the video at the rate of 30 frames per second.
2. **Blob Detection:** Object detection and recognition test bed are built so as to result in higher performance.
3. **Preprocessing:** Preprocessing is done on each identified blob to have a constant resolution (256*256).
4. **Feature Extraction:** Various features such as appearance features and geometric features are considered.
5. **Classification:** Based on classifiers such as SVM (Support Vector Machine), AdaBoost, etc. and the classification performance will be evaluated using precision, recall, error rate and accuracy.

A Frame-Based Decision Pooling Method for Video Classification (Mohanty & Sethi, 2013)

Here, they split a given video into individual frames (as many as there are in the video), and extract features from the individual, such frames, and each frame is then classified into one of the five classes by an SVM classifier trained on a large image database. The features that were used to train the proposed model were of three types viz. Color features, structural features and textural features. Here 500 training images from the Corel image dataset are used for training. The features that were extracted from these images were:

- 18 color features using RGB histograms.
- 258 texture features using entropy of image, FFT, range of gray scale image.
- 6 structural features from Canny's operator.

An SVM classifier was then trained on these 282 features from the 500 training images. To do feature extraction, it is important to have a constant resolution for all the training images. Choosing 256 by 256 as a constant resolution for all the images which helps to extract the features easily.

Color Features Using RGB Histograms

For color features, statistical properties of normalized RGB histograms are used. The RGB color space can be defined as a unit cube with red, green, and blue axes and a normalized RGB color space is one in which the RGB color cube has unit dimensions, thus, all values of red, green, and blue are assumed to be in the range [0, 1]. An RGB histogram can be thought of as a type of bar graph, where each bar represents a particular color of the RGB color space being used. 'Bars' in a color histogram are referred to as bins and are represented on the x-axis, and y-axis represents the number of pixels in each bin. Consider 12 bins each for red, green and blue.

First order histogram is given in Equation 1 (Mohanty & Sethi, 2013)

$$P(g) = \frac{N(g)}{M} \tag{1}$$

where

M = Total number of pixels in an image,
$N(g)$ = Number of pixels at grey level.

The six color features are extracted based on P (g) i-e, first histogram probability.

1. **Mean:** Mean is the representative for brightness. It is given in Equation 2 and Equation 3 (Mohanty & Sethi, 2013)

$$g' = \sum_{0}^{L-1} g P(g) \tag{2}$$

$$g' = \sum_{r} \sum_{c} \frac{I(r,c)}{M} \tag{3}$$

where

L = Total intensity level,
r, c represents row and column.

2. **Standard Deviation:** Measure of contrast. It is calculated using Equation 4 (Mohanty & Sethi, 2013)

$$\sigma = \sqrt{\sum_{K=0}^{L-1} P(g)(g - g')^2} \tag{4}$$

3. **Skew and Skew:** Measure of asymmetry about the mean in the intensity level distribution. It is given in Equation 5 and Equation 6 (Mohanty & Sethi, 2013)

$$Skew = \frac{1}{\sigma_{g^3}} \sum_{K=0}^{L-1} P(g)(g - g')^3 \tag{5}$$

$$Skew' = \frac{g' - \left\{\left(P(g)\right)\right\}_{max}}{\sigma_g} \tag{6}$$

4. **Energy:** It is an indication of distribution of intensity levels. It is calculated using Equation 7 (Mohanty & Sethi, 2013)

$$Energy = \sum_0^{L-1} P(g)^2 \tag{7}$$

5. **Entropy:** It gives us the number of bits needed to code the image data. It is given in Equation 8 (Mohanty & Sethi, 2013)

$$Entropy = -\sum_0^{L-1} P\left(g\right) \log_2 P(g) \tag{8}$$

Hence, 6 color features for R, 6 color features for G and 6 color features for B. Total of 18 (6*3) color features for RGB. Thus, we have six color features for each histogram, and, hence, accounting for the three histograms (R, G, B), we have a total of 18 color features.

Structural Features Using Edge Detection

Here edge detection is the process of finding and locating sharp discontinuities in an image, which are abrupt changes in pixel intensity which characterize boundaries of objects in a scene. In our work, we made use of a very efficient edge detection technique. Canny's edge detection algorithm is used.

In the first step of edge detection, 2-D spatial gradient measurement on an image is computed. At each point in the output, pixel values indicate the estimated absolute magnitude of the spatial gradient of the input image at particular point. The convolution masks using are Equation 9 and Equation 10 (Mohanty & Sethi, 2013)

$$G_x = \begin{matrix} -1 & 0 & +1 \\ -2 & 0 & +2 \\ -1 & 0 & +1 \end{matrix} \tag{9}$$

$$G_y = \begin{matrix} +1 & +2 & +1 \\ 0 & 0 & 0 \\ -1 & -2 & -1 \end{matrix} \tag{10}$$

Let us refer to the convolution results as G_{xx} and G_{yy}. Thus, $G_{xx} = G_x*I$, and $G_{yy} = G_y*I$, where I is the gray scale image. These are then used to find the magnitude and orientation, given in Equation 11 and Equation 12 (Mohanty & Sethi, 2013)

$$|G| = |G_{xx}| + |G_{yy}|$$ (11)

$$\theta = \tan^{(-1)}\left(G_y y / G_{(xx)}\right)$$ (12)

The angle, obtained in the above step, is the edge orientation that has to be resolved into one of four directions (0 degree, 45 degrees, 90 degrees and 135 degrees). We then use the six statistical features as described in the section on 'color features', to get 6 structure features.

Texture Features

Texture can be described as an attribute representing the spatial arrangement of the gray levels of the pixels in a region of a digital image.

In our model, we used the entropy of FFT of an image converted into gray scale. This gave us one texture feature to work with. Another texture feature that we used was entropy of the gray scale image. In addition to these entropy features, we used 'range' of the gray scale matrix. This gave us additional 256 features. The number of features obtained from the range function is equal to no. of columns in the gray scale image. So, we obtain 256 features from range in addition to the two entropy features. Hence, total number of texture features extracted was 258.

Training the SVM Classifier

Representing each image as a feature vector of '282'elements, we can consider every image as a point in a 282-D space. And to classify these points, we employ a support vector machine (SVM). A Support Vector Machine constructs a hyper-plane or set of hyper-planes in a high or infinite dimensional space, which can be used for classification as shown in Figure 5.

Dataset Used:

1. Corel dataset.

Advantages:

1. When a video cannot be classified distinctly into one particular class, it can still be classified under broader sub-class.
2. When this method is implemented, say in a search engine, this would enable users to search using vague keywords.

Disadvantages:

Figure 5. Classification model using SVM
Source: Mohanty & Sethi, 2013.

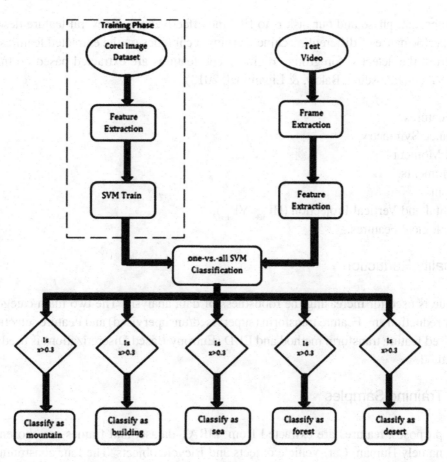

1. This paper discusses the object classification only on the five fixed classes (mountains, sea, forest, buildings, and deserts) which give accurate results only when the classes are more specific because the main focus is on structural features by using canny operator.

Multi Class Object Classification in Video Surveillance Systems (Elhoseiny, Bakry, & Elgammal, 2013)

In this, detection and segmentation is done on the dynamic objects through a motion detection module.

Object Detection

In surveillance systems, monitoring and detection are important. For stationary camera, the background modeling is the most efficient approach having low false alarm and high accuracy. Here dynamic objects are detected by using non-parametric kernel density Estimation (KDE) approach as shown in Figure 7. For this step, the outcome is the series of connected foreground areas. These are further segmented into blobs.

Feature Extraction

This is the important phase and our task is to find the effectiveness of several feature descriptors and evaluate the performance of different machine learning techniques on the extracted features. This stage takes input from the detected foreground regions. The features are extracted based on the following methods described (Elhoseiny, Bakry, & Elgammal, 2013):

1. HOG Features,
2. Luminance Symmetry,
3. Central Moments,
4. ART Moments,
5. Cumulants,
6. Horizontal and Vertical Projection ($HP_{i,+}$, $VP_{+,j}$),
7. Morphological Features.

Dimensionality Reduction

This technique is used for increasing the robustness of data analysis. The two main categories for dimensionality reduction are: Feature transform (supervised/unsupervised) and Feature selection. PCA, as an unsupervised feature transform method and EBD (Entropy Based Discretization) is used as a feature selection method.

Extracting Training Samples

For training purposes, features are extracted from VIRAT dataset and frames are chosen to classify five classes' namely Human, Car, Vehicle objects and Bicycle objects. The four constraints described in (Elhoseiny, Bakry, & Elgammal, 2013) were considered in training phase, while only the first three are considered on the test videos.

Object Classification

For training, we have a list of pairs $\left\{(xi, yi)\right\}_{i=1}^{N}$ where $x_i \in R^d$ is the feature vector, and $y_i \in \{1, 2, ...,K\}$ is the sample label. Let $X = [x1, x2, ...,x_N]^t$ is $N \times d$ matrix and $Y = [y1, y2, ..., y_N]^t$ is N dimensional column vector.

Dataset Used:

1. Virat dataset.

Advantages:

1. 71.4% accuracy was achieved based on HOG features.
2. SVM and AdaBoost classification techniques performed well for recognizing objects.
3. Geometric features perform significantly better in surveillance systems.

Figure 6. Experimental framework for multiclass object classification
Source: Elhoseiny, Bakry, & Elgammal, 2013.

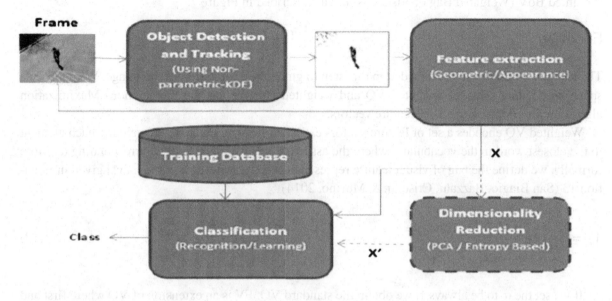

Disadvantages:

1. In Surveillance Systems, the experiments shows that using appearance features like HOG features did not perform well and is less discriminating for recognizing object classes.
2. PCA technique violates the real time requirements of surveillance systems.
3. The effect of changing the weak classifiers count for the training and test accuracy for every category significantly increases the processing time by using AdaBoost classification.

Weighted Bag of Visual Words for Object Recognition (San Biagio, Bazzani, Cristani, & Murino, 2014)

Visual object recognition is one of the most studied problems in computer vision. Bag of Visual words (BoV) is one of the most successful and important strategies for object recognition, used to represent an image as a vector of counts using a learned vocabulary.

Feature Extraction

Here grid of pixel locations with spacing of 4 pixels in both x, y directions is defined on the image. Around these pixel locations, patches of different sizes (12×12, 18×18, 24×24, 30×30 pixels) are extracted. On each patch, we calculate the SIFT descriptor, generating a set of local descriptors for each image.

Codebook Creation

In this step, the local descriptors are used to generate a codebook with K words, by GMM (Gaussian Mixture Model) clustering. This result in a weighted histogram, where the height of each bin depends on the number of associated codes retrieved in the image, and on their related weights. This is called a Weighted BoV (Weighted Bag of Visual words) as described in Figure 7.

Encoding

The salience of each patch is included in this step to guide the exploration of the image. We apply two specific encoding schemes weighted VQ and weighted FV. We use the Expectation- Maximization algorithm for GMM on the SIFT feature vectors.

Weighted VQ encodes a set of feature vectors extracted from an image by associating each element to the closest word in the vocabulary, where the association is weighted by the corresponding α_i. More formally, we define the bag of visual feature representation as a vector $v = [v_1, v_2 . . . v_K]$ given in Equation 13 (San Biagio, Bazzani, Cristani, & Murino, 2014)

$$V_k = \sum_{i=1}^{NI} \alpha_i \delta \left(x_i, \mu_k \right) \tag{13}$$

If we set the α_i to be always 1, we obtain the standard VQ. FV is an extension of VQ where first and second order statistics are also considered.

Spatial Pyramid Matching

In this step, the image is partitioned into increasingly finer spatial sub-regions and Weighted BoV is computed from each sub-region, following a spatial pyramid scheme. Typically, $2r \times 2r$ sub-regions, with $r = \{0, 1, 2\}$ are used.

Figure 7. Object recognition pipeline
Source: San Biagio, Bazzani, Cristani, & Murino, 2014.

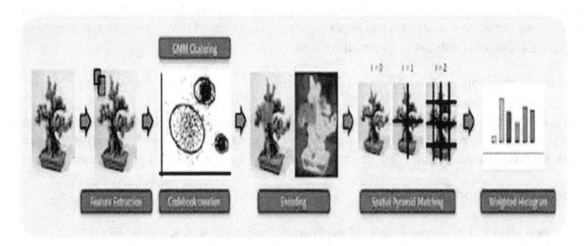

Weighted Histogram

All the Weighted BoV extracted from each sub region is pooled and concatenated together, generating the final Weighted BoV representation of the image. Finally, classification is done using one-vs-all linear SVM.

Datasets Used:

1. Caltech-101, Caltech-256, VOC2007 datasets.

Advantages:

1. Weighting the BoV representation is always beneficial for both VQ an FV.
2. This technique gives the robustness to spatial translations of features, the efficiency to compute it, and its competitive performance in different image categorization tasks.
3. This bag of visual words method performs well with both dense and sparse patches on challenging recognition datasets.

Disadvantages:

1. Some pairs of images of Caltech-256 are correctly classified using the weighted BoV approach but miss-classified with the classic BoV.
2. Some examples are miss-classified using the weighted BoV representation but classified correctly using the classic BoV.

Color Texture Classification with Color Histograms and Local Binary Patterns (Pietikainen, Maenpaa, & Viertola, 2002)

This paper describes about the approaches proposed for color texture discrimination, are of two types:

1. Color and texture information are processed separately, and
2. Spatial interactions of pixels both within and between color bands are considered and the current approaches to color texture analysis are, processing color and texture processing information separately and those that considering color and texture as a joint phenomenon.

Approaches to Texture Analysis

1. The use of joint color texture features has been a popular approach to color texture analysis. Statistics derived from co-occurrence matrices and difference histograms were considered as texture descriptors.
2. In some approaches, the spatial interactions within bands are considered. For example, Caelli and Reye proposed a method that extracts features using three multiscale isotropic filters from three spectral channels.
3. Paschos compared the effectiveness of different color spaces when Gabor features computed separately for each channel were used as color texture descriptors.

4. Another way of analyzing color textures is to process luminance and chrominance components separately by dividing the color signal.

5. The texture features are extracted based on the discrete cosine transform from a gray level image; while measures derived from color histograms were used for color description is described by Tan and Kittler.

6. Dubuisson Jolly and Gupta proposed a method for aerial image segmentation, in which likelihoods are computed independently in color and texture spaces. Then, the final segmentation is obtained by evaluating the certainty with which each color or texture classifier alone would make a decision.

7. The texture features are computed based on Local Binary Patterns and color percentile features derived from color histograms, and combined these for classifying defects in parquet slabs described by Kyllonen and Pietikoinen.

8. The psychophysical studies of Poirson and Wandell suggest that color and pattern information are processed separately.

9. Mojsilovic suggest that the overall perception of color patterns is formed through the interaction of a luminance, chrominance and achromatic pattern components.

In this paper, histograms of LBP patterns are used for classification with LBP distributions, a log-likelihood dissimilarity measure was chosen. Simple variations of the basic LBP operator were also used. In the case of an RGB image, the local threshold can be taken from three different color channels. The neighborhood to be threshold can also be taken from three channels, which consists of total of nine different combinations. Since all of these produce one feature distribution, we ended up with nine of them. The histograms were concatenated into a single distribution containing $256*9 = 2304$ bins. Gray scale Gabor features are extracted by using City-block distance scaled with the standard deviations. For color, texture Gabor features, a squared Euclidean distance scaled with feature variances was used.

Datasets Used:

1. Vision texture, Outex texture datasets.

Advantages:

1. It is possible to achieve real time performance by considering simple texture measures like the LBP even in very demanding tasks.

2. There are applications where color and texture need to be used for maximum performance.

Drawbacks:

1. RGB and Ohta histograms are severely degraded when the illumination source is not kept constant.

2. Ohta intensity component performs clearly better than color histograms. 3D RGB and 3D Ohta histograms beat the texture features in the constant illumination case.

Integrated Feature Extraction for Image Retrieval
(Poorani, Prathiba, Ravindran, 2013)

In this paper, Color feature is extracted by Color Histogram and Color Descriptor. The Color histogram specifies the color pixel distribution in an image. Color histogram uses two types of color space that are RGB, HSV. Color descriptor consists of color expectancy, color variance and color skewness.

1. Color expectancy is the average or mean of intensity in image (Mean is the representative for brightness).
2. Color variance is the square root of the standard deviation (SD is the measure of contrast).
3. Color skewness is a measure of the asymmetry of the probability distribution of a real valued random variable. Two types of skewness are Positive skewness and Negative skewness (Skew and skew' are the measures of asymmetry about the mean in the intensity level distribution).

Texture feature is extracted by using texture analysis and texture filter methods. The characteristics provide the details about the texture of an image.

1. A local variation in the gray level co-occurrence matrix is considered as contrast.
2. The joint probability occurrence of the specified pixel pairs is measured as correlation.
3. Energy gives the sum of squared elements in the gray level co-occurrence matrix also known as uniformity or the angular second moment.
4. The closeness of the distribution of elements in the GLCM to the gray level co-occurrence matrix diagonal is measured as homogeneity.
5. Comparing to Texture analysis statistics and Texture filter characteristics, texture analysis method provides the shape feature characteristics.
6. Texture filter method provides the Range filter which calculates the local range of an image.
7. Standard deviation filter which calculates the local deviation of an image.
8. Entropy filter calculates the local entropy of a gray scale image. Entropy is a statistical measure of randomness.

Shape is a very powerful feature. Here, shape features are extracted by Edge detection method and Hu moment Invariant. In Edge detection method, we considered two methods that are Sobel and canny. Canny operator is described in (Mohanty & Sethi, 2013).

The Sobel operator is the magnitude of the gradient computed using Equation 14 (Poorani, Prathiba, & Ravindran, (2013)

$$M = \sqrt{S_x^2 + S_y^2} \tag{14}$$

where the partial derivatives are computed using Equation 15 and Equation 16 (Poorani, Prathiba, & Ravindran, (2013)

$$S_x = \left(a_2 + ca_3 + a_4\right) - \left(a_0 + ca_1 + a_6\right) \tag{15}$$

$$S_x = \left(a_2 + ca_3 + a_4\right) - \left(a_6 + ca_5 + a_4\right) \qquad (16)$$

where c=2 which is a constant. Like the other gradient operators, S_x and S_y can be implemented using convolution masks.

Hu derived expressions are derived from algebraic invariants by applying to the moment generating function under a rotation transformation. They consist of groups of nonlinear centralized moment expressions given in Equation 17 (Poorani, Prathiba, Ravindran, 2013)

$$\mu_{pq} = \int\limits_{-\infty}^{\infty} \int\limits_{-\infty}^{\infty} \left(x - x'\right)^p \left(y - y'\right)^q f\left(x, y\right) dx dy \qquad (17)$$

This gives a set of rotation moment invariants, which can also be used for scale, position, and rotation invariant.

Similarity measure is done by Euclidean distance. Euclidean distance is the most common metric for measuring the distance between two vectors, and is given by the square root of the sum of the squares of the differences between vector components. Euclidean Distance is calculated by using Equation 18 (Poorani, Prathiba, Ravindran, 2013)

$$\|D\| = \sqrt{\sum_{i=1}^{n} \left(p_i - q_i\right)\left(q_i - p_i\right)} \qquad (18)$$

Datasets Used:

1.   Wang dataset.

Advantages:

1.   The canny edge detection provides more details than the Sobel method.
2.   The values computed using the Hu moment method are proved to be invariants to the image scale, rotation, and reflection except the seventh one, whose sign is changed by reflection.

Disadvantages:

1.   Texture filter method does not bring the shape detail for the given object.

Histograms of Pattern Sets for Image Classification and Object Recognition (Voravuthikunchai, Cremilleux, & Jurie, 2014)

This paper proposed a method that relies on:

1.   Multiple random projections of the input space, followed by local binarization of projected histograms encoded as sets of items, and

2. The representation of images as HoPS (Histograms of Pattern Sets).

Their method has two steps:

1. Real-valued histograms are turned into lists of binary items, and
2. Image representations are computed from histograms of mined patterns. This method discovers dependencies between image features, and encodes them through Histograms of Pattern Sets (HoPS).

Datasets Used:

1. Pascal VOC2007, KTH TIPS2a, Oxford-Flowers17, Daimler Classification Dataset.

Advantages:

1. Texture recognition approach improves the performance of the baseline with more than 5%.
2. Combining texture and shape features (LBP and HOG features) together gives better results.

Disadvantages:

1. PCA does not work well when reducing the dimensionality of input histograms.
2. The proposed HoPS approach improves baseline with the reduction of about 30% of the false positive rate.

Occlusion Handling in Videos Object Tracking: A Survey (Materka & Strzelecki, 1998)

The paper mainly describes about the problems related to occlusion and the methods to overcome occlusion during object tracking in a real time video monitoring.

Occlusion is of two kinds.

1. Self-Occlusion (one object occludes another),
2. Inter Object Occlusion (structure in the background occludes the tracked object in the foreground).

Many challenges still remains while tracking objects, this can arise due to the object motion, changing appearance patterns of objects and the scene, non-rigid object structures and most significantly is handling occlusion of tracked object. Handling occlusion in single object environment is a straight forward task but handling occlusion in multi object environment becomes more complicated. Now-a-days inter object occlusion is a challenging problem especially when the targets are identical.

The severity of occlusion can be explained in three categories. During non-occlusion, the tracked object appears as a single blob having all tracking features. This can be avoided by using template matching and mean shift algorithms. When partial occlusion is happened, some of the key features of the tracked object are hidden from the camera during tracking and can be avoided by using mean shift method. Full occlusion happens when the tracking is completely invisible while knowing that object has not left the area of the view of the camera can be avoided using Kalman filter technique. These occlusions can be handled with the help of optimal camera placement method (camera is mounted on the ceiling of a room, which is, when a birds-eye view of the scene is available).

Dataset Used:

1. PETS, ETISEO dataset

Depth Aided Tracking Multiple Objects under Occlusion (Chen, Feris, Zhai, Brown, & Hampapur, 2012)

This paper, mainly describes about the tracking method aiming at detecting objects and maintaining their label or identification over the time. The key factors of this method are to use depth information and different strategies to track objects under various occlusion scenarios.

Depth Estimation

1. Block matching algorithms (color image, depth image)
2. Foreground segmentation and Shadow cancellation(for static image - absolute difference method, dynamic object - GMM)
3. Blob detection, extraction
4. Occlusion detection
5. Object tracking (without occlusion, under occlusion)

Dataset Used:

1. Any video dataset.

Disadvantage:

1. This approach works only under indoor environments and moving object velocity should be low.

Drawbacks of Existing Methods

1. Object classification is done only on the fixed classes and gives accurate results only when the classes are more specific.
2. HOG features did not perform well and is less discriminating for recognizing object classes.
3. RGB and Ohta histograms are severely degraded when the illumination source is not kept constant.
4. Most of the existing methods work only under indoor environments.

The proposed approach considers the training dataset consisting of various types of image samples and the features are extracted from those samples, which is used to train the classifier. To classify the objects, video is divided into frames and features viz. Color features using RGB and HSV histograms, Structure features using HoG, DHoG, Harris, Prewitt, LoG operators and Texture features using LBP, Fourier and Wavelet transforms are extracted. Additionally BoV (Bag of Visual Words) is used for improving the classification performance and accuracy. SVM, Bagging, Boosting, J48 classifiers is used for classification.

PROPOSED SYSTEM

This section presents the detailed functionality of the proposed system to perform moving object classification in a video sequence. Our proposed method gives solution to the drawbacks mentioned before.

Architecture

Figure 8 presents architecture of the proposed system.

Step 1: Collect the image samples of various types (human, 2-wheeler, 3-wheeler, 4-wheeler and other objects) at different views.

Step 2: Preprocessing: Each image sample is preprocessed because it is important to have constant resolution (say 256*256) to extract the features easily.

Figure 8. Proposed method for moving object classification

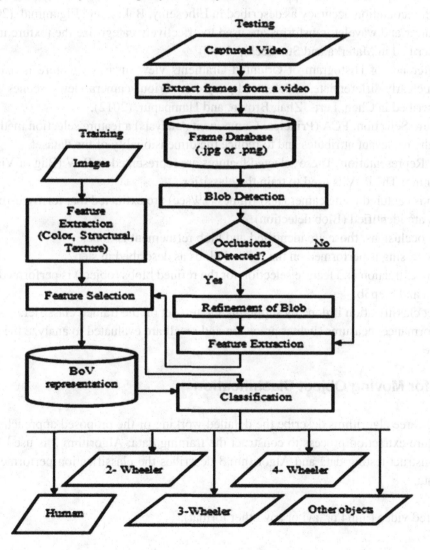

Step 3a: These samples are represented as Bag of Visual words (BoV) by extracting various features such as Color, Structure, and Texture.

1. **Feature Extraction:** To extract color features, statistical properties of RGB histogram or HSV histogram will be used, since HSV histogram provide more efficiency. The reason is Red channel use 256 bins similarly other two channels uses 256 bins. In HSV histogram, hue use 16 bins, Saturation use 8 bins and value use 1 bin as described in Poorani, Prathiba, and Ravindran, (2013).

2. Using Local Binary Patterns in addition to color measurements increases the overall classification accuracy of the LBP operator in a 3x3 neighborhood which is better than that of the much larger Gabor filters as given in Pietikainen, Maenpaa, and Viertola, (2002).

3. We extend the work of Mohanty and Sethi, (2013) by including the features based on corner detection (Harris operator) instead of computing only the edge detection to increase the performance. By considering other operators such as LoG (Laplacian of Gaussian) and Prewitt we can calculate more features to classify an image only to a single class.

4. In this step we combine Appearance features like HOG features with geometric features for high recognition accuracy as described in Elhoseiny, Bakry, and Elgammal, (2013).

5. Fourier and wavelet transforms are used to effectively categorize the texture in real time as described in Materka and Strzelecki, (1998).

6. Difference of Histogram of Oriented Gradients view invariant feature is extracted which can clearly differentiate vehicles and people in various camera views, scenes with shadows described in Chen, Feris, Zhai, Brown, and Hampapur, (2012).

Step 3b: Feature Selection: PCA (Principal Component Analysis) a feature selection method is used to discard the irrelevant attributes and to reduce the dimensionality of the dataset.

Step 3c: BoV Representation: These selected features are represented as BoV (Bag of Visual Words).

Step 3d: Training: The BoV is used to train the classifier.

Step 4: Video is captured using camera and frames (29/sec) are extracted that forms testing images.

Step 5: Blobs are identified (Blob detection).

Step 6: If any occlusions, those are identified and blob refinement is done.

Step 7: Preprocessing is performed on the refined blobs as described in step2.

Step 8: Feature extraction and feature selection for the refined blobs (object) is performed as described in Step3a and Step3b.

Step 9: Object classification is done by supplying the test set on the trained classifier.

Step 10: Performance measures such as precision and recall are evaluated to analyze the accuracy and error rate.

Algorithm for Moving Object Classification

The following three algorithms describe the detailed working of the proposed approach. Algorithm 1 states the feature extraction process to construct the training data; Algorithm 2 is used for extracting features to construct testing data and Algorithm 3 describes the classification performed on training and testing data.

Input: Captured video (.mp4 or .avi or any other format).

Output: An object classified into a particular class (Humans, 2-wheeler {bicycle, bike}, 3-wheeler {auto}, 4-wheeler {bus, car}, other objects {bag, lunchbox etc.}).

Algorithm 1: Constructing Feature Set for Training the Classifier

Input: Sample images 1…n (n-total no of images)
Output: Features (train dataset)

```
//Feature Extraction
for 1: n do
//Color Features
```

$$P(g) = N(g) / M ;$$

$$g' = \sum_{0}^{L-1} gP(g) ;$$

$$\sigma = \sqrt{\sum_{K=0}^{L-1} P(g)\left(g - g'\right)^2} ;$$

$$Skew = \frac{1}{\sigma_{g^3}} \sum_{K=0}^{L-1} P(g)\left(g - g'\right)^3 ;$$

$$Skew' = \frac{g' - \left\{\left(P(g)\right)\right\}_{\max}}{\sigma_g}$$

$$Energy = \sum_{0}^{L-1} P(g)^2 ;$$

$$Entropy = -\sum_{0}^{L-1} P(g) \log_2 P(g) ;$$

```
//Structural Features
```

$$E\left(u,v\right) = \left[u,v\right]M\begin{bmatrix} u \\ v \end{bmatrix};$$

```
LoG(x,y) = -1/πσ⁴ [1- ((x²+y²)/2 σ²)]e⁻⁽⁽ˣ²⁺ʸ²⁾/² σ²⁾ ;
HoG=HOGDescriptor(win_size,block_size, threshold,levels);
//Texture Features
```

$$LBP = \sum_{p=0}^{p-1} 2p_s\left(i_p - i_c\right);$$

$$f\left(x\right) = \int_{-\infty}^{\infty} F\left(k\right)e^{2\pi ikx}dk;$$

```
 Implement feature selection and construct BoV;
write(f1); //f1 represents a file of train dataset
end
```

Algorithm 2: Constructing Testing Data

Input: Video (.mp4 or .avi)
Output: Features Extracted (test dataset)

```
if (video) then
Extract frames (1…m) from video;
Perform blob detection;
end
if(occluded objects) then
        Blob refinement;
end
//Feature Extraction
for 1: y do  //total no of blobs identified
//Color Features
```

$$P\left(g\right) = N\left(g\right)/M;$$

$$g = \sum_{0}^{L-1} gP(g);$$

$$g' = \sum_{0}^{L-1} gP(g);$$

$$\sigma = \sqrt{\sum_{K=0}^{L-1} P(g)\left(g - g'\right)^2};$$

$$Skew = \frac{1}{\sigma_{g^3}} \sum_{K=0}^{L-1} P(g)\left(g - g'\right)^3;$$

$$Skew' = \frac{g' - \left\{\left(P(g)\right)\right\}_{\max}}{\sigma_g};$$

$$Energy = \sum_{0}^{L-1} P(g)^2;$$

$$Entropy = -\sum_{0}^{L-1} P\left(g\right)\log_2 P\left(g\right);$$

```
                         //
              Structural Features
```

$$E\left(u, v\right) = \left[u, v\right] M \begin{bmatrix} u \\ v \end{bmatrix};$$

```
    LoG(x,y) = -1/πσ⁴ [1- ((x²+y²)/2 σ²)]e⁻((x2 +y2)/2 σ2) ;
    HoG=HOGDescriptor(win_size,block_size, threshold,levels);
//Texture Features
```

$$LBP = \sum_{p=0}^{p-1} 2p_s\left(i_p - i_c\right);$$

$$f(x) = \int\limits_{-\infty}^{\infty} F(k) e^{2\pi i k x} dk \, ;$$

```
        Implement feature selection;
  write(f2); //f2  represents a file of  test dataset
end
```

Algorithm 3: Classification

Input: Training and Testing datasets
Output: Assigns a class label C $\in$ C$_1$, C$_2$, C$_3$, C$_4$, C$_5$

```
Train the classifier using f1; //f1 - training data
Upload f2 for testing; //f2 - testing data
for 1…i do   //i - no of rows (data samples) in f1
 for 1…j do //j - no of rows (data samples) in f2
        if (features in f1 is nearly ∈features in f2) then
                Assigns (1…m) ∈ C;
 end
end
```

Implementation and Results

OpenCV and Weka tool are used for implementing moving object classification algorithm. OpenCV requires setup the connection between OpenCV library and Microsoft Visual Studio2010. A GUI is designed for human intervention in order to analyze the process. We provide security to the GUI application by having a login page before performing the entire process. By this only the authorized user can access the application. After the successful login, page is redirected to welcome page, which defines the steps (modules) need to be done in an application. Each step is designed with AWT Button component, on clicking the button it performs necessary action and outputs are saved in the ".txt" file. An alert box is provided which indicates an ERROR_MESSAGE for every module if something goes wrong with the GUI.

Data Sets Used

Training

To train the classifier, we have taken the different samples of human, 2-wheeler, 3-wheeler, 4-wheeler and other objects at different views. Total 133 samples are taken to form the training dataset as shown in Figure 9.

Figure 9. Training dataset

Testing

A real time video between Hyderabad to Vijayawada highway is considered as an input. This video is taken as it includes all the objects (human, 2-wheeler, 3-wheeler, 4-wheeler, and other objects) and it has dynamic background as shown in Figure 10.

Figure 10. Video considered for testing

Performance Measures

Performance is calculated for the different classification algorithms (SVM, AdaBoost, J48, and Bagging) depending on the parameters given below:

- **Confusion Matrix:** The confusion matrix is also known as contingency table or error matrix. It is a specific table that allows the visualization of the classification performance of the proposed algorithm as mentioned in the Table 1. In the confusion matrix, 'a' and 'b' are class labels and
  - TP indicates True Positives,
  - FP indicates False Positives,
  - TN indicates True Negatives,
  - FN indicates False Negatives.

Precision: It is the ratio of the number of matched objects retrieved to the total number of irrelevant and relevant matched objects retrieved and it is defined as in Equation 19 (han & Kamber, 2006)

$$Precision = \frac{tp}{tp + fp} * 100\% \tag{19}$$

where

tp = True positives, is the number of relevant matched objects retrieved.
fp = False positive, is the number of irrelevant matched objects retrieved.

- **Recall:** It is the ratio of the number of relevant matched objects retrieved to the total number of relevant matched objects present. It is defined as in Equation 20 (han & Kamber, 2006)

$$Recall = \frac{tp}{tp + fn} * 100\% \tag{20}$$

where

tp = True positive, is the number of relevant matched objects retrieved.
fn = False negatives, is the number of relevant matched objects not retrieved.

- **Error Rate:** It measures the average magnitude of the error.

Table 1. Confusion matrix 1

	A	B
actual a=0	TP	FN
actual b=1	FN	TP

Source: han & Kamber, (2006).

- **Accuracy:** The accuracy is measured using all the prediction values that are correct that is TP (true positives), FP (false positive), TN (true negative) and FN (false negative). It is defined as in Equation 21 (han & Kamber, 2006)

$$accuracy = \frac{tp + tn}{tp + tn + fp + fn} * 100\% \qquad (21)$$

Comparison

The Table 2 and Table 3 present the different classification results for National Highway video.

Classification is done using Weka tool, by using different classifiers such as SVM, AdaBoost, Bagging and J48. To perform classification, total 9 blobs are considered for testing, 7 objects are correctly classified and 2 objects are not classified by implementing our proposed approach using SVM classifier. This miss classification is due to the objects are not included in the training dataset. Even we predicted that the two objects will not be classified and hence our prediction is true in the case of miss classification. From the above results, it clearly show that our proposed approach using SVM classifier works better than the existing methods because proposed approach combines the different techniques such as edge detection operators, HoG, LBP and wavelet transforms whereas existing methodology uses only one of the above mentioned techniques.

Table 2. SVM and Adaboost classifiers comparison for existing and proposed method

Category	SVM Classification				AdaBoost Classification			
	[1]	[2]	[6]	Proposed	[1]	[2]	[6]	Proposed
Total instances	9	9	9	9	9	9	9	9
Correctly classified	1	3	6	7	3	5	5	3
Incorrectly classified	8	6	3	2	6	4	4	6
Error rate	0.30	0.29	0.25	0.25	0.28	0.26	0.25	0.25
Precision	0.12	0.66	0.75	0.83	0.11	0.43	0.43	0.11
Recall	0.11	0.33	0.66	0.77	0.33	0.55	0.55	0.33

Table 3. Bagging and J48 classifiers comparison for existing and proposed method

Category	Bagging Classification				J48 Classification			
	[1]	[2]	[6]	Proposed	[1]	[2]	[6]	Proposed
Total instances	9	9	9	9	9	9	9	9
Correctly classified	1	5	7	3	4	5	7	7
Incorrectly classified	8	4	2	6	5	4	2	2
Error rate	0.30	0.26	0.21	0.25	0.24	0.26	0.10	0.25
Precision	0.12	0.43	0.85	0.11	0.22	0.43	0.85	0.83
Recall	0.11	0.55	0.77	0.33	0.44	0.55	0.77	0.77

From the Table 2 and 3, precision for proposed approach is 0.83 which is better than 0.12, recall is 0.77 better than 0.11. When error rate is considered, the existing methods have higher error rate of 0.30 than 0.25 for proposed approach.

The Table 4 presents the individual class label results for National Highway video.

In the Table 4, accuracy is calculated based on the Equation (21). For the class "Human", accuracy is 100%, "Vehicles" class, accuracy is 71% and for "Other objects" class accuracy is 100%. Total overall accuracy for the proposed approach is 78%.

All the frames are stored in the frame database. We can view the frames by clicking on "view" button shown in Figure 11.

The Figure 12 shows the area of the blob which is detected.

The output can be seen by clicking on view button on GUI. The Figure 13 indicates the blobs that are identified.

All the features are saved in .csv file as training.csv and testing.csv shown in Figure 14 and Figure 15.

Both the training and testing csv files are converted into .arff (Attribute Relation File Format) which can be used for classification purpose. Arff files are given to different classifiers and the classification results are obtained.

Table 4. Individual class label results for proposed approach

Classification Algorithm	People Detection		Vehicle Detection		Other Objects		Accuracy
	TP	FP	TP	TP	TP	FP	
SVM	1	0	5	2	1	0	78%
J48	1	0	5	2	1	0	78%
AdaBoost	1	0	1	6	1	0	40%
Bagging	1	0	1	6	1	0	40%

Figure 11. Sample frames extracted from video

Figure 12. Blob detection

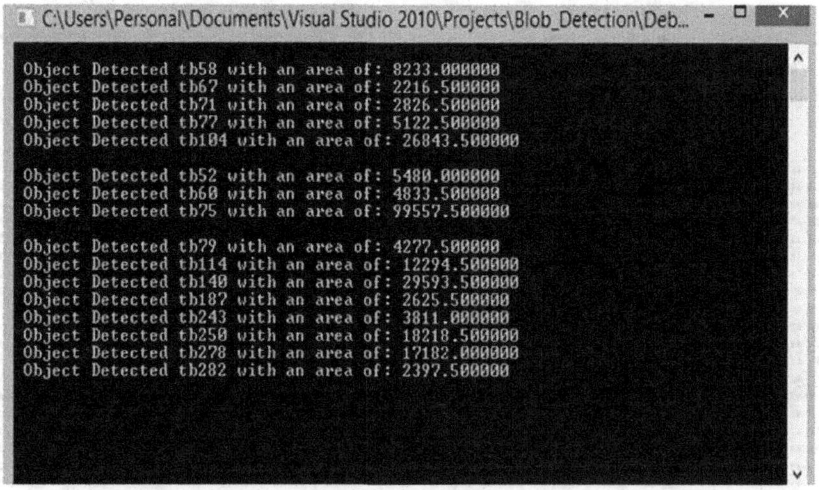

```
C:\Users\Personal\Documents\Visual Studio 2010\Projects\Blob_Detection\Deb...  —  ☐  ✕

Object Detected tb58 with an area of: 8233.000000
Object Detected tb67 with an area of: 2216.500000
Object Detected tb71 with an area of: 2826.500000
Object Detected tb77 with an area of: 5122.500000
Object Detected tb104 with an area of: 26843.500000

Object Detected tb52 with an area of: 5480.000000
Object Detected tb60 with an area of: 4833.500000
Object Detected tb75 with an area of: 99557.500000

Object Detected tb79 with an area of: 4277.500000
Object Detected tb114 with an area of: 12294.500000
Object Detected tb140 with an area of: 29593.500000
Object Detected tb187 with an area of: 2625.500000
Object Detected tb243 with an area of: 3811.000000
Object Detected tb250 with an area of: 18218.500000
Object Detected tb278 with an area of: 17182.000000
Object Detected tb282 with an area of: 2397.500000
```

Figure 13. Blobs that are identified

CONCLUSION

Main purpose of moving object classification is to classify the different types of objects that are present in a video while monitoring the video in the important areas. Our proposed approach uses frame extraction from a video, blob detection to identify the different objects in a frame based on area of the blob with a resolution of 256*256, feature extraction to know the color, structure and texture information of the object, classifying the object depending on the information.

To perform classification, video is taken and frames are extracted from a real time video at the rate of 29 frames per second. Blob detection is done on the frames by using parameters such as area, circularity and each blob is indicated by a rectangle. During blob detection, if the occluded area is less than 20%, blob is refined otherwise the blob is discarded. Preprocessing is done on each blob to get resolution of 256*256. RGB and HSV color features are extracted from each blob and a total of 36 color features are obtained. Total of 144 structural features are extracted using Canny, Prewitt, LoG, Harris, and HoG operators. 102 Texture features are extracted using LBP, Fourier and Haar wavelet transform. These features

Figure 14. CSV file indicating features of train dataset

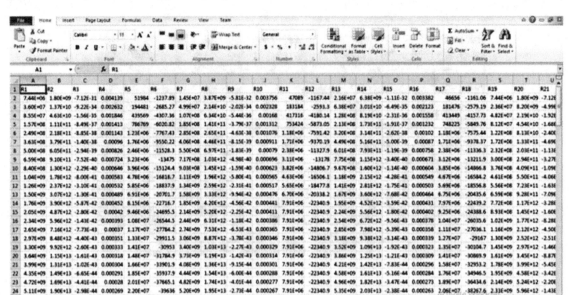

Figure 15. CSV file indicating features of test dataset

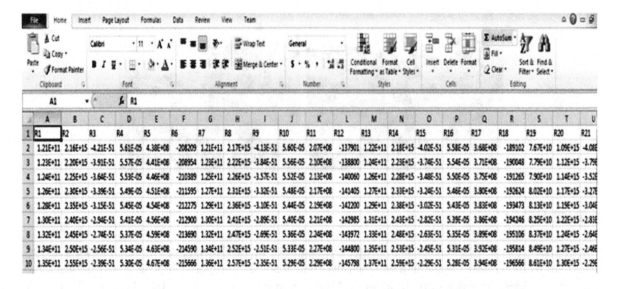

are computationally inexpensive and all these features are saved in CSV file format, later converted to ARFF file. Training and testing is done by using ARFF files. Classification is done using Weka tool, by using different classifiers such as SVM, AdaBoost, Bagging and J48. The performance of proposed system is mainly calculated based on precision, recall, error rate and accuracy. Total 9 blobs are considered for testing, 7 blobs are vehicles, 1 blob is human and another 1 blob is a tree (other object). Out of 9 blobs, 7 blobs are correctly classified and 2 blobs are not classified as they are not in the part of training dataset which is considered. Our results proved that SVM classifier works better compared to remaining classifiers. SVM classifier works well in outdoor environments. We have considered 133 samples for training and the performance can be improved by considering 300-400 samples as training dataset.

Our future work is to apply the proposed method on challenging surveillance videos with occluded object classes. This work can be extended by calculating Zernike moments.

REFERENCES

Chen, Feris, Zhai, Brown, & Hampapur. (2012). *An Integrated System for Moving Object Classification in Surveillance Videos*. Academic Press.

Elhoseiny, Bakry, & Elgammal. (2013). Multi Class Object Classification in Video Surveillance Systems. In *Proceedings of CVPR 2013*, (pp. 788-793).

Han & Kamber. (2006). *Data Mining: Concepts and Techniques* (2nd ed.). Morgan Kaufmann Publishers.

Kaehler, A., & Bradski, G. (2013). *Learning OpenCV – Computer Vision in C++ with OpenCV Library* (1st ed.). O'Reilly Media.

Lee, B. Y., View, L. H., Cheah, W. S., & Wang, Y. C. (2014). *Occlusion Handling in Videos Object Tracking: A Survey*. IOP Conference.

Materka & Strzelecki. (1998). Texture Analysis Methods – A Review. Institute of Electronics, COST B11 report.

Mohanty & Sethi. (2013). A Frame Based Decision Pooling Method for Video Classification. *Annual IEEE India Conference (INDICON)*.

Patel & Patel. (2013). Illumination Invariant Moving Object Detection. *International Journal of Computer and Electrical Engineering, 5*(1).

Pepik, Stark, Gehler, & Schiele. (2013). *Occlusion Patterns for Object Detection*. CVPR 2013.

Pietikainen, Maenpaa, & Viertola. (2002). *Color Texture Classification with Color Histograms and Local Binary Patterns*. Machine Vision Group.

Poorani, M., Prathiba, T., & Ravindran, G. (2013). Integrated Feature Extraction for Image Retrieval. *IJCSMC, 2*(2), 28–35.

San Biagio, Bazzani, Cristani, & Murino. (2014). *Weighted Bag of Visual Words for Object Recognition*. Academic Press.

Voravuthikunchai, W., Cremilleux, B., & Jurie, F. (2014). Histograms of Pattern Sets for Image Classification and Object Recognition. *CVP*.

Chapter 5
Image Registration Techniques and Frameworks:
A Review

Sayan Chakraborty
Bengal College of Engineering and Technology, India

Prasenjit Maji
Bengal College of Engineering and Technology, India

Prasenjit Kumar Patra
Bengal College of Engineering and Technology, India

Amira S. Ashour
Tanta University, Egypt

Nilanjan Dey
Techno India College of Technology, India

ABSTRACT

Image registration allude to transforming one image with reference to another (geometrically alignment of reference and sensed images) i.e. the process of overlaying images of the same scene, seized by assorted sensors, from different viewpoints at variant time. Virtually all large image evaluating or mining systems require image registration, as an intermediate step. Over the years, a broad range of techniques has been flourished for various types of data and problems. These approaches are classified according to their nature mainly as area-based and feature-based and on four basic tread of image registration procedure namely feature detection, feature matching, mapping function design, and image transformation and resampling. The current chapter highlights the cogitation effect of four different registration techniques, namely Affine transformation based registration, Rigid transformation based registration, B-splines registration, and Demons registration. It provides a comparative study among all of these registration techniques as well as different frameworks involved in registration process.

DOI: 10.4018/978-1-5225-1022-2.ch005

INTRODUCTION

Image registration (Lucas & Kanade 1981) is a significant process in several domains including computer vision, medical imaging, biological imaging and brain mapping, military automatic target recognition, and compiling/analyzing satellites' images and data. It is the act for transforming different sets of data into one coordinate system. This process involves designating one image as a reference image (fixed image), while applying geometric transformations to the other images, thus they align with the reference image. A geometric transformation maps locations in one image to new locations in another image. Consequently, the key step for the image registration process is to determine the correct geometric transformation parameters. Image registration allows the common features comparison in different images (Irani & Peleg 1991; Li et al., 2005). Data may be multiple photographs or data from different sensors; times; depths; or viewpoints. Registration (Christensen & He, 2001) is necessary in order to be able to compare or integrate the data obtained from these different measurements.

There are various algorithms that employed for image registration. The registration techniques can be categorized based on different parameters as follows. The registration methods categories based on image alignment are:

1. Intensity-based method that compares the images' intensity patterns using correlation metrics. The registration (Hong & Zhang, 2005) is performed for the entire images or for a part of that image (sub-images), where for sub-images registration the centers of corresponding sub-images are treated as corresponding feature points.
2. Feature-based method that treats the image features (lines, points and contours) as a parameter to find the correspondence between a numbers of especially distinct points in images (Siu & Lau, 2005; Wahed *et al.*, 2013). Knowing the correspondence between a numbers of points in images, a geometrical transformation is then determined to map the target image to the reference images, thereby establishing point-by-point correspondence between the reference and target images.

The registration methods categories based on the domain are:

1. Frequency domain method: it determines the transformation parameters for the images registration while working in the transform domain. Such method uses transformations, such as translation, rotation, and scaling. Then, apply phase correlation method to a pair of images, which produces a third image that contains a single peak.
2. Spatial domain method: it operates in the image domain to match the intensity patterns or features in the images. The corresponding control points (CP) are chosen from the images. When the CP number exceeds the minimum requirements to define the appropriate transformation model, iterative algorithms, such as the RANSAC (Random sample consensus) can be used to robustly estimate the parameters of a particular transformation type (e.g. affine) for the images registration.

The registration methods categories based on the source of images are:

1. Single methods or mono modal methods: such methods tend to register images (Ashburner & Friston, 2007; Wang *et al.*, 2011) in the same modality acquired by the same scanner/sensor type.

2.   Multi-modality methods: multi-modality registration methods tended to register images acquired by different scanner/sensor types.

The registration methods categories based on the level of automation are:

1.   Automatic methods: do not allow any user interaction and perform all registration steps automatically (Smeets *et al.*, 2012).
2.   Interactive methods: reduce the user bias by performing certain key operations while still relying on the user to guide the registration (Janiczek *et al.*, 2005).
3.   Manual methods: provide tools to align the images manually.
4.   Semi-automatic methods: perform more of the registration steps automatically but depend on the user to verify the correctness of a registration (Ho *et al.*, 2007).

The registration methods categories based on the used transformation models to relate the target image space to the reference image space, which are:

1.   Linear transformations methods: include rotation, scaling, translation, and other affine transforms. Linear transformations are global in nature, thus they cannot model local geometric differences between images.
2.   'Elastic' or 'Non-rigid' transformations: These transformations (Li *et al.*, 2013) are capable of locally warping the target image to align with the reference image. Non-rigid transformations include radial basis functions (thin-plate or surface splines, multi-quadrics, and compactly-supported transformations), physical continuum models (viscous fluids), and large deformation models.

Consequently, the main contribution of the current study is to analyze four different image registration algorithms, namely:

1.   Rigid registration (Araki *et al.*, 2015),
2.   Affine registration (Araki *et al.*, 2015),
3.   B-Splines registration, and
4.   Demons registration.

Rigid transformation (Lakshmanan *et al.*, 2013; Xie *et al.*, 2004) was first proposed by Bottema and Roth. It is mainly depends on scaling, rotation or combination of these two techniques. Affine registration is mainly based on affine transformation, which is similar to rigid transformation. However, the affine registration is also applied in shear mapping. Affine transformations (Tustison *et al.*, 2007) were first described by Berger. In the year 2000, Boor proposed a new transformation method based on splines. Later, this technique was further revised and modified into B-Splines registration. B-Splines registration uses unique features like control points, weighting function which leads to spline curves generation and grid build up. Demons algorithm (Hansen *et al.*, 2008; Bai & Brady, 2008) proposed by Thirion that uses displacement deviator between two images for transformation. Previously, lot of research has been done in the area of image registration which has used these registration techniques. To our best knowledge, no previous work has compared the effect of these four image registration techniques and their effect on a video content. Comparison among various methods based on their properties is described in the current work.

LITERATURE REVIEW

Image registration domain attracts many researchers. A brief literature review to significant image registration work is described in this section. Lucas and Kanade (1981) proposed an iterative image registration technique with an application to stereo vision proposed a new image registration technique that makes use of the spatial intensity gradient of the images to find a good match using a type of Newton-Raphson iteration. Irani and Peleg (1991) proposed improving resolution using image registration. They proposed the approach of back projection used in tomography. The improved resolution was given for gray level and color images when there is an unknown image displacement. Li *et al.* (2005) introduced a contour-based approach to multi-sensor image registration proposed that they present two contour based method which use region boundaries and other strong edges as matching primitives.

In 2001, Christensen and He (2001) worked on a consistent nonlinear elastic image registration method. In this work, the authors described a new bidirectional image registration algorithm that estimates a consistent set of nonlinear forward and reverse transformations between two N dimensional images. Later, Hong, and Zhang (2005) proposed an image registration technique for high resolution remote sensing image in hilly area. An automated image registration technique that based on the combination of feature-based and area-based matching was proposed. Wavelet-based feature extraction technique and relaxation-based image matching technique are employed in this research.

In the same year, Siu and Lau (2005) proposed image registration based on IBR (Image-Based Rendering). The requirements of an image registration technique for reducing the spatial sampling rate and based on those requirements were analyzed. A novel image registration technique to automatically recover the geometric proxy from reference images was presented using chain-code correlation shape similarity criteria. Recently, Wahed *et al.* (2013) proposed automatic image registration technique of remote sensing images. An image registration technique of multi-view, multi-temporal and multispectral remote sensing images was proposed. Initially, a preprocessing step was performed by applying median filtering to enhance the images. Secondly, the steerable pyramid transform was adopted to produce multi-resolution levels of reference and sensed images. Apart from the mentioned work, much work has been done in the field of image registration over the year.

In the following section, different image registration techniques are discussed. Followed a comparative study of image registration framework is included. Afterward, a comparative study among the recent works in image registration field is presented. Finally, the conclusion of the chapter is depicted.

IMAGE REGISTRATION TECHNIQUES

Since, there is various image registration techniques according to the rigid transformation of matrices or based on the non-rigid transformation operation. Thus, in the current work, these techniques are discussed and grouped into four main methods.

Rigid Registration

Rigid registration is mainly based on rigid transformation, which includes operations on matrices such as rotation, translation, reflection. Ashburner and Friston (2007) used rigid registration technique to discuss the analysis of functional brain images and their statistical mapping. The authors used different modality images of PET (Positron Emission Tomography) and MRI (Magnetic resonance imaging) scans. Afterward, both of the brain image set using rigid registration technique were mapped. In 2011, Wang *et al.* used two-dimensional rigid registration in order to discuss the geometric topological inference algorithm (GTI). The rigid registration method was tested on synthetic (Wang *et al.*, 2011) data sets and real SAR (Synthetic aperture radar) images. Later, Smeets *et al.* (2012) introduced a novel feature-based piecewise rigid registration that was applied on 2-D medical images. In this work, points of interest in the medical images were obtained and compared with the results of SIFT (Scale Invariant Feature Transform) algorithm. The multiple rigid motions (Smeets *et al.*, 2012) were sampled and clustered using the mean shift algorithm. Rigid registration being one of the most widely used techniques for transformation as well as image registration. It provides one of the simplest frameworks for images transformation of images. Since, it involves basic transformation operations; hence it is now considered to be one of the least popular technique among the medical image registration field.

Non-Rigid Affine Registration

Affine registration is one of the oldest non-rigid transformation based technique. Affine transformation is a combination of rigid and non-rigid transformation. It includes translation, compositions, scaling, similarity transformation, rotation, shear mapping and reflection. In (Janiczek *et al.*, 2005) a novel affine registration on magnetic resonance (MR) images was proposed. The active contours were used to analyze the registration errors in first-pass MR analysis.

Ho *et al.* (2007) applied affine registration to match 2-D point sets. Two step algorithms were used to reduce the orthogonal cases and the unknown rotation. This algorithm was involved to compute the roots of low degree polynomial with coefficients which was applied on synthetic 2D point sets. In addition, this algorithm was used as feature matching algorithm of those point sets. Li *et al.* (2013) used the affine registration and applied optimization of mutual information. A stochastic gradient approximation technique was used to tune the parameters of affine registration, which led to optimize the mutual information of two image frames. Later in 2013, on medical registration, affine transformation was also applied by Lakshmanan *et al.* (2013) applied the affine transformation on the medical registration. This novel work included registration of brain MRI images using affine transformation. Thus, the affine registration was introduced to medical field much later than rigid registration, although in recent works it turned out to be one of the finest non-rigid registration method compared to rigid one.

B-Splines Registration

Affine is considered to be one of the registration methods that based on non-rigid transformation. Based on the B-Splines curves and the transformation using B-Splines curve, B-Splines registration was introduced. B-Splines registration plays a major role on the neighboring points or control points. These control points lead to mesh creation. Dense transformation fields between control points help the interpolation in B-Splines. Equation 1 is used for the cubic function ($Q(u)$) which builds the B-Spline curves (Araki *et al.*, 2015).

$$Q(u) = a_3 u^3 + a_2 u^2 + a_1 u^1 + a_0 \tag{1}$$

where, a_0, a_1, a_2 and a_3 are parameters. The value of u varies from 0 to 1. Thus, to generate each control point the following expression is used:

$$b_0(u) = \frac{1}{6} u^3$$

$$b_1(u) = \frac{1}{6}[1 + 3u + 3u^2 - 3u^3]$$

$$b_2(u) = \frac{1}{6}[4 - 6u^2 + 3u^3]$$

$$b_3(u) = \frac{1}{6}[1 - 3u + 3u^2 - 3u^3] \tag{2}$$

Here, $b_0(u), b_1(u), b_2(u)$ and $b_3(u)$ are known as the basis function that controls the generation of the CPs.

Xie and Farin (2004) proposed an image registration technique using hierarchical B-Splines. An application using image registration to match the images at increasing level of detail was presented. This registration was applied on brain MRI images. Nicholas *et al.* (2007) improved the Free-Form Deformation (FFD) B-Spine image registration technique. In this work, assertion of straight forward gradient learning in suboptimal in certain cases was done in order to remedy the sub optimality present in free form deformation. The Thirion's demons registration was taken as reference to solve the sub optimality issue of the FFD B-splines registration. Parameterization of B-Splines image registration was introduced in Hansen *et al.*, (2008). The control points' measurements were refined using this method. The registration cost was minimized as well as the parameters of registration were minimized to make the registration process faster and efficient. Bai & Brady (2008) applied B-splines registration on PET images in order to observe respiratory motion correction. In this B-Splines registration process Markov random field was used to regularize the deformation in B-Splines. The algorithm was later applied on PET images which were aligned to effectively suppress the noise present in the images. Lijuan *et al.* (2012) presented highly accurate B-splines image registration. In this algorithm, a novel B-splines based approach was presented in order to dense image registration which was based on deformation model of B-Splines.

Demons Registration

Demons registration is one of the most popular non-rigid registration techniques because of its linear computational complexity as well as the easier implementation feature. It successfully solves the diffusion problems present in registration by estimating force vectors that helps the deformation to align properly and kernels of Gaussian convolution smoothens the force vectors. Araki *et al.* (2015) calculated the displacement vector $D_{i,j}$ between reference image (a) and target image (b) using the following expression:

$$D_{i,j}^{n} = D_{i,j}^{n-1} - \frac{(b_{i,j}^{n-1} - a_{i,j}^{0})\nabla a_{i,j}^{0}}{|\nabla a_{i,j}^{0}|^{2} + |(b_{i,j}^{n-1} - a_{i,j}^{0})|^{2'}} \tag{3}$$

for i, j = 1,2,3,....., N using the following initial conditions:

$$D_{i,j}^{0} = 0$$
$$b_{i,j}^{0} = \hat{b}_{i,j} \tag{4}$$
$$a_{i,j}^{0} = \hat{a}_{i,j}$$

Where, $\hat{b}_{i,j}$ and $\hat{a}_{i,j}$ refer to the intensities of the original target and reference images at the corresponding pixel (i,j) and N is any positive integer.

Lu *et al.* (2010) proposed demons registration based on point wise mutual information. A proposed a diffeomorphic demons registration technique was used, where the standard demons metric was replaced by point wise mutual information. The authors compared the B-Splines registration and demons registration in order to establish the superiority of their method. Later, Freiman *et al.* (2011) proposed an affine based demons registration technique, where they combined the smoothening kernels of Thirion's demons with affine transformation. The method was applied on the CT images to demonstrate the algorithm's robustness. Mishra *et al.* (2012) modified the demons registration to reduce the time complexity of the process. In this paper, it was focused on reducing the runtime of the demons registration process instead of focusing on the alignment errors present in demons algorithm. Mishra *et al.* (2015) introduced another modified version of the demons algorithm. Modifications to the smoothing kernels were performed. By tweaking smoothing kernels [16], it was obtained faster registration technique with improved accuracy during the registration process. In the same year Mishra *et al.* used demons registration on the VLSI (Very Large Scale Integration). The modified demons algorithm was synthesized in Virtex6-xc6vlx760-2-ff1760 and they observed better frequency during the process.

IMAGE REGISTRATION FRAMEWORKS

During registration, capturing devices can be the same or different. According to modality (image capturing devices) image registration frameworks can be divided into two subsections, namely:

1. Monomodal Framework, and
2. Multimodal framework.

Monomodal Frameworks

During image registration process, the multiple image frames are processed, which are usually captured using the same device. For example, if an image of bouncing ball which is captured by a camera is registered by any of the image registration processes, then it is considered to be monomodal registration.

Typically, during monomodal registration, a video is broken into multiple image frames and those frames are further registered using image registration.

Sassi *et al.* (2008) used SSIM (structure similarity index) in order to register MR images using monomodal registration. The luminance and contrast component comparison assisted to develop this technique, which was applied on the MR images. Al-Azzawi *et al.* (2010) used non-subsampled contourlet transform and mutual information to execute monomodal registration on MR images. Efficiency of multi-resolution representation was used to extract salient edges from medical images. The MR images were decomposed using contourlet transform and the mutual information based image registration was done. Ghaffari and Fatemizadeh (2013) used correntropy measure for mono-modal image registration. In this work, correntropy measure which is a measurement between two random variables based on information theoritic learning was used to register using monomodal registration. Mutual information and SSD (Systems, Signals and Devices) both were used during registration. Ghaffari and Fatemizadeh (2013) proposed a monomodal registration based on sparse based similarity measure. A combined SSD, CC, MI and CR similarity measures in order to structure the image transformation matrix was done.

Multimodal Frameworks

Multimodal registration refers to the image registration process, where the images are captured using different devices are processed. For example, if the same medical image is captured using MRI and CT scan, then using the image transformation matrix of the MRI to CT scan can be used as multimodal registration technique. This process is very much effective in medical field as the transformation matrix that can help to obtain different type of scanned images. Makrogiannis *et al.* (2007) proposed multimodal image registration based fusion methodology which was applied to drug discovery research. In this work, multimodal registration and fusion of PET and MRI was done. An optimization of this registration technique using genetic algorithm was proposed. Reducindo *et al.* (2010) introduced multimodal registration based on particle filter. This registration technique was based on Bayesian estimation theory especially on particle filters. This proposed method was used to reduce the runtime of the multimodal registration technique. Pradeepa and Vennila (2012) proposed multimodal registration using mutual information. The optimization of CT and MRI images were done by down sampling and image registration based on mutual information. Recently, Hernandez *et al.* (2015) used multimodal registration on multiple retinal images. This method was based on line structures. As the retinal images had several modality-invariant features, hence multimodal registration was needed to stabilize the framework. The registration process was based on salient line structures and was aligned to minimize the chamfer distance. Figure 1 demonstrated the registration process according to the used modality.

COMPARATIVE STUDY

An explanation and comparison between different image registration techniques and algorithms are conducted in the current work. Generally, image registration has been done based on the mutual information of multiple images. Nevertheless, the main focus of the current study is related to techniques based on transformation algorithms. In addition, the image registration frameworks are introduced according to the used modality. A comparative study is depicted in Table 1 for the most important image registration techniques.

Figure 1. Registration process according to the modality

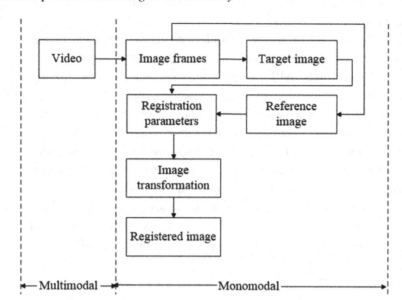

Table 1. Image registration review in tabular form

Authors/Year	Advantages	Disadvantages
Lucas and Kanade (1981)	An Iterative Image Registration Technique with an Application to Stereo Vision proposed a new image registration technique that makes use of the spatial intensity gradient of the images to find a good match using a type of Newton-Raphson iteration.	Optimization of the parameters was not performed in this work. Moreover, the time complexity was not discussed.
Irani and Shmuel Peleg (1991)	Image registration technique was proposed by improving resolution by image registration. The authors proposed the approach of back projection used in tomography. The improved resolution was given for grey level and colour images when there was an unknown image displacement.	Only the resolution was improved, while other mutual information among images was not considered. Modality of image registration was not involved.
Li and Mitra (1995)	Contour-Based approach to multi-sensor image registration proposed that they present two contour based method, which use region boundaries and other strong edges as matching primitives. A chain-code correlation shape similarity criteria was included.	Only correlation was used as similarity index, while other SSIM parameters were not used. Time complexity was higher than rigid registration.
Christensen and He (2001)	This paper proposed Consistent Nonlinear Elastic Image Registration. A new bidirectional image registration algorithm was described to estimate a consistent set of nonlinear forward and reverse transformations between two N dimensional images.	Registration of color images and modality of images were not discussed in this method.
Hong and Zhang (2005)	A new automated image registration technique was proposed based on the combination of feature-based and area-based matching. Wavelet-based feature extraction technique and relaxation-based image matching technique were employed. Local distortions caused by terrain relief can be greatly reduced in this procedure. The IKONOS and QuickBird data were conducted to evaluate this technique.	Matching techniques were used to apply transformation, rather than using existing registration algorithm. Distortion was less but transformation was not better.
Siu and Lau (2005)	Image registration based on IBR was employed. The requirements of an image registration technique was analysed for reducing the spatial sampling rate. Based on those requirements, a novel image registration technique to automatically recover the geometric proxy from reference images was used.	Image enhancement prior to registration was not discussed. Parameters of registration were not optimized to make the method robust.
Wahed and El-karim (2013)	Automatic image registration technique of remote sensing images was suggested. In this work, it was proposed an image registration technique of multi-view, multi-temporal and multispectral remote sensing images. Firstly, a pre-processing step was performed by applying median filtering to enhance the images. Secondly, the Steerable Pyramid Transform was adopted to produce multi-resolution levels of reference and sensed images.	Deduction of runtime of registration process was not discussed. Registration process was not verified on monomodal images.

From the preceding survey, it is recommended in future further optimization of various image registration techniques and their parameters can open a wide research area. Reducing runtime of the process can also play a huge part in future work.

CONCLUSION

Aligning images or transforming one image with reference to other can be said as image registration techniques. Image registration has different goals and is applicable to various fields. It is one of the most researched topics in the recent era. As the image registration frameworks and techniques were introduced, more focus was done on the optimization of the frameworks and stabilization of the techniques. Reducing time and space complexity has been marked as one of the major issue in image registration. Due to less transformation, rigid algorithm might have the smallest time complexity among other available algorithms, but as it is having minimal effect during transformation, thus it is neglected in the medical field. Medical imaging mostly trusts multimodal demons and B-splines techniques as they have the most stabilized and robust framework. Mutual information based methods are mostly used according to line structures extracted from images. However, being a less effective method makes the mutual information based registration less popular among the researchers. Generally, it is concluded that image registration is still pilot in nature.

REFERENCES

Al-Azzawi, N. A., Sakim, H. A. M., & Abdullah, W. A. K. W. (2010). MR image monomodal registration based on the nonsubsampled contourlet transform and mutual information.*2010 International Conference on Computer Applications and Industrial Electronics (ICCAIE)*, (pp. 481 - 485). doi:10.1109/ICCAIE.2010.5735128

Araki, T., Ikeda, N., Dey, N., Chakraborty, S., Saba, L., Kumar, D., & Suri, J. S. et al. (2015). A comparative approach of four different image registration techniques for quantitative assessment of coronary artery calcium lesions using intravascular ultrasound. *Computer Methods and Programs in Biomedicine*, *118*(2), 158–172. doi:10.1016/j.cmpb.2014.11.006 PMID:25523233

Ashburner, J. T., & Friston, K. J. (2007). *Rigid body registration. In Statistical Parametric Mapping: The Analysis of Functional Brain Images* (pp. 49–62). Academic Press. doi:10.1016/B978-012372560-8/50004-8

Bai, W., & Brady, M. (2008). Regularized B-spline deformable registration for respiratory motion correction in PET images.*2008 IEEE Nuclear Science Symposium Conference Record.*

Christensen, G. E., & He, J. (2001). Consistent nonlinear elastic image registration. *MM-BI, A01*, 1–5.

Freiman, M., Voss, S. D., & Warfield, S. K. (2011). Demons registration with local affine adaptive regularization: application to registration of abdominal structures.*2011 IEEE International Symposium on Biomedical Imaging: From Nano to Macro.* doi:10.1109/ISBI.2011.5872621

Ghaffari, A., & Fatemizadeh, E. (2013). Mono-modal image registration via correntropy measure. *2013 8th Iranian Conference on Machine Vision and Image Processing* (MVIP), (pp. 223 - 226). doi:10.1109/IranianMVIP.2013.6779983

Ghaffari, A., & Fatemizadeh, E. (2013). Sparse based similarity measure for mono-modal image registration. *2013 8th Iranian Conference on Machine Vision and Image Processing* (MVIP), (pp. 462 - 466).

Hansen, M. S., Larsen, R., Glocker, B., & Navab, R. (2008). Adaptive parametrization of multivariate B-splines for image registration. *IEEE Conference on Computer Vision and Pattern Recognition.* doi:10.1109/CVPR.2008.4587760

Hernandez, M., Medioni, G., Hu, Z., & Sadda, H. (2015). Multimodal Registration of Multiple Retinal Images Based on Line Structures.*2015 IEEE Winter Conference on Applications of Computer Vision,* (pp. 907 - 914). doi:10.1109/WACV.2015.125

Ho, J., Yang, M. H., Rangarajan, A., & Vemuri, B. (2007). A New Affine Registration Algorithm for Matching 2D Point Sets.*IEEE Workshop on Applications of Computer Vision.* doi:10.1109/WACV.2007.6

Hong, G., & Zhang, Y. (2005). The Image Registration Technique for High Resolution Remote Sensing Image in Hilly Area. *International Society of Photogrammetry and Remote Sensing Symposium.*

Irani, M., & Peleg, S. (1991). Improving resolution by image registration. *CVGIP: Graphical Models and Image Proc., 53,* 231–239.

Janiczek, R. L., Gilliam, A. D., Antkowiak, P., Acton, S. T., & Epstein, F. H. (2005). Automated Affine Registration of First-Pass Magnetic Resonance Images.*Conference Record of the Thirty-Ninth Asilomar Conference on Signals, Systems and Computers.* doi:10.1109/ACSSC.2005.1599747

Lakshmanan, A. G., Swarnambiga, A., Vasuki, S., & Raja, A. A. (2013). Affine based image registration applied to MRI brain.*2013 International Conference on Information Communication and Embedded Systems (ICICES).* doi:10.1109/ICICES.2013.6508186

Li, H., Manjunath, B. S., & Mitra, S. K. (1995). A contour based approach to multisensor image registration. *IEEE Transactions on Image Processing, 4*(3), 320–334. doi:10.1109/83.366480 PMID:18289982

Li, Q., Sato, I., & Murakami, I. (2007). Affine registration of multimodality images by optimization of mutual information using a stochastic gradient approximation technique.*2007 IEEE International Geoscience and Remote Sensing Symposium.* doi:10.1109/IGARSS.2007.4422814

Lijuan, Z., Dongming, L., Junnan, W., & Hui, Z. (2012). High-accuracy image registration algorithm using B-splines. *2012 2nd International Conference on Computer Science and Network Technology (ICCSNT),* (pp. 279 - 283). doi:10.1109/ICCSNT.2012.6525938

Lu, H., Reyes, M., Šerifović, A., Weber, S., Sakurai, Y., Yamagata, H., & Cattin, P. C. (2010). Multi-modal diffeomorphic demons registration based on point-wise mutual information.*2010 IEEE International Symposium on Biomedical Imaging: From Nano to Macro,* (pp. 372 – 375). doi:10.1109/ISBI.2010.5490333

Lucas, B., & Kanade, T. (1981). An iterative image registration technique with an application to stereo vision.*Proc. DARPA Image Understanding Workshop,* (pp. 121–130).

Makrogiannis, S., Wellen, J., Wu, Y., Bloy, L., & Sarkar, S. K. (2007). A Multimodal Image Registration and Fusion Methodology Applied to Drug Discovery Research. *IEEE 9th Workshop on Multimedia Signal Processing.* doi:10.1109/MMSP.2007.4412883

Mishra, A., Mondal, P., & Banerjee, S. (2012). Modified Demons deformation algorithm for non-rigid image registration. *2012 4th International Conference on Intelligent Human Computer Interaction* (IHCI), (pp. 1 - 5). doi:10.1109/IHCI.2012.6481800

Mishra, A., Mondal, P., & Banerjee, S. (2015). VLSI-Assisted Nonrigid Registration Using Modified Demons Algorithm. *IEEE Transactions on Very Large Scale Integration (VLSI) Systems, 23*(12), 2913–2921.

Mishra, B., Pati, U. C., & Sinha, U. (2015). Modified demons registration for highly deformed medical images.*2015 Third International Conference on Image Information Processing (ICIIP)*, (pp. 152 - 156). doi:10.1109/ICIIP.2015.7414757

Pradeepa, P., & Vennila, I. (2012). A multimodal image registration using mutual information.*2012 International Conference on Advances in Engineering, Science and Management (ICAESM)*, (pp. 474 - 477).

Reducindo, I., Arce-Santana, E. R., Campos-Delgado, D. U., & Alba, A. (2010). Evaluation of multimodal medical image registration based on Particle Filter. *2010 7th International Conference onElectrical Engineering Computing Science and Automatic Control* (CCE), (pp. 406 - 411). doi:10.1109/ICEEE.2010.5608648

Sassi, O. B., Delleji, T., Taleb-Ahmed, A., Feki, I., & Hamida, A. B. (2008). *MR Image Monomodal Registration Using Structure Similarity Index. 2008 First Workshops on Image Processing Theory* (pp. 1–5). Tools and Applications.

Siu, A., & Lau, E. (2005). Image Registration for Image-Based Rendering. *IEEE Transactions on Image Processing, 14*(1), 241–252. doi:10.1109/TIP.2004.840690 PMID:15700529

Smeets, D., Keustermans, J., Hermans, J., Vandermeulen, D., & Suetens, P. (2012). Feature-based piecewise rigid registration in 2-D medical images. *2012 9th IEEE International Symposium on Biomedical Imaging* (ISBI).

Tustison, N. J., Avants, B. A., & Gee, J. C. (2007). Improved FFD B-Spline Image Registration. *2007 IEEE 11th International Conference on Computer Vision.*

Wahed, M., El-tawel, G. S., & El-karim, A. G. (2013). Automatic Image Registration Technique of Remote Sensing Images. *International Journal (Toronto, Ont.), 4*, 177–187.

Wang, W., Liu, L., Jiang, Y., & Kuang, G. (2011). Point-based rigid registration using Geometric Topological Inference algorithm. *2011 3rd International Asia-Pacific Conference on Synthetic Aperture Radar* (APSAR), (pp. 1-3).

Xie, Z., & Farin, G. E. (2004). Image registration using hierarchical B-splines. *IEEE Transactions on Visualization and Computer Graphics, 10*(1), 85–94. doi:10.1109/TVCG.2004.1260760 PMID:15382700

ADDITIONAL READING

Chakraborty, S., Dey, N., Nath, S., Roy, S., & Acharjee, S. (2014*).Effects of rigid, affine, b-splines and demons registration on video content: A review.2014 International Conference on Control, Instrumentation, Communication and Computational Technologies (ICCICCT)*, pp. 497 - 502. doi:10.1109/ICCICCT.2014.6993013

de Boor, C. (1978). *A Practical Guide to Splines*. Springer-Verlag. doi:10.1007/978-1-4612-6333-3

Thirion, J. P. (1998). Image mat ching as a diffusion process: An analogy with Maxwell's demons. *Medical Image Analysis*, 2(3), 243–260. doi:10.1016/S1361-8415(98)80022-4 PMID:9873902

KEY TERMS AND DEFINITIONS

Affine Registration: Is the registration process that combines rotation, scaling, shear mapping, and translation.

B-Splines Registration: The B-Splines curve helps to register images.

Image Registration: Is the process of registering one image with reference to other.

Image Transformation: Refers to transformation of images according to some transformation matrix.

Rigid Registration: Is the registration process that combines rotation and scaling.

Section 2
Feature Detectors and Descriptors

Feature detectors and descriptors have a vital role in numerous applications including video camera calibrations, object recognition, biometrics, medical applications and image/video retrieval. Extract point correspondences "Interest points" between two similar scenes, objects, images/video shots is their main task. This section outlines the different feature detectors and descriptors types that involved in various applications.

Chapter 6

An Overview of Steganography:
"Hiding in Plain Sight"

Al Hussien Seddik Saad
Minia University, Egypt

Abdelmgeid Amin Ali
Minia University, Egypt

ABSTRACT

Nowadays, due to the increasing need for providing secrecy in an open environment such as the internet, data hiding has been widely used. Steganography is one of the most important data hiding techniques which hides the existence of the secret message in cover files or carriers such as video, images, audio or text files. In this chapter; steganography will be introduced, some historical events will be listed, steganography system requirements, categories, classifications, cover files will be discussed focusing on image and video files, steganography system evaluation, attacks, applications will be explained in details and finally last section concludes the chapter.

INTRODUCTION

Steganography is a Greek word which means covered writing that comes from two words stegauw (steganos) and grafein (graphein). It is a strategy in which secret message is concealed in any other information or cover file in such a way that the cover file does not change significantly and appears same as the original. So, it attempts to hide the existence of the communication. Another definition for steganography is "hiding in plain sight." In which the secret message is out in the open but goes undetected because the existence of it is a secret. In short, the communication is taking place in front of someone's eyes, but unless they're the sender or receiver the message goes unnoticed (Jack & Russ, 2007).

DOI: 10.4018/978-1-5225-1022-2.ch006

STEGOSYSTEM MODEL

StegoSystem components are:

- **Cover File (Carrier):** Is the host file into which the secret message is hidden. It is also known as innocent file or original file. (Archana, Antony, & Kaliyamurthie, 2013) (Shatha, Baker & Ahmed, 2013).
- **Secret Message (Payload):** Is the massage that will be hidden within the carrier file. It can be plaintext, an image, or anything that can be represented as a bit stream (Shatha, Baker, & Ahmed, 2013) (Tayana, 2012).
- **Stegofile (Stego-Object):** Is the resultant file obtained after hiding the payload into the carrier file.
- **Stegokey:** Is an optional password that may be used to encode the secret message to increase the level of security (Shatha, Baker, & Ahmed, 2013) (Jack & Russ, 2007).

Actually, steganography is comprised of an embedding algorithm which concerned with embedding a secret message within a carrier file, and it is the most carefully constructed process, and an extraction algorithm which is much simpler process as it is an inverse of the embedding algorithm where the payload is extracted at the end.

The entire process of steganography can be presented in Figure 1. The inputs are passed through the encoder to embed the payload (secret message) within the carrier file. The system may require a key which is also used at the extraction phase. The resulting output from the system encoder is the Stegofile. This file is then sent over some communications channel along with the key (if used). Both the Stegofile and the key are then fed into the decoder where the secret message is extracted (Deepa & Umarani, 2013) (Saeed, Kabirul, &Baharul, 2013).

Figure 1.

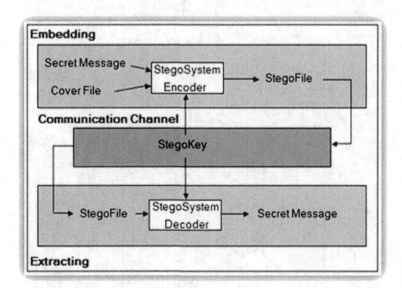

STEGANOGRAPHY HISTORY

This section will give some interesting insights and context into steganography by showcasing some of the prominent people, events, and methods used throughout history.

1. The Egyptians

The Egyptians, by their use of hieroglyphics, are considered the first to use encryption. Hieroglyphic uses characters in the form of pictures and they can be read as a symbol, picture or a sound, see figure 2. Although hieroglyphics are not thought of as a form of secret writing in these days, some hieroglyphics were written in such a way that only few people could read them properly. This can be considered one of the first instances of secret writing (Greg, 2004).

2. The Greeks

The Greeks used steganography when they carved their secret messages into tablets then covered them with wax to hide these messages (Leezaben, 2010). Also when a Greek named Histaiaeus shaved the head of his messenger, wrote the secret message on his bare scalp, as shown in Figure 3, and then waited for the hair to grow. When his messenger reached the destination his head was shaved and the message read properly (Rohit, 2012) (Viraj, 2010).

Figure 2.

Figure 3.

3. The Chinese

The Chinese used steganography to communicate secretly in an inventive way without being noticed. Where the Moon Festival was drawing near, they made a special cakes called "moon cakes", see Figure 4. Typed into each cake was a message says "attack". On the night, the cakes were dispersed, the messages read, and they attacked and overthrew the government successfully (Jack & Russ, 2007).

4. Gaspar Schott

Gaspar Schott in his book Schola Steganographya, as shown in Figure 5, described a method of encoding secret message by matching letters to musical notes. This "music" would never be pleasant if listened, but to the normal eye it would appear to be normal sheets of music (Jack & Russ, 2007).

5. Girolamo Cardano

The Cardano Grille system may be something that is already familiar. The basics are that each recipient has a piece of paper with some holes which called the "grille", when this grille is placed over the message (cover message) the holes line up with words and reveal the secret message.

Figure 4.

Figure 5.

THE PRISONERS' PROBLEM

To illustrate how steganography works, let's look at the most commonly used example, which was described by G. Simmons in what he called "The Prisoners' Problem". It goes like this; Alice and Bob have been arrested and placed in different cells. The goal was to develop an escape plan. The snag is that their only way to communicate is through the warden, Wendy. Wendy won't allow them to communicate in code (encryption) and if she should notice anything suspicious, they will be put in solitary confine-

Figure 6.

FOR LEVERAGE TRY
ONLY CEREALS
LISTED IN SOCIAL
LOVERS EDITIONS

FLEE AT ONCE ALL IS DISCOVERED

ment. So they must communicate in such a way that won't arouse suspicion. Thus, steganography must be used (Harsh, Tushar, Gyanendra, & Sunil, 2012) (Seifedine & Sara, 2013).

The scenario goes on to explain that a smart way of doing this would be to hide the message in another message or picture. Bob could draw a picture of a blue cow in a green pasture, and then ask the warden to pass it along to Alice. Wendy would of course check it before passing it on, and thinking it is just a piece of art, would hand it to Alice not knowing that the colors in the picture conveyed a message to her (Harsh, Tushar, Gyanendra, & Sunil, 2012) (Leezaben, 2010) (Giacomo, 2009).

The Prisoner's Problem can be applied to a lot of situations where steganography might be used to communicate. Alice and Bob are the two parties (sender and receiver), who want to communicate, and the warden is the eavesdropper (Jack & Russ, 2007).

STEGANOGRAPHY REQUIREMENTS

Actually, there are some factors should be considered when designing a steganography system:

- **Invisibility / Imperceptibility (Undetectability):** Is the ability to be unnoticed. Imperceptibility means how it is difficult to determine the existence of an embedded message or payload. (Chandrakant, Ashish, Bhupesh, Keerti, & Kaushal, 2013) (Shikha & Sumit, 2013) (Nagham, Abid, Badlishah, & Osamah, 2012).
- **Capacity (Payload):** Is the maximum amount of data that can be embedded safely in a carrier (Shikha & Sumit, 2013) (Nagham, Abid, Badlishah, & Osamah, 2012).
- **Robustness:** Is the algorithm resistance to either malicious or unintentional changes to the stego-file (Chandrakant, Ashish, Bhupesh, Keerti, & Kaushal, 2013) (Nagham, Abid, Badlishah, & Osamah, 2012).

- **Security:** Means that no targeted attacks can detect or view the hidden message unless they know the embedding algorithm (Hemalatha, Dinesh, Renuka, & Priya, 2013).

Consequently, as shown in Figure 7 (Nagham, Abid, Badlishah, Dheiaa, & Lubna, 2013), although steganography systems should achieve a balance among these three requirements, *steganography systems do not need to be robust*, i.e. robustness is not an issue or a top priority for steganography systems, but they should satisfy *high steganography capacity* and *secret data imperceptibility* (Adel, 2010).

CATEGORIES OF STEGANOGRAPHY

There are three categories of steganography: pure steganography, secret key steganography, and public key steganography (Mustafa, ElGamal, ElAlmi, & Ahmed, 2011).

Pure Steganography

No exchange of information required between the sender and the receiver and depends on secret through obscurity. This means that the used algorithm is not known, and therefore the level of testing is also unknown (Bhavna, Shrikant, & Anant, 2013) (Chinchu & Iwin, 2013) (Sravanthi, Sunitha, Riyazoddin, & Janga, 2013).

Secret Key Steganography

The exchange of a secret key is required between sender and receiver before communication. It hides the payload in the carrier by using a secret key. Only the parties who know the key can extract the payload.

Unlike Pure steganography, Secret Key steganography exchanges a key, which makes it more susceptible to interception (Mustafa, ElGamal, ElAlmi, & Ahmed, 2011) (Bhavna, Shrikant, & Anant, 2013) (Sravanthi, Sunitha, Riyazoddin, & Janga, 2013).

Figure 7.

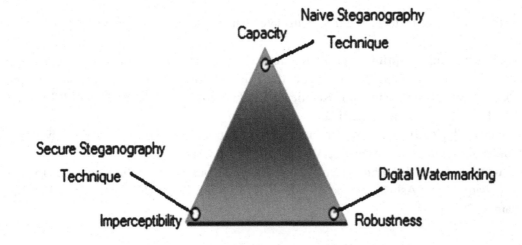

Public Key Steganography

Public Key steganography is the steganography system that uses a public key and a private key to secure the communication between the sender and the receiver. The sender will use the public key during the encoding process and only the private key can decode the secret message (Mustafa, ElGamal, ElAlmi, & Ahmed, 2011) (Bhavna, Shrikant, & Anant, 2013) (Chinchu & Iwin, 2013) (Sravanthi, Sunitha, Riyazoddin, & Janga, 2013).

STEGANOGRAPHY CLASSIFICATION

Steganography is classified into three major categories, as shown in Figure 8 (Cengage, 2010):

LINGUISTIC STEGANOGRAPHY

Linguistic steganography is any form of steganography that uses language in the cover (Souvik, Indradip & Gautam, 2011). It hides the payload in the carrier in several ways. A number of forms of linguistic steganography will be covered next (Cengage, 2010).

1. Grilles (Cardano's Grille)

Named for its inventor, Girolamo Cardano (1501–1576), and it has been discussed before in the steganography history section.

2. Type Spacing and Offsetting

Type spacing or offsetting is a way of distorting the text in a message to hide additional data. To encode a payload using this method, all one would have to do is to adjust the positions of specific letters. The letters that have been adjusted indicate the secret message (Jack & Russ, 2007).

Figure 8.

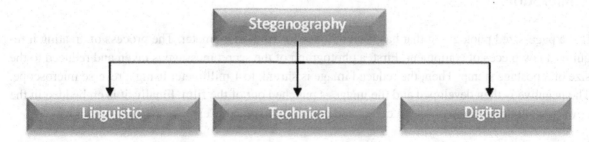

3. Newspaper Code

During the Victorian era, newspapers could be sent for free; the poorer classes of the time made use of this and invented the newspaper code to communicate secretly. Holes were poked above the required letters in the newspaper so that when the dots were transferred and written together the secret message would be extracted (Jack & Russ, 2007).

4. Null Ciphers

In the null cipher, the payload is "camouflaged" in a larger cover message. It is also known as an open code (Jack & Russ, 2007).

An example of a message containing null ciphers is

Fishing freshwater bends and saltwater coasts reward anyone feeling stressed. Resourceful anglers usually find masterful leapers fun and admit swordfish rank overwhelming any day.

By taking the 3rd letter in every word, the following message appears:

Send lawyers guns and money

TECHNICAL STEGANOGRAPHY

Technical steganography is a little broader in scope because it does not necessarily deal with the written words even though it communicates information. It is the method where a device, tool, chemical or physical methods is used to hide the payload. It can include the following methods: (Cengage, 2010)

1. Invisible Ink

Invisible inks are liquids that are colorless (invisible) and to become visible they require heat, light, or special chemicals. There are several types of liquids that can be used as invisible inks such as lemon juice, milk and even urine (Shawn, 2007).

2. Microdots

It is a page-sized photograph that has been reduced to 1 mm in diameter. The process of creating it requires a few pieces of equipment. First, a photograph of the secret message is taken and reduced to the size of a postage stamp. Then, the reduced image is shrunk to 1 millimeter using a reverse microscope. The negative is then developed and the image is punched out of the film. Finally, it is embedded in the cover text, under a stamp or over a colon, and cemented in place (Cengage, 2010).

DIGITAL STEGANOGRAPHY

In this section we will move forward into the digital age of steganography. The following techniques are the most commonly used in digital steganography:

1. Injection

This technique works as follows; the secret message is inserted directly into a carrier file, which could be a video, image, audio or text file. The drawback of the injection technique is that, the size of the carrier file increases, making it easy to be detected. To overcome this drawback the original cover file (carrier) should be deleted once the stego-file has been created (Cengage, 2010) (Rajanikanth, 2009) (Jiri, 2011).

2. Substitution System

It replaces redundant or unneeded bits of a carrier with the bits of the payload (Vanmathi & Prabu, 2013). Several steganography tools use Least-Significant Bit (LSB) method of encoding the secret message (Shatha, Baker, & Ahmed, 2013) (Saeed, Kabirul & Baharul, 2013). LSB works like this: in a digital cover (image, video, or audio file), there is a tremendous amount of redundant space; it is the space that the steganography algorithm will use to hide the secret message (Rajanikanth, 2009) (Adel, 2010), for example, the following string of bytes represents part of a cover, an image or a video frame:

10000100 10000110 10001001 10001101

01111001 01100101 01001010 00100110

Each byte is comprised of eight bits; these bits make up a color value in an image. Now, the bits that make up the byte go from left to right in order of importance to the color value they are representing, see Figure 9. For example, changing the first bit in the first string from a 1 (10000100) to a 0 (00000100) will drastically change the color, as opposed to changing the last number from a 0 (10000100) to a 1 (10000101). It is the last bit that is considered the least significant, because changing its value has little effect on the color the byte is representing.

Figure 9.

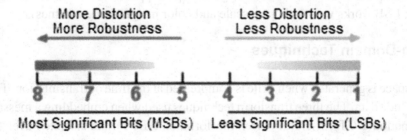

Let's see how a substitution system in steganography can be used to hide a message. Using this string of bytes:

10000100 10000110 1000100110001101

01111001 01100101 01001010 00100110

Suppose that the secret message is the number of a locker in a bus terminal, locker number 213:

$_{10}$ is represented as $(11010101)_2$. (213)

Now, using the LSB method, 213 will be embedded into the cover image. The embedding will be done one bit at a time.

10000100: 0 is replaced by a *1*, the first bit in the message.

10000110: 0 is replaced by a *1*, the second bit in the message.

10001001: 1 is replaced by a *0*, the third bit in the message.

10001101: 1 is left alone because it corresponds to the *1* in the message.

01111001: 1 is replaced by a *0*, the fifth bit in the message.

01100101: 1 is left alone because it corresponds to the *1* in the message.

01001010: 0 is left alone because it corresponds to the *0* in the message.

00100110: 0 is replaced by a *1*, the eighth bit in the message.

So, of the eight bytes of information, only five have been altered, and the message has been embedded. Now, while this example deals with only 8 bytes of data, imagine the amount of redundant information in a cover image that is 500 kilobytes or 1 megabyte. Within all those 1's and 0's are a lot of LSBs that can be changed with little or no noticeable difference to the cover image.

The LSB technique is commonly used in steganography applications because the algorithm is quick and easy to use; LSB works well with gray-scale and color images (video frames).

3. Transform-Domain Techniques

A transformed space is generated when a file is compressed at the time of transmission. This transformed space is used to hide data. The three transform techniques used when embedding a message are: discrete cosine transformation (DCT), discrete Fourier transformation (DFT), and discrete wavelet transformation (DWT). These techniques embed the secret data in the cover at the time of the transmission process. The transformation can either be applied to an entire carrier file or to its subparts. The embedding process is performed by modifying the coefficients, which are selected based on the protection required. The hidden data in the transform domain is present in more robust areas, and it is highly resistant to signal processing (Cengage, 2010).

4. File Generation

Instead of selecting a cover to embed a secret message, this technique generates a new carrier solely for the purpose of hiding data. As illustrated in "The Prisoners' Problem," Bob creates a picture of something not suspicious that can be passed to Alice; this picture is the cover that provides the mechanism for conveying the message, which in that example was the color of a cow (Cengage, 2010) (Adel, 2010).

5. Statistical Method

In this method, one bit of information is embedded in the cover file, creating a statistical change. This change is indicated as a one, while no change is indicated as a zero. This method is based on the ability of the receiver to differentiate between modified and unmodified cover (Cengage, 2010) (Adel, 2010).

6. Distortion Technique

This technique changes the carrier in order to embed the secret message. The encoder performs a sequence of modifications to the carrier that corresponds to the payload. The payload (secret message) is extracted by comparing the distorted cover with the original one. The drawback of this technique is that the decoder needs access to the original carrier file (Cengage, 2010) (Adel, 2010).

STEGANOGRAPHY COVER FILES (CARRIERS)

The various techniques used in steganography are applied differently depending on the type of cover file used to hide the secret message. Digital cover files (carriers) are such as: text files, audio files, image files and video files (Cengage, 2010) (Gayathri & V.Kalpana, 2013).

TEXT FILES (TEXT STEGANOGRAPHY)

Text file as a carrier is the oldest technique used (Gayathri & V.Kalpana, 2013). Text steganography can be classified into three types: (Souvik, Indradip, & Gautam, 2011)

1. Format-based.
2. Linguistic method.
3. Random and statistical generations.

Format-Based Methods

They change or use the formatting of the cover-text to hide secret message. They do not change any word or sentence, so it does not harm the "value" or the meaning of the text. A format-based text steganography method is such as "open space method". In this method extra white spaces are added into the text to hide the secret message. These spaces can be inserted after the end of each word, sentence or paragraph. A single space represents "0" and two consecutive spaces represent "1". Although a short message can be hidden in a document, this method can be applied to almost all types of text without being noticed (Souvik, Indradip, & Gautam, 2011).

Open-space methods can be categorized in the following three ways:

- **Inter-Sentence Spacing:** This method works by inserting one or two spaces after every terminating character (Cengage, 2010).
- **End-of-Line Spacing:** Secret message is placed at the end of a line in the form of spaces (Cengage, 2010).
- **Inter-Word Spacing:** This method works by adding single space or double spaces between words (Cengage, 2010).

While, *linguistic-based methods* deals with the linguistic properties of text, such as the following methods:

- **Syntactic Steganography:** This method manipulates punctuation to embed data, such as the following example:
  ◦ Laptop, PC
  ◦ Laptop PC

In the second phrase, the punctuation mark has been removed. This mark can be used to hide the secret message (Cengage, 2010).

- **Semantic Steganography:** This method involves changing the words themselves. It assigns two synonyms primary and secondary values for each word. When decoded, the primary value represents "1" and the secondary value represents "0" (Cengage, 2010).
- While *Random and statistical generation methods* generate cover-text according to the statistical properties of language. They use example grammars to produce cover-text in a certain natural language (Souvik, Indradip & Gautam, 2011).

AUDIO FILES (AUDIO STEGANOGRAPHY)

Message embedding in an audio file can be done by techniques such as LSB or using inaudible frequencies that are over 20,000 Hz which can't be heard by the human ear, so it can hide a secret message. Information can also be hidden using musical tones with a substitution scheme. For example, tone A could represent 1, and tone F could represent 0. By using the substitution technique a simple musical piece can be composed with a secret message, or an existing piece can be used to hide a secret message (Cengage, 2010) (Gayathri & V.Kalpana, 2013). Next are some audio steganography techniques:

1. Low-Bit Encoding in Audio Files

Digital steganography is based on the fact of artifacts; audio files, contains redundant information. Compression techniques such as MP3 remove parts of the redundancy, allowing the files to be compressed. By using steganography, the computer forensic investigator can replace some of the redundant information with other data. Low-bit encoding replaces the LSB of information in each sampling point with a coded binary string. The low-bit method encodes large amounts of hidden data into an audio signal at the expense of producing significant noise in the upper frequency range (Cengage, 2010) (Indradip, Souvik, & Gautam, 2013).

2. Echo Data Hiding

With echo hiding, information is embedded by adding an echo into the discrete audio signal. To hide the data successfully, three parameters of the echo need to be altered: decay rate, amplitude and offset (delay time) from the original signal. The echo is not easily resolved as all the three parameters are set below the human audible threshold limit (Souvik, Indradip, & Gautam, 2011).

IMAGE FILES (IMAGE STEGANOGRAPHY)

As Duncan Sellars explains; "To a computer, an image is an array of numbers that represent light intensities at various points or pixels that make up the images data" (Babita & Ayushi, 2013) (Kaustubh, 2012).

Pixel (Picture Element)

It is the smallest controlled unit of an image, see Figure 10. Each pixel has its own address which corresponds to its coordinates (Sudipta, 2010). Pixels are displayed row by row (horizontally) (Deepa & Umarani, 2013).

Bit Depth / Bits Per Pixel (bpp)

It is the number of bits required to represent the color (tone) of a pixel in an image. Higher bit depth gives a broader range of colors (Sudipta, 2010) (Tayana, 2012).

IMAGE TYPES

According to colors, images can be categorized into:

Figure 10.

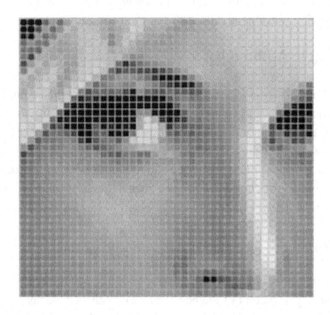

1. Binary Image

It is also called 1-bit monochrome image, it has two colors only white and black, as shown in Figure 11. That is for every pixel in the image, there are two possible values either 0 or 1. Where 0 represents black color and 1 represent white color. In the field of image processing, these images might be sometimes the output of image segmentation or image threshold (Atallah, 2012).

2. Grayscale Image

In a grayscale images, everything is black, white and shades of grey, as shown in Figure 12. Every pixel in this image contains 8 bits. So we have $2^8 = 0$ to 255, total 256 color shades. Take an example of "00000000" which represents the value of 0 means it is black and "11111111" represents the value of 255 which means it is white, while "00110101" represents a shade of gray (Leezaben, 2010) (Atallah, 2012) (Bhavna, Shrikant & Anant, 2013). So, each pixel is represented by one byte (a value between: 0 to 255) - for example, a dark pixel value might be 30, and a bright one might be 200 (Nian & Mark, 2004).

3. RGB Color Image

In RGB color image, see Figure 13, every pixel is represented by 24 bits (3 bytes). That can be divided into chunks of 8 bits for red, 8 bits for green and 8 bits for blue. There are 256 possible color shades of red, 256 possible color shades of green and 256 color shades of blue. So, according to the intensity of these colors we have 2^{24} possible color shades or 16,777,216 colors (Atallah, 2012) (Bhavna, Shrikant & Anant, 2013).

Figure 11.

Figure 12.

Figure 13.

IMAGE STEGANOGRAPHY

Since the advent of internet and digital technology, data hiding has shifted its paradigm from text to other forms such as videos and images (Naidele, 2012). Images are the most widely used cover files because of a lot of reasons such as: they are easy to find, have higher degree of distortion tolerance over other type of files (Archana, Antony, & Kaliyamurthie, 2013) and also they contain a lot of redundancy.

Redundancy or redundant information is the bits that provide accuracy far greater than necessary for the object's use and display. The redundant bits are also those bits that can be altered without the alteration being detected easily (Sneha & Sanyam, 2013) (Tayana, 2012). Image steganography means hiding a secret message within an image in such a way that others can't detect the presence of the hidden data.

Actually, hiding a message into an image requires two files. The first is the image file that will hold the secret message, called the cover image. The second file is the secret message (Payload); the information to be hidden. When combined, the cover image and the payload, they make a stego-image (Mamta & Parvinder, 2013).

In fact, RGB images are the most suitable covers because they contain a lot of information that help in embedding the payload with a bit change in the image resolution which does not affect the image quality making the message more secure (Babita & Ayushi, 2013).

An important note is that, the cover image must be larger than the payload, so that the small is embedded inside the large, where larger carriers are the best secured data transmission and less distortion (Mohammed, 2012).

IMAGE STEGANOGRAPHY TECHNIQUES

There are a number of techniques that hide the secret message in the cover image, all of which have corresponding strong and weak points. Here are some of them:

1. Transform Domain Techniques

These techniques are embedding data in the frequency domain of a cover image, i.e. secret message is embedded in the image after it has been transformed (as discussed before in the digital steganography section). These techniques are much stronger robustness than embedding principles that operate in the spatial domain (Nagham, Abid, Badlishah, Dheiaa, & Lubna, 2013) (Poornima & Iswarya, 2013).

2. Spatial Domain Technique

Also known as substitution techniques; are the techniques that embeds the secret message (payload) within the pixels of the cover image. One of the ways to do so is to hide secret message in the LSB of the cover image (Nagham, Abid, Badlishah, Dheiaa & Lubna, 2013) (Manoj, Sivasankaran, Vikram, & Bharatha, 2013) (Tatiana, 2013) (Firas, 2013).

3. Spread Spectrum

The Spread Spectrum Image Steganography (SSIS) provides the ability to hide a significant amount of secret message bits within digital images without being detected. During the encoding phase, the secret message is converted into pseudo-noise which is then added with the cover image to produce the stego-image. To retrieve the secret message from the stego-image, first the image is filtered to find the noise and then the message is extracted from this noise (Nagham, Abid, Badlishah, Dheiaa, & Lubna, 2013).

4. Distortion Techniques

These techniques require knowledge of the original cover image during the extraction process where the decoder functions check for differences between the original cover image and the distorted cover image in order to extract the payload. The encoder, on the other hand, adds a sequence of changes to the cover image. So, information is described as being stored by signal distortion (Nagham, Abid, Badlishah, & Osamah, 2012) (Nagham, Abid, Badlishah, Dheiaa, & Lubna, 2013) (Vanmathi & Prabu, 2013).

5. Statistical Methods

These techniques modify the statistical properties of an image. This modification is typically small, and it is thereby able to take advantage of the human weakness in detecting luminance variation (Indradip, Souvik, & Gautam, 2013) (Vanmathi & Prabu, 2013).

They exploit the existence of a "1-bit"; this process is done by modifying the cover image to make a statistical characteristics change if a "1" is transmitted, otherwise it is left unchanged. To send multiple bits, the image is broken into sub-images, each corresponding to a single bit of the message (Nagham, Abid, Badlishah, & Osamah, 2012).

6. Image Generation Technique

In this technique, a message is converted into image elements and then collected to form a stego-image. Stego-image can't be broken by rotating or scaling the image, parts of the payload may be destroyed because of cropping, but it is still possible to recover the rest of the message (Nagham, Abid, Badlishah, Dheiaa, & Lubna, 2013).

7. Image Element Modification Techniques

Some techniques do not try to hide message using the actual elements of the image. Instead, they modify the image elements in undetectable ways, for example, by modifying the color of the eye or color of the hair of someone in a photograph. These modifications can then be used to embed the secret message. In addition, this information will survive scaling, rotations and lossy compression (Nagham, Abid, Badlishah, & Osamah, 2012) (Nagham, Abid, Badlishah, Dheiaa, & Lubna, 2013).

VIDEO FILES (VIDEO STEGANOGRAPHY)

Video Steganography refers to using video files as carriers to hide the payload by using some embedding algorithms. But before applying any embedding algorithm on video it must be divided into video frames, see Figure 14 (Deepa & Umarani, 2015). Actually, video file consists of a set of frames (images) that are played at certain frame rates based on the video standards. Video quality depends on the following: (fps) which is the number of frames (images) per second, number of pixels in a frame, and size of the frame (image size). The fps parameter is almost between 24 and 30 fps (i.e. between 24 and 30 images per second) in many common video formats, however, the other two parameters varies depending on video standards. Each video frame or image, as discussed before, is a collection of pixels that are arranged horizontally (Chandel & Jain, 2016) (Deepa & Umarani, 2015).

So, almost all Image steganography researches done can be extended to be used in video steganography. As an example, LSB method of image steganography, can be applied on least significant bit of the cover video frames to hide the bits of the payload (Sahu & Mitra) (Choudry & Wanjari, 2015).

Advantages of video steganography are such as: video streams have high degree of redundancy; which means more data could be hidden (Khan & Gorde, 2015), thus they are considered as good covers for data hiding. Also, generally speaking, video steganography is the extension of image steganography, so, a video file can simply be viewed as a sequence of images, yielding video data hiding similar to image data hiding. Moreover, videos provide new dimensions for data hiding such as hiding messages in motion components. Furthermore, as shown in Figure 14, the audio components contained in video files can also be utilized for data hiding (audio steganography). Finally, video content is dynamic, lower chances of detection of the hidden data compared with images (Sadek, Khalifa, & Mostafa, 2014)

However, there are many disadvantages using videos as carriers such as: attacks; in addition to the image attacks that can be applied on the separate frames of video, there are much more attacks for videos such as lossy compression, change of frame rate, formats interchanging, addition or deletion of frames during video processing that may destroy part or the whole secret message (Sadek, Khalifa, & Mostafa, 2014). Also, the large size of a video files is not regularly communicated over normal transmission channels (Tayana, 2012).

Focusing on video steganography techniques, it can be classified into a number of ways; one way is the classification that based on the domain of embedding, i.e. spatial domain techniques and transform domain techniques (as discussed in image steganography section) see Figure 15 (Sadek, Khalifa, & Mostafa, 2014).

Figure 14.

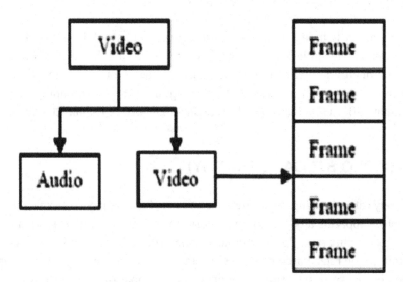

Figure 15.

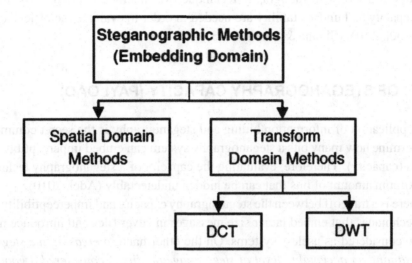

1. Spatial Domain Techniques (Substitution-Based Techniques)

As explained before, they replace redundant data of the cover video frames with the required secret message. Their advantages are the simplicity and the high embedding capacity relative to other techniques. These techniques are such as Least Significant Bit (LSB) technique, which is one of the oldest and most famous substitution-based techniques (Sadek, Khalifa, & Mostafa, 2014).

2. Transform Domain Techniques

The drawback of substitution techniques is that it is not robust to cover modification such as format change. So, the hidden data can be destroyed easily. Transform domain techniques are more complex than substitution techniques, but they have higher robustness than them. Basically, any transform-domain technique consists of at least the following steps: the cover is transformed into the frequency domain first by using one of these transformations; Discrete Fourier Transform (DFT), Discrete Wavelet Transform (DWT) or Discrete Cosine Transform (DCT), then payload is hidden in some or all of the transformed coefficients, and finally these coefficients are transformed back to the original form of the cover (Sadek, Khalifa, & Mostafa, 2014).

STEGANOGRAPHY SYSTEMS EVALUATION

In order to make a decision of which steganography system is better than another, an evaluation scheme is needed. Actually, there is no standard test or measure is available to evaluate the performance or the effectiveness of steganography systems. However, there are some guidelines and general procedures that can be considered when evaluating or designing steganography systems (Adel, 2010).

The amount of hidden information and the difficulty of detection of stego-files are the two most important aspects of any steganography system. Therefore, measuring these two characteristics will determine the superiority of a steganography technique over another. Consequently, the measures of steganography capacity and undetectability are needed in order to evaluate the efficiency of steganography systems (Adel, 2010) (Shikha & Sumit, 2013).

EVALUATION OF STEGANOGRAPHY CAPACITY (PAYLOAD)

Since the main application of information hiding and steganography is the secret communication, it is important to determine how many bits a steganography system can embed imperceptibly in comparison to other methods (capacity). Therefore, evaluating the capacity of a steganography technique means to find out the maximum amount of bits that can be hidden undetectably (Adel, 2010).

As known, there is a tradeoff between the steganography capacity and imperceptibility. Nevertheless, steganography techniques that embed larger size messages in cover files and introduce more distortion to stego-files are considered as useless systems. On the other hand, *increasing the steganography capacity and maintaining an acceptable level of stego image quality is considered a good contribution*. Additionally, *improving the stego image quality while maintaining the steganography capacity is also considered a significant contribution* (Adel, 2010).

EVALUATION OF IMPERCEPTIBILITY

Methods or techniques that can evaluate the undetectability or imperceptibility of steganography systems are different depending on the type of carrier used for hiding the secret message. For example, image quality represents an indication for the imperceptibility of image based steganography, while file size may reveal the presence of hidden message within a text file and therefore lead to its detection (Adel, 2010).

Two types of perceptibility can be distinguished and evaluated in signal processing systems which are fidelity and quality. Fidelity is the perceptual similarity between signals before and after processing. However, quality is an absolute measure of the goodness of a signal. For example, a grayscale, low resolution image can be used (low quality) for data hiding. The stego image looks identical to the cover image, so it has high fidelity, but it is also has low quality. For image based steganography, the fidelity is defined as the perceptual similarity between the original cover image and the stego image (Adel, 2010).

Therefore, both versions of the image before and after embedding are required for the fidelity evaluation. However, attackers, and most likely recipients, do not have access to the unmodified original cover image. Additionally, steganography systems must avoid attracting the attention of anyone not involved in the secret communication process and therefore stego images must have very good quality. Therefore, quality is the major concern for most steganography techniques in order to avoid any suspicion and therefore detection.

EVALUATING THE QUALITY OF DIGITAL IMAGES

The usage of image compression or processing technologies has increased significantly during last few decades. Furthermore, evaluating and measuring the quality of processed images still represent a significant issue in many image processing applications, such as digital image steganography. Thus, image quality represents a key factor in most applications and assessing the perceived quality of digital images is very important (Adel, 2010).

Actually, there are two primary ways to measure image quality: objective quality methods which are automated methods and subjective quality methods which are human based methods. The objective methods measure the physical aspects of images and psychological issues while the subjective methods are psychologically based methods. Additionally, subjective methods use humans to evaluate the quality of images. For example, humans can be asked to compare a modified image (stego-image) with its original version (cover image) in order to know amount of degradation (Adel, 2010).

1. Objective Quality Evaluation (Automated Evaluation)

In order to get a faster and cheaper measure of image quality, designing an evaluation metrics for image quality that can automatically predict the quality of the perceived image is the main goal of objective image quality assessment research. Thus, the assessment algorithms designed for objective image quality evaluation should be in close agreement with subjective human evaluation (Adel, 2010).

Objective image quality evaluation metrics are classified into three generic categories according to the availability of the unmodified or original image (reference). These categories are: full-reference (FR), no-reference (NR), and reduced-reference (RR) image quality assessment. The full reference means that the original image (cover image) and the test image (stego-image) are available. However, the no-reference means that only the test image (stego-image) is available. On the other hand, the reduced-reference means that the test image (stego-image) and some information about the original image (cover image) are available (Adel, 2010).

In the literature, the peak signal-to-noise ratio metric (PSNR) has shown the best advantage almost overall objective image quality metrics under different image distortion environments and strict testing conditions. Indeed, PSNR and the MSE (mean squared error) metrics are the most commonly used measures to evaluate the quality of image steganography, coding, etc. (Adel, 2010).

PSNR and MSE

Since the computing of these two metrics is very easy and fast, they are the most commonly used full-reference (FR) metrics for the evaluation of objective image quality (Adel, 2010). PSNR measures the similarity between two images; it determines the degradation in the stego image with respect to the cover image. PSNR values over 36 dB are acceptable in terms of degradation, which means no significant degradation is observed by HVS, which also means that the two images are undistinguishable (Saeed, Kabirul &Baharul, 2013). While the MSE measures the statistical difference in the pixel values between the original (cover) and the reconstructed (stego) image.

Moreover, PSNR and MSE are defined as shown in Equation 1 and Equation 2: (Adel, 2010) (Saeed, Kabirul, & Baharul, 2013)

$$MSE = \left(\frac{1}{MN}\right) \sum_{i=1}^{M} \sum_{j=1}^{N} \left(X_{ij} - X'_{ij}\right)^2 \qquad (1)$$

$$PSNR = 10 log_{10} \frac{I^2}{MSE} \qquad (2)$$

where, X_{ij} is the ith row and the jth column pixel in the original (cover) image, X'_{ij} is the ith row and the jth column pixel in the reconstructed (stego) image, M and N are the height and the width of the image, I is the dynamic range of pixel values, or the maximum value that a pixel can take, for 8-bit images: I = 255. However, the MSE and PSNR for color images is defined as shown in Equation 3 and Equation 4 (Saeed, Kabirul, & Baharul, 2013).

$$MSE_{AVG} = \frac{MSE_R + MSE_G + MSE_B}{3} \qquad (3)$$

$$PSNR = 10 log_{10} \frac{I^2}{MSE_{AVG}} \qquad (4)$$

where, MSE_R, MSE_G, and MSE_B are the MSE of red, green, and blue components respectively, as shown in Equation 3. Thus, the best image quality can be found when the MSE value is very small or going to be zero since the difference between the original and reconstructed image is negligible. *However, PSNR values between 30 and 40 can be considered as typical values.* Moreover, the higher the PSNR value of a stego image, the better the degree of hidden message imperceptibility. For example, it is difficult for the human visual system to recognize any difference between a grayscale cover image and its stego image if the PSNR value exceeds 36 dB (Adel, 2010).

2. Subjective Quality Evaluation (Human Evaluation)

In this kind of evaluation, humans are asked to evaluate the visual quality of some images. However, the visual sensitivity differs from someone to another and it also changes over time in anyone. Therefore, different viewers will behave differently. Nevertheless, almost all objective image quality measures do not perfectly reflect the impression of humans. Hence, the subjective quality measure represents a true performance benchmark for image processing tools. Unlike objective quality measures, subjective measures represent the most reliable method to determine the actual image quality since human beings are the ultimate proposed receivers in most applications. Furthermore, it has been stated that the subjective test is the best method to evaluate the quality of images surely and reliably (Adel, 2010).

STEGANOGRAPHY ATTACKS

Steganography attackers are the interceptors of stego-files in the communication channels in order to detect embedded messages in these files. In the "Prisoners' Problem", Simmons represented these attackers by warden's mediating the communication between the two prisoners.

There are three general types of attacks; the first technique deals with protecting the steganography system against secret message detection in a *passive attack,* while the second one protects the message against detection and modification by an *active attack.* However, the third technique protects the system against the forgery of a *malicious attack.* The next three subsections explain these three attacks in details (Adel, 2010).

1. Passive Attack

Passive attackers just observe the communication without any interference. Therefore, if the attacker is restricted from modifying the contents of stego-files during communication, he is called a passive attacker. It has only the right to prevent the message from being delivered. Therefore, the communication between two parties will be blocked (Adel, 2010).

2. Active Attack

If the attacker modifies the contents of stego-files during communication, then the attacker is called an active attacker. An active attack is thus the process of altering (modifying) stego-files during the communication process to prevent secret communication. In these attacks, the attacker can capture and modify a stego-file sent from the sender and then forward this modified file to the receiver (Adel, 2010).

Therefore, protecting secret messages against active attacks is another problem. However, steganography systems which resist these types of attacks and keep the secret message readable at the receiver are known as robust systems. These robust systems are most likely to be used for watermarking rather than for steganography methods (Adel, 2010).

3. Malicious Attack

If the attacker fakes messages or acts as one of the communication partners (sender or receiver) during the communication process, he is called a malicious attacker. In the malicious attack, the attacker may try to remove the hidden message, impersonate one of the communicating parties, or trick them. Therefore, in this kind of attacks, the attacker can pass his own message to the receiver as if it is sent by the sender (Adel, 2010).

However, this is the most difficult attack among them because the attacker needs to know more information about the communication process such as the encoding algorithm and the secret key (if used). So, these kinds of attacks are considered infrequently because they are difficult to be applied and easy to be detected by the actual receiver (Adel, 2010).

STEGANOGRAPHY APPLICATIONS

Steganography is used in areas where there is a need to protect or secure the information to be transmitted. Following are some applications of steganography:

Medical Records

In medical records, image steganography is used to avoid any mix-up of patients' records. Every patient has his own EPR (electronic patient record), which contains examinations and other medical records stored in it (Cengage, 2010). Also, it can be used to hide classified patient information in X-ray and scan images of the patient. This is securely associate patient records with their own X-rays and scans.

ATM (Automated Teller Machine)

Image steganography could be used to embed secure information like customer name, account information and key presses in ATM camera feeds.

Military and Intelligence Services

Military and intelligence agents require secret communications. Even if the message is encrypted, the detection of a signal may rapidly lead to an attack on the sender. Steganography either video or images can be used to keep these signals hidden (Tayana, 2012).

Sensitive Information

Steganography can also be used for storing information without the desire of communicating it to anyone else. Sensitive information such as banking details, private images and passwords can be embedded in a cover object such as video or image that is then stored on personal computer (Tayana, 2012).

Video Error Correction

Data hiding in video (video steganography) can be used for video error correction during transmission or for transmitting additional data (e.g. subtitles) without requiring larger bandwidth (Sadek, Khalifa, & Mostafa, 2014).

Surveillance System

Video steganography can be used for hiding data in a video captured by a surveillance system. That is, in order to protect the privacy of authorized people, their images are extracted from the surveillance video and embedded in its background (Sadek, Khalifa, & Mostafa, 2014).

CONCLUSION

In this chapter, an overview of steganography has been presented. First of all, steganography has been defined and introduced. Then, setgo-system model has been discussed including the components and the two algorithms (embedding and extraction). Also, some of the events, and methods used steganography throughout history have been listed including Egyptians, Greeks and Chinese. Moreover, the prisoner's problem has been explained to make it easy to understand how steganography works. Furthermore, steganography system requirements have been discussed to allow anyone to know the needs of any steganography system. Also, a comparison has been made among the three categories of steganography. In addition to the previous sections, a steganography classification has been discussed including linguistic, technical and digital steganography to remove any misunderstanding among the three classifications especially between technical and digital steganography. Also, the most commonly used carriers which are (text, audio, image and video files) have been introduced and explained how to be used with examples. The next section was steganography evaluation, which is one of the most important sections in this chapter, it allows anyone to judge or decide whether the system (either his own system or a previous one) is efficient enough to be used in the field of steganography or not. The penultimate section was steganography attacks; when designing a steganography system it is important to know the attacks that the system will face, because of this reason, the three types of attacks have been explained in details in this section. Finally, if steganography has no applications that make use of it, it will be useless, so, the section that precedes the conclusion listed some uses or applications that are related to steganography, some of them are cover dependent such as medical records and video error correction and the other some will work with any cover file.

REFERENCES

Adel, A. (2010). *Steganography-Based Secret and Reliable Communications: Improving Steganography Capacity and Imperceptibility* (Doctor of Philosophy Thesis). Department of Information Systems and Computing, Brunel University.

Archana, S., Judice, A. A., & Kaliyamurthie, K. P. (2013). A Novel Approach on Image Steganography Methods for Optimum Hiding Capacity. *International Journal of Engineering and Computer Science, 2*(2), 378 - 385.

Atallah, M. A. (2012). A New Method in Image Steganography with Improved Image Quality. *Applied Mathematical Sciences Journal, 6*(79), 3907–3915.

Babita & Ayushi. (2013). Secure Image Steganography Algorithm using RGB Image Format and Encryption Technique. *International Journal of Computer Science & Engineering Technology, 4*(06), 758–762.

Bharti, C., & Shaily, J.(2016). Video Steganography: A Survey. *IOSR Journal of Computer Engineering, 18*(1).

Bhavna, S., Shrikant, B., & Anant, G. K. (2013). Biometric Feature Based Steganography Scheme- An Approach Based On LSB Technique And Huffman Coding. *International Journal of Innovative Research and Studies, 2*(8), 247–257.

Chandrakant, B., Ashish, K. D., Bhupesh, K. P., Keerti, Y., & Kaushal, K. S. (2012). A new steganography technique: Image hiding in Mobile application. *International Journal of Advanced Computer and Mathematical Sciences, 3*(4), 556–562.

Chinchu, E. A., & Iwin, Th. J. (2013). An Analysis of Various Steganography Algorithms. *International Journal of Advanced Research in Electronics and Communication Engineering, 2*(2), 116–123.

Deepa, S., & Umarani, R. (2013). A Study on Digital Image Steganography. *International Journal of Advanced Research in Computer Science and Software Engineering, 3*(1), 54–57.

Deepa, S., & Umarani, R. (2015). A Prototype for Secure Information using Video Steganography. *International Journal of Advanced Research in Computer and Communication Engineering, 4*(8).

Firas, A. J. (2013). A Novel Steganography Algorithm for Hiding Text in Image using Five Modulus Method. *International Journal of Computers and Applications, 72*(17), 39–44.

Gayathri, C., & Kalpana, V. (2013). Study on Image Steganography Techniques. *International Journal of Engineering and Technology, 5*(2), 572–577.

Giacomo, C. (2009). *New techniques for steganography and steganalysis in the pixel domain* (Ph.D. Thesis). Dipartimento di Ingegneria dell'Informazione, Universita Degli Studi di Siena.

Greg, K. (2004). *Investigator's Guide to Steganography*. Auerbach Publications.

Harsh, P., Tushar, S., Gyanendra, O., & Sunil, C. (2012). Information Hiding in an Image File: Steganography. *International Journal of Computer Science and Information Technologies, 3*(3), 4216–4217.

Hemalatha, S., Dinesh, U. A., Renuka, A., & Priya, R. K. (2013). A Secure and High Capacity Image Steganography Technique. *Signal & Image Processing: An International Journal, 4*(1), 83 - 89.

Indradip, B., Souvik, B., & Gautam, S. (2013). Study and Analysis of Steganography with Pixel Factor Mapping (PFM) Method. *International Journal of Application or Innovation in Engineering & Management, 2*(8), 268–266.

Jack, W., & Russ, R. (2007). *Techno Security's Guide to Managing Risks for IT Managers, Auditors, and Investigators*. Elsevier Inc.

Jiri, H. (2011). *Artificial Intelligence Applied on Cryptoanalysis Aimed on Revealing Weaknesses of Modern Cryptology and Computer Security* (Dissertation thesis). Department of Informatics and Artificial Intelligence, Zlin Faculty of Applied Informatics, Tomas Bata University.

Kaustubh, C. (2012). Image Steganography and Global Terrorism. *Global Security Studies, 3*(4), 115–135.

Kedar, N. C., & Aakash, W. (2015). A Survey Paper on Video Steganography. *International Journal of Computer Science and Information Technologies, 6*(3).

Leezaben, A. P. (2010). *Steganography Using Cylinder Insertion Algorithm and Mobile Based Stealth Steganography* (Master of Science Thesis). San Diego State University, San Diego, CA.

Mamta, J., & Parvinder, S. S. (2013). Information Hiding using Improved LSB Steganography and Feature Detection Technique. *International Journal of Engineering and Advanced Technology, 2*(4), 275 - 279.

Manoj, G., Senthur, T., Sivasankaran, M., Vikram, M., & Bharatha, S. G. (2013). AES Based Steganography. *International Journal of Application or Innovation in Engineering & Management, 2*(1), 382 - 389.

Mennatallah, M. S., Amal, S. K., & Mostafa, G. M. (2014). *Video steganography: A comprehensive review*. Springer.

Mohammed, N. H. A. (2012). Text Realization Image Steganography. *International Journal of Engineering*, *6*(1), 1–9.

Mustafa, A. E., ElGamal, A.M.F., ElAlmi, M.E., & Ahmed, B. D. (2011). A Proposed Algorithm For Steganography In Digital Image Based on Least Significant Bit. *Research Journal Specific Education*, (21), 752 - 767.

Nagham, H., Abid, Y., Badlishah, A., & Osamah, M. (2012). Image Steganography Techniques: An Overview. *International Journal of Computer Science and Security*, *6*(3), 168–187.

Nagham, H., Abid, Y., Badlishah, R., Dheiaa, N., & Lubna, K. (2013). Steganography in image files: A survey. *Australian Journal of Basic and Applied Sciences*, *7*(1), 35–55.

Nagham, H., Abid, Y. R., Badlishah, A., & Osamah, M. (2012). Image Steganography Techniques: An Overview. *International Journal of Computer Science and Security*, *6*(3), 168–187.

Naidele, K. M. (2012). *Stealthy Plaintext. Master's Projects, Master's Theses and Graduate Research.* San Jose State (SJSU) University.

Nishi, K., & Kanchan, S. G. (2015). Video Steganography by Using Statistical Key Frame Extraction Method and LSB Technique. *International Journal of Innovative Research in Science, Engineering and Technology*, *4*(10).

Poornima, R., & Iswarya, R. J. (2013). An Overview Of Digital Image Steganography. *International Journal of Computer Science & Engineering Survey*, *4*(1), 23–31. doi:10.5121/ijcses.2013.4102

Rajanikanth, R. K. (2009). *A High Capacity Data-Hiding Scheme in LSB-Based Image Steganography* (Master of Science Thesis). Graduate Faculty of the University of Akron, Akron, OH.

Rohit, G. (2012). Comparison Of Lsb & Msb Based Steganography In Gray-Scale Images. *International Journal of Engineering Research & Technology*, *1*(8), 1–6.

Seifedine, K., & Sara, N. (2013). New Generating Technique for Image Steganography. *Lecture Notes on Software Engineering*, *1*(2), 190–193.

Shatha A. B., & Ahmed, S. N. (2013). Steganography in Mobile Phone over Bluetooth. *International Journal of Information Technology and Business Management*, *16*(1), 111-117.

Shawn, D. D. (2007). *An Overview of Steganography*. James Madison University.

Shikha, S., & Sumit, B. (2013). Image Steganography: A Review. *International Journal of Emerging Technology and Advanced Engineering*, *3*(1), 707–710.

Sneha, A., & Sanyam, A. (2013). A New Approach for Image Steganography using Edge Detection Method. *International Journal of Innovative Research in Computer and Communication Engineering*, *1*(3), 626–629.

Souvik, B., Indradip, B., & Gautam, S. (2011). A Survey of Steganography and Steganalysis Technique in Image, Text, Audio and Video as Cover Carrier. *Journal of Global Research in Computer Science*, *2*(4), 1–16.

Sravanthi, G.S., Sunitha, D. B., Riyazoddin, S.M., & Janga, R. M. (2012). A Spatial Domain Image Steganography Technique Based on Plane Bit Substitution Method. *Global Journal of Computer Science and Technology Graphics & Vision, 12*(15).

Sudipta, K. (2010). *An Enhanced, Secure and Comprehensive Data Hiding Approach Using 24 Bit Color Images* (Master of Technology thesis). Faculty Council for UG and PG Studies in Engineering & Technology, Jadavpur University, Kolkata, India.

Tatiana, E. (2013). *Reversible Data Hiding in Digital Images* (Master's thesis). Tampere University of Technology.

Tayana, M. (2012). *Image Steganography Applications for Secure Communication* (Master of Science Thesis). Faculty of Engineering, Built Environment and Information Technology, University of Pretoria, Pretoria, South Africa.

Uma, S., & Saurabh, M. (n.d.). A Secure Data Hiding Technique Using Video Steganography. *International Journal of Computer Science & Communication Networks, 5*(5), 348-357.

Vanmathi, C., & Prabu, S. (2013). A Survey of State of the Art techniques of Steganography. *IACSIT International Journal of Engineering and Technology, 5*(1), 376–379.

Viraj, S. G. (2010). *Steganography Using Cone Insertion Algorithm and Mobile Based Stealth Steganography* (Master of Science Thesis). San Diego State University, San Diego, CA.

Ze-Nian, L., & Mark, S. D. (2004). *Fundamentals of Multimedia*. Pearson Education International.

KEY TERMS AND DEFINITIONS

Cover File (Carrier): Is the original file that will hold the secret message.

Least Significant Bit (LSB): Is the most commonly known steganography algorithm that replaces least significant cover bits with secret message bits.

Pure Steganography: The most famous steganography technique as no information should be exchanged between sender and receiver.

Robustness: Is the ability of the stego-file to keep the secret message inside after it has been modified.

Secret Message (Payload): Is the important information that needs to be hidden or secured.

Stegofile (Stego-Object): Is the cover file that contains the important information.

Stegokey: Is a password that is used in some types of steganography in order to increase the security.

Chapter 7
Design of Reconfigurable Architectures for Steganography System

Sathish Shet
JSS Academy of Technological Education, India

A. R. Aswath
Dayananda Sagar College of Engineering, India

M. C. Hanumantharaju
BMS Institute of Technology and Management, India

Xiao-Zhi Gao
Aalto University School of Electrical Engineering, Finland

ABSTRACT

The most crucial task in real-time processing of image or video steganography algorithms is to reduce the computational delay and increase the throughput of a steganography embedding and extraction system. This problem is effectively addressed by implementing steganography hiding and extraction methods in reconfigurable hardware. This chapter presents a new high-speed reconfigurable architectures that have been designed for Least Significant Bit (LSB) and multi-bit based image steganography algorithm that suits Field Programmable Gate Arrays (FPGAs) or Application Specific Integrated Circuits (ASIC) implementation. Typical architectures of LSB steganography comprises secret message length finder, message hider, extractor, etc. The architectures may be realized either by using traditional hardware description languages (HDL) such as VHDL or Verilog. The designed architectures are synthesizable in FPGAs since the modules are RTL compliant. Before the FPGA/ASIC implementation, it is convenient to validate the steganography system in software to verify the concepts intended to implement.

INTRODUCTION

Steganography is the art as well as the skill that deals with smacking secret data in an image, video, audio, etc. Steganography may be stated as the method of smacking a furtive information inside a larger one such that someone may not know the presence of concealed information (Kipper, 2003). The main difference between steganography and cryptography is that the steganography scheme hides the secret

DOI: 10.4018/978-1-5225-1022-2.ch007

message inside the cover medium whereas the cryptography encrypts the secret message before transmitting it to the destination. Also, the breaking of steganography is known as steganalysis and breaking of cryptography is labeled as cryptanalysis. In image steganography, secret information is hidden inside the lower bits of image pixels, such that Human Visual System (HVS) is unable to visualize the presence of it. The Steganography system deals with the implementation of embedding and extraction hardware that hides and extracts secret information from the cover image. The original archives is denoted as cover media. The secret message encapsulated in cover media is referred to as stego image. Hiding or extraction process is protected by secret code to protect it from hackers.

Hardware implementation of steganography algorithms can be classified into several categories based on the architectural approach used; these are General Purpose Processor (GPP), dedicated Digital Signal Processors (DSPs), ASICs and FPGAs implementations. The choice of a particular approach is based upon the flexibility, memory requirement, throughput and cost etc. The image steganography algorithms implementation on GPP possess a great deal of flexibility compared with other implementations. In the recent years, processing speed of GPP has been increased significantly. However, the technology has approached the upper limit. The GPP are limited in performance since their instruction sets do not seem to suit for fast processing of high resolution of image enhancement algorithms. In addition, the instructions in GPP are executed in sequence, eventually, the throughput of the system decreases. DSPs (Hines, 2004) have been employed for enhancement of images provides some improvement over the GPPs. Only marginal improvement has been achieved since parallelism and pipelining incorporated in the design are inadequate. The hardware implementation of image enhancement techniques has been strongly influenced by the evolution of Very Large Scale Integration (VLSI) technology. The complex computational tasks in image processing such as 2D convolution, filtering, smoothing and contrast stretching operations encountered in image enhancement techniques are well suited for VLSI implementations. In addition, VLSI based hardware implementation exploits pipelining and massive parallel processing operations, resulting in increased throughput of the image enhancement system.

The image steganography methods implemented in ASIC increases the performance compared with other realizations. However, ASIC implementations require a large time to market and initial investments are high. The ASIC designs are best suited for relatively high volume productions. In the recent years, FPGAs has become an attractive choice of solution for hardware implementation of image steganography algorithms, especially when high throughputs are the needs of the hour. In an image steganography application, for rapid prototyping of new algorithms developed and to dynamically reconfigure, FPGAs are the right choice. FPGAs provides an optimal blend of flexibility of a GPPs and high speed processing that can be achieved using ASICs. An image enhancement architecture design for FPGA/ASIC technology can exploit fully the data and I/O parallelism for image enhancement application. Further, FPGA implementation of image enhancement techniques is cost effective for low volume productions. Therefore, in this chapter, LSB based image steganography algorithms are developed and implemented in FPGAs.

Steganography algorithms are highly essential to secure the data from hackers. Embedding secret message in image or video requires comparatively less delay to ensure real-time results. However, the computation complexity and hence the delay increases with increase in the payload. Therefore, the hardware implementation of steganography algorithms is the best choice that is preferred for this type of application. Although numerous researchers have been proposed software based steganography solutions, throughput achieved from these schemes are inadequate. The top design module invokes several other sub modules developed in the form of tree structure. The test bench has been developed to pass the stimulus for the top design and analyze the output results. A Matlab program was written in order to acquire a standard picture in "tif" format and convert into a text format. The text file generated by Matlab program is in raster scan order and can be easily used by Modelsim for simulation. The Verilog design of top module was run in Modelsim, to get the processed picture in "txt" format. These obtained "txt" file were converted back into "tif" format using another Matlab program. This program automatically displays both the original picture as well as stego picture. Subsequently, the design is synthesized, place and routed, and implemented on FPGA device using Synopsis Synplify and Xilinx ISE tools, respectively. The hardware implementation of steganography algorithms occupy less area in the silicon chip and is capable of embedding and extracting data in real-time. The cost, speed of operation, chip area and power consumed by the image or video steganography algorithms are relatively small compared to their software implementation counterparts.

BACKGROUND

The image steganography methods are broadly categorized into three types, namely,

1. **Spatial Domain Techniques:** Some spatial steganography schemes are available in the literature. These methods modify few bits in image pixel for hiding furtive information. Among various spatial steganography schemes, LSB/multi-bits based steganography is one of the simplest schemes that hides a secret message in the LSBs or other lower significant bits of the pixel without noticeable changes. Modifications in LSB or lower nibble of the pixels are indiscernible to the human eye. Embedding of message bits can be done either in ordered basis or randomly. LSB or lower nibble bit replacement techniques and matrix embedding are some of the spatial domain transformation methods. The main advantages in spatial domain bit modification schemes are: no original image degradation and enormous hiding capacity. However, this scheme has lower robustness and secret data hidden be destroyed by simple steganalysis.

2. **Masking and Filtering Approaches:** Masking and Filtering is a steganography technique that can be used on gray level images. Masking and filtering schemes are similar to placing watermarks on a printed image. These methods embed the information in the major areas of the image than just hiding it into the noise level. The masking and filtering techniques are superior to LSB replacement concerning the image compression. However, these methods apply to gray scale images.

3. **Transform Domain Methods:** The frequency domain hiding methods store the secret message in the coefficients of the transformed image that leads to massive information hiding capacity and is more robust against attacks. Currently, most of the strong steganography systems developed using transform domain approach have an improvement over spatial domain LSB or multi-bit hiding algorithms. As these schemes hide information in areas of the image that are less exposed to compression, cropping, and image processing. Transform domain methods do not depend on the image format, and they may outstrip lossless as well as lossy format conversions.

Transform domain techniques available in the literature are as follows:

- Discrete Fourier Transformation (DFT)
- Discrete Cosine Transformation (DCT)
- Discrete Wavelet Transformation (DWT)

4. **Distortion Techniques:** This method hides information through signal distortion and measures deviation from the original cover in the decoding process. Distortion techniques need evidence of the original cover image during the decoding process. However, decoder verifies the difference between original cover image and distorted cover image to restore the secret information. A stego image is formed by applying a sequence of variations on the cover image. This sequence of changes is used to match the secret message required to transmit. The message is encoded into the pixels chosen randomly. If the stego image is dissimilar from the cover image at the known message pixel, the message bit is a logic '1'. Otherwise, the message bit is a logic '0'.

As it is evident from the conventional steganography schemes presented that the complexity of techniques varies from one method to another. It may be noted that for most of the approaches, as the capacity of the message is increased, embedding and extraction process also increases. Although, numerous researchers have developed steganography methods in software, increase in embedding capacity reduces the throughput of the system. Therefore, this chapter presents the highly efficient, and fast LSB and multi-bit based steganography scheme to overcome the limitation of the conventional steganography methods.

In recent years, design of reconfigurable architectures were used in diverse field of image processing applications such as pre-processing (Genevese, 2015), segmentation (Hemalatha, 2015), motion estimation (Shah, 2016) etc. The design of hardware for image steganography algorithms and achieving real-time results is a challenging task since the hiding and extraction schemes are computationally complex. In this method, novel reconfigurable architectures were developed and implemented in FPGA device. The FPGAs are programmable devices that can be programmed by the designers as many times as possible. Before illustrating the developed reconfigurable architectures, contributions made by the other researchers are presented. Hussain and Hussain (2013) conducted a detailed survey of image steganography techniques.

In an earlier work, Tseng at al. (2002) remarkably developed steganography based on weight matrix for binary images. As is evident from the method that the r-bit secret message is embed into 2r-1 bit carrier data by changing at most two bits. With a little improvement of embedding process Fan et al. (2013) revealed that it is possible to increase the maximum number of secret bits that can be embedded into a block of size m×n to [$\log_2(2mn)$] by changing at most one pixel. Although, the work presented by Li is capable of achieving good performance by improving imperceptibility, embedding rate and embedding efficiency, visual inspection of experimental result reveals that this approach is pertinent to gray images and may not be suited for color images.

Mohd et al. (2012) proposed an FPGA implementation of LSB steganography technique on Altera family of Cyclone II FPGA. The design utilizes Nios embedded processor as well as specialized logic to perform steganographic steps. The scheme balances the trade-offs such as imperceptibility, quality and capacity. A new 2/3 LSB scheme presented in this work provides good image quality and facilitates simple memory access. However, the architecture developed is implemented on Nios processor this limits the efficiency of the algorithm compared to the design implemented on FPGA based on hardware description language based coding.

Among various steganographic algorithms, Crandall's (1998) work on matrix coding appears to be more representative compared with other techniques. Further to increase the embedding rate and efficiency, Fan et al. (2011) presented an extended matrix coding. The extended matrix-coding algorithm is capable of producing remarkable performance while using the binary logo as a secret message. However, regarding cost of symbol bits, the performance of the extended matrix coding is inferior to the matrix coding when the secret message is a stochastic signal. This drawback has been overcome by the researcher Fan et al. (2013) in his work on the mathematical analysis of extended matrix coding for steganography. In this work, authors have been carried out detailed mathematical analysis of extended matrix coding. Besides, this researcher has proved theoretically that extended matrix coding is not suitable for the applications that need random secret messages.

Amirtharajan and Rayappan have proposed an intelligent chaotic embedding scheme to enhance the quality of the stego image (2012). In this work, Adaptive Random (AR) k-bit embedding approach is presented to improve the quality of stego image. Firstly, the cover image is separated into non-overlapping blocks of same sizes. This follows embedding of encrypted confidential data into each block using four dissimilar random walks. An efficient random walk that provides minimum degradation for a particular block is detected and is fixed for that block. Also, AR embedding method is combined with Adaptive Random Inverted Pattern (ARIP) technique to enhance the visual quality of stego image. Although this scheme uses the spatial domain to hide furtive information into the cover image, that is quite easier, compared to the frequency domain and compressed domain methods, experimental results presented reveals that this scheme has limited embedding capacity.

Wang et al. (2013) proposed a high capacity lossless reversible data hiding method for Joint Photographic Expert Group (JPEG) images. This approach modified quantization table and quantized Discrete Cosine Transform (DCT) coefficients to store secret information. Reducing particular quantization table values and modifying corresponding DCT coefficients creates an option for embedding. This scheme is capable of embedding large volume of data with lower distortion in the stego image. Authors claim that proposed method is superior concerning image quality, hiding capacity and file size compared to steganographic schemes suggested by Chang et al. (2007) and Xuan et al. (2007). However, the experimental results presented shows that the reconstructed images are not satisfactory.

Upadhyay et al. (2012) present detailed survey and analysis of hardware cryptographic and steganographic methods implemented on a reconfigurable device such as FPGA. Gómez-Hernández et al. (2008) shows hardware architecture of steganographic context technique suitable for FPGA implementation. The context technique utilizes the noisy regions as well as rapid gray level changes in an image where the information is difficult to detect. The procedure to find these areas in an image is highly regular and complex from a computation point of view. Therefore, authors have been employed FPGA device to implement the design. The design is capable of processing pixels at the rate of 61.5 Mbps. As it is evident from the results presented that the algorithm has not optimized from the hardware perspective. This scheme also lags in employing new techniques such as pipelining and parallel processing to increase the throughput of the system.

Rajgopalan et al. (2014) proposed the FPGA implementation of the steganographic algorithm that randomizes the volume of data embedded in each pixel, according to the same MSBs. Three separate techniques have been proposed based on indicator selection among RGB channels. It has been shown that increase in image size increases the message embedding time and number of logic elements. The throughput achieved in this scheme for a color image size of 256×256 pixels is 5.898 ms or 2,94,900 clock cycles of 50 MHz. Although the experimental results presented for various color images guarantees that this approach works well for medium to large size images, inadequate hardware implementation details reveal that achieving a high throughput is tough.

As it can be seen from the review of existing methods, most of the schemes offer satisfactory results in Matlab simulation but at the cost of increased computational complexity. This problem is effectively addressed in this chapter by implementing the steganography technique in reconfigurable FPGA device. High degrees of pipelining and parallel processing techniques are incorporated at the coding level increases, the system throughput.

MAIN FOCUS OF THE CHAPTER

Among the available steganography techniques, the simplest one is the LSB steganography. Advanced steganography schemes exploit the LSB method for embedding and extraction. Therefore, the present authors proposed a design of reconfigurable architectures suited for FPGA/ASIC implementation. Although, the steganography algorithms are simulated in software's using high-level programming languages such as C, C++, Java, Matlab, etc., throughput achieved from these schemes is inadequate. Therefore, the hardware implementation of steganography algorithms is the right choice to achieve high system performances.

Least Significant Bit (LSB) Steganography Technique

In LSB steganography, LSB or, the eighth bit of a pixel in an image is modified to a bit of a secret message. Digital images are mainly of two types:

1. 24-bit images (Color images) and
2. 8-bit images (Gray images).

In 24-bit images, we can embed three bits of information in each pixel, one in each LSB position of the three eight bit values. Increment or decrement of the pixel value by changing the LSB does not alter the visual appearance of the image. The resultant stego image appears almost similar to that of the cover image. In 8-bit images, one bit (or multi-bit utilizing lower nibble of the cover image pixel) information can be hidden. The cover image may be the gray or color image, or the stego image is obtained by applying LSB algorithm cover images. The secret message hidden inside a picture is extracted from the stego-image by using the reverse process. If the LSB of the pixel value of cover image C(i, j) is equal to the message bit 'm' of secret message to be embedded, C(i, j) remain unchanged; if not, set the LSB of C(i, j) to m. The secret message embedding procedure summarized as follows:

$$s(i,j) = \begin{cases} C(i,j) - 1, & if\, LSB[C(i,j)] = 1\ and\ m = 0 \\ C(i,j), & if\, LSB[C(i,j)] = m \\ C(i,j) + 1, & if\ LSB[C(i,j)] = 0\ and\ m = 1 \end{cases} \quad (1)$$

where $LSB[C(i,j)]$ stands for the LSB of cover image $C(i,j)$, m is the secret message bit to be embedded and $s(i,j)$ is the stego image.

The LSB hiding process for a gray level cover image is illustrated in Figure 1. It is well known that each pixel of the gray cover image consists of one byte of data with either '1' or '0'. For instance, if it is required to hide a secret message "hi" (104 and 105) in three pixels of a gray level cover image (8-bit).

Figure 1. LSB hiding process for gray scale cover image

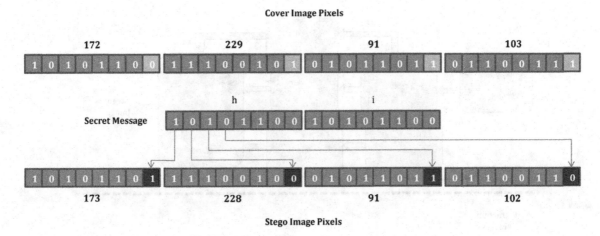

LSB Steganography Algorithm for Hardware Implementation

The methodology presented in this chapter for LSB steganography is a two-step approach, namely, embedding and extraction process. The embedding process consists of secret message length finder, LSB extractor, embedding algorithm module, etc. The designed system reads the secret message in the form of text, converts into American Standard Code for Information Interchange (ASCII) format. Further, the ASCII data is transformed into the binary format before it is embedded into the cover image. The gray or color form of the cover image is read at the same time as the secret message is accepted. In the next step, LSB bit is extracted from each pixel of the cover image. The core part of the embedding process is embedding algorithm. This module accepts three inputs concurrently, namely, the length of the secret message, binary pattern of the secret message and LSB of the cover image. The secret message is hidden in the cover image as long as the length of the secret message does not exceed its limit. Once, the length of secret data exceeds its limit, then embedding process stops. The process of embedding a secret message in a cover image is referred as steganography and the image produced a stego image. Conversely, the extraction module performs the withdrawal of secret message from the stego image. Based on the length of the secret message, the secret message is extracted from the stego image pixels. Extracted LSBs are equivalent to the binary form of secret data as long as the length of the secret message is within the limit. The LSBs removed from the stego pixels up to the length of the secret message constitutes the secret message. This LSB data is divided to the size of character length, and then the message is displayed. The algorithm to embed and extract the secret message in/from the cover image is illustrated in Figure 2.

The embedding and extraction process presented in Figure 2 are mainly suitable for FPGA/ASIC implementation. As is evident from the Matlab (Ver. R2014a) simulation, conventional LSB steganography is computationally complex in time. The time complexity in software implementation of embedding and extraction of steganography algorithms is resolved by the development of reconfigurable architectures followed by FPGA/ASIC implementation. The embedding process comprises secret message length finder, LSB extractor, message hiding module. The extractor part includes LSB finder, length matching module and message extractor. The developed modules have high degrees of pipelining and parallel processing that enormously improves the throughput of the system. The pictorial illustration of embedding and extraction process are presented in Figures 3 through 7, respectively.

Figure 2. Embedding and extraction technique suitable for fpga implementation

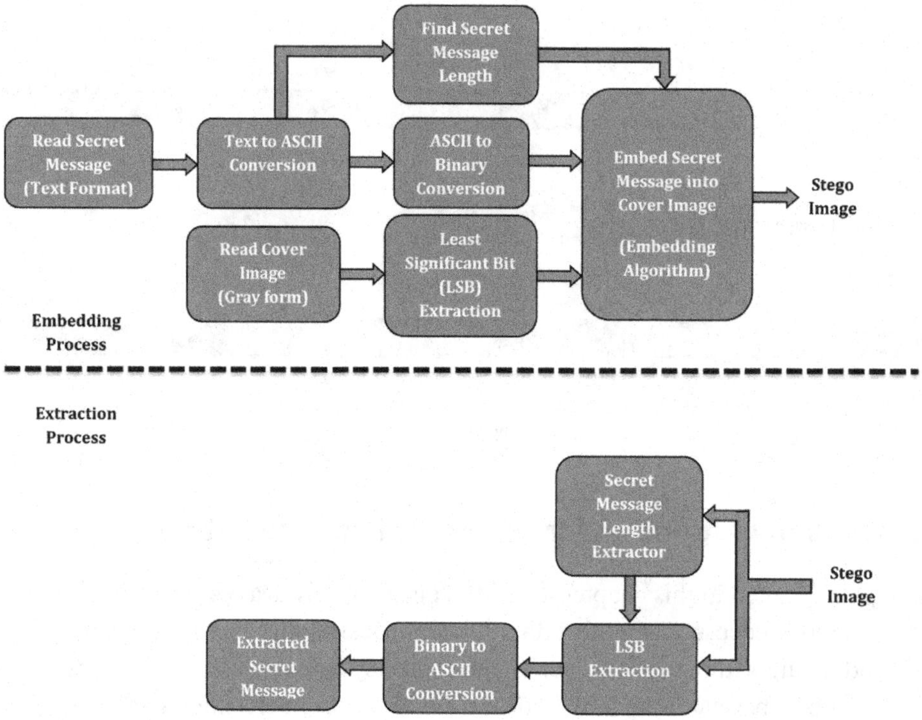

Figure 3. Sample secret message represented in ASCII and binary formats

Sample Secret Message

ATM PIN 22345714

ASCII Version of the Secret Message

65	84	77	32	80	73	78	32
50	50	51	52	53	55	49	52

Binary Version of the Secret Message

0	1	0	0	0	0	0	1
0	1	0	1	0	1	0	0
0	1	0	0	1	1	0	1
0	0	1	0	0	0	0	0
0	1	0	1	0	0	0	0
0	1	0	0	1	0	0	1
0	1	0	0	1	1	1	0
0	0	1	0	0	0	0	0
0	0	1	1	0	0	1	0
0	0	1	1	0	0	1	0
0	0	1	1	0	0	1	1
0	0	1	1	0	1	0	0
0	0	1	1	0	1	0	1
0	0	1	1	0	1	1	1
0	0	1	1	0	0	0	1
0	0	1	1	0	1	0	0

Length of the Secret Message: **128 bits**

Figure 4. Cover image represented in matrix and binary formats

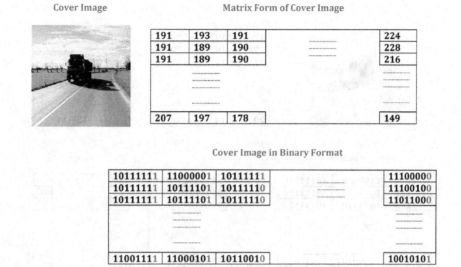

Figure 5. LSB based embedding strategy using equation (1)

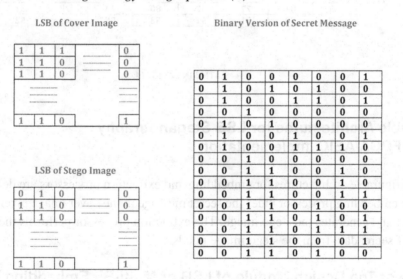

Figure 6. Stego image in matrix and graphics format

Matrix Form of Stego Image

190	193	190		224
191	189	190		228
191	189	190		216
207	197	178		149

Stego Image

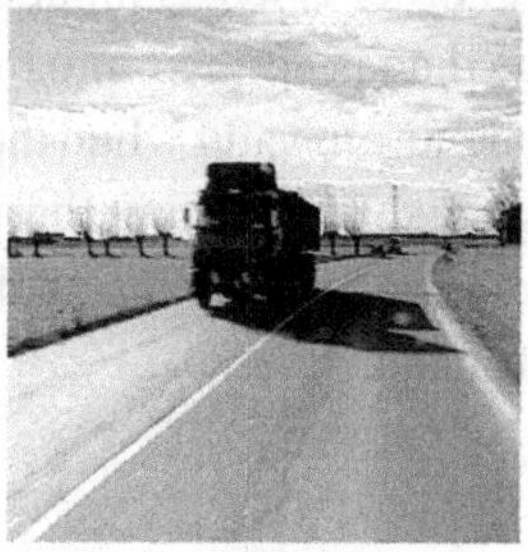

Figure 7. Secret message extraction from the stego image

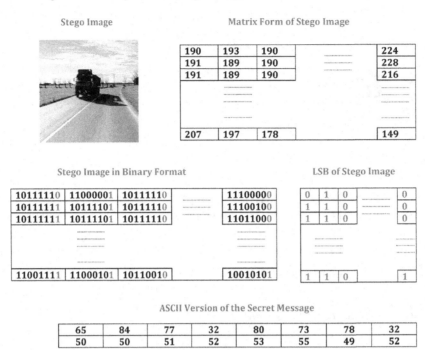

Extracted Secret Message

ATM PIN 22345714

Reconfigurable Architectures for LSB Steganography Suitable for FPGA/ASIC Implementation

Design of reconfigurable architectures for embedding and extraction processors are dealt in the following section. The embedding processor hides the secret message in the cover image pixels either by using LSB or multi-bit and generates the stego image. The extraction processor performs the reverse process of withdrawal of secret data from the stego image pixels.

Architecture of Top Design Module of LSB or Multi-Bit Embedding Processor

The top level architecture of LSB or Multi-bit embedding processor comprises message length finder, LSB or multi-bit extractor, and secret message hiding module. The pixels are read from the host system in raster scan order is fed as input to the embedding processor through the signal pixel_in [7:0]. The secret message data is also fed into the system after converting into binary format. These pixels and secret data components are valid when the 'start' signal is asserted. After processing the pixels and secret data, desired stego pixels are output through stego_out with their corresponding valid signal, stego_valid. The top module architecture and signal description of LSB steganography processor are shown in Figure 8 and Table 1, respectively.

Figure 8. Top level architecture of LSB steganography

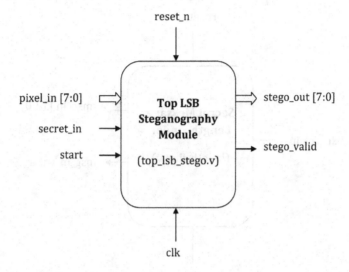

Table 1. Signal description of top level module of LSB steganography

Signals Description	Signals Description
clk	Global clock signal
reset_n	Active low system reset
start	Signal to initiate secret message embedding
pixel_in [7: 0]	Pixel input of cover image
secret_in	Secret message input
stego_out [7: 0]	Stego pixel output
stego_valid	Validity signal of stego pixel

Architecture of Secret Message Length Finder

The secret message length finder accepts secret data through secret_in signal after binary conversion. After processing, the module generates size of the message through the signal msg_ln [12:0] on arrival of corresponding valid signal. The signal description and reconfigurable architecture of the secret message length finder are presented in Figure 9 and Table 2, respectively.

Architecture of LSB Module

The LSB of the cover image pixel is extracted by feeding cover image pixel through pixel_in signal. The output of the module is LSB of the cover image pixel, generated through lsb_out on asserting lsb_valid signal. The signal description of LSB module and architecture are shown in Figure 10 and Table 3, respectively.

Figure 9. Architecture of secret message length finder

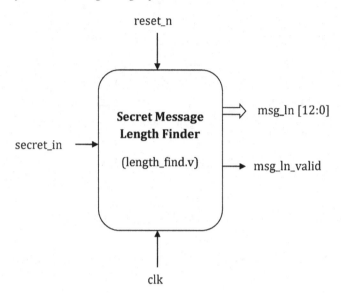

Table 2. Signal description of the secret message length finder

Signals Description	Signals Description
clk	Global clock signal
reset_n	Active low system reset
secret_in	Secret message input in binary format
msg_ln [12:0]	Secret message length output
msg_ln_valid	Validity of secret message length

EXPERIMENTAL RESULTS AND DISCUSSIONS

This section presents experimental results of embedding and extraction algorithms for LSB/multi-bits steganography implemented in reconfigurable FPGA device. The algorithms are simulated successfully at 50 MHz clock operation and verified in Matlab and C programs prior to the development of hardware architectures. This is due to the fact that this approach provides assurance to the hardware implementation and also assists in validation of concepts developed. Further, architectures are designed to suit for FPGA/ASIC implementation. The architectures presented in earlier section with its signal diagram and description assists in writing Verilog Register Transfer Level (RTL) description. The RTL style adapted in this research work is generally used in industry since it best suited for synthesis or netlist generation. Prior to logic synthesis, the code has been compiled and simulated by the authors in Modelsim HDL simulator and found to be working satisfactorily. Also, the coding strategy incorporates parallel and pipelining processing schemes for each modules to increase the throughput of a system. This approach achieves enormous improvement in the performance compared to others schemes.

Figure 11 shows the waveform for LSB/Multi-bit steganography of top module. The 'reset_n' signal is active low, used to reset the steganography processor after 20 ns. This signal is followed by the start

Figure 10. Module for LSB extraction from the cover image pixel

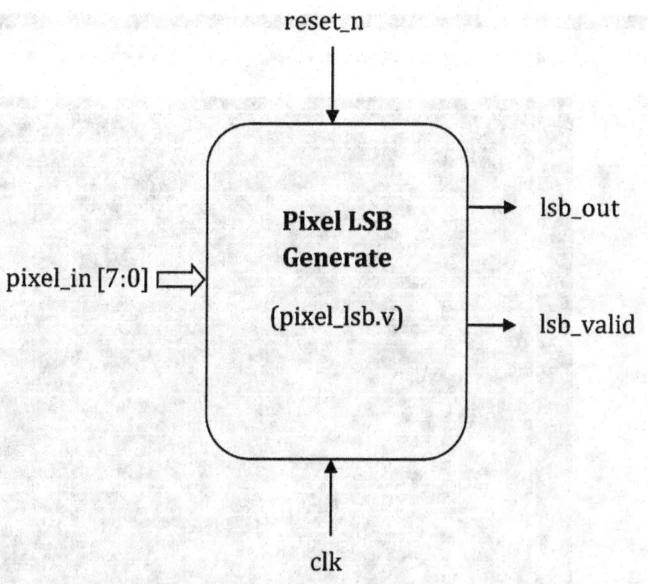

Table 3. Signal description of LSB extraction module

Signals Description	Signals Description
clk	Global clock signal
reset_n	Active low system reset
pixel_in [7: 0]	Pixel input of cover image
lsb_out	LSB output of cover image pixel
lsb_valid	Validity of LSB bit of cover image pixel

signal that starts the embedding process. Prior to the assignment of start signal, the system needs to ensure that the secret data arrives in binary format and cover image pixels are fetched in the raster scan order from the host.

The waveforms for message length finder presented in Figure 12 shows that the process starts at 60 ns after asserting the 'reset_n' signal. This module performs cumulative addition of random binary data.

In this method, the size of the binary message is 1344 bits. The duration of clock pulse is 20 ns. Therefore, the message length starts at 60 ns and ends at 26,940 ns. As is evident from this data that the latency of message length finder latency 26,880 ns. End of the message length finder waveform is shown in Figure 13.

Generation of LSB output or multi-bit output from the cover image pixels are shown in Figure 14. The resolution of the cover image considered in this work is 256×256 pixels. LSB or multi-bit processor starts at 26,980 ns. This is due to fact that LSB module is simple logical 'AND' operation. Therefore, LSB module require one clock cycle delay. In order to match the delay of message length finder and LSB module, LSB module is buffered up to 26,980 ns. Next, the LSB module produces output in every clock cycle. The last LSB of cover image pixel appears at 13,37,700 ns.

Figure 11. Waveforms for LSB/multi-bit steganography: validity of the input data

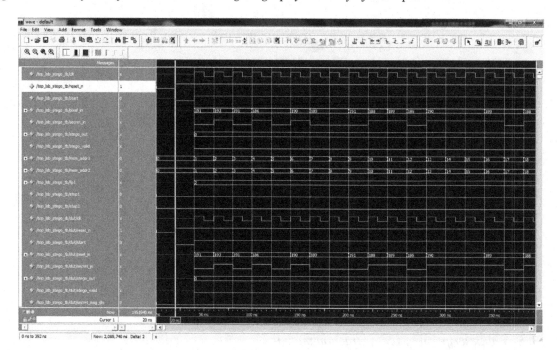

Figure 12. Waveforms for message length finder: process starts at 60 ns

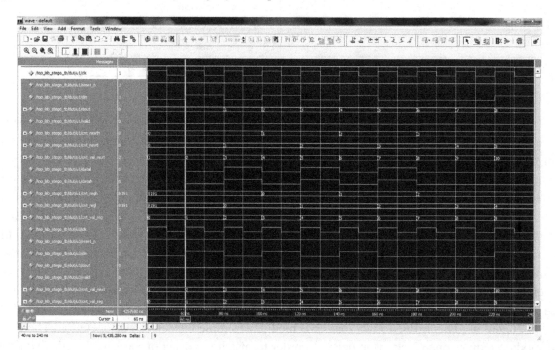

Figure 13. Waveforms for message length finder: process starts at 26,940 ns

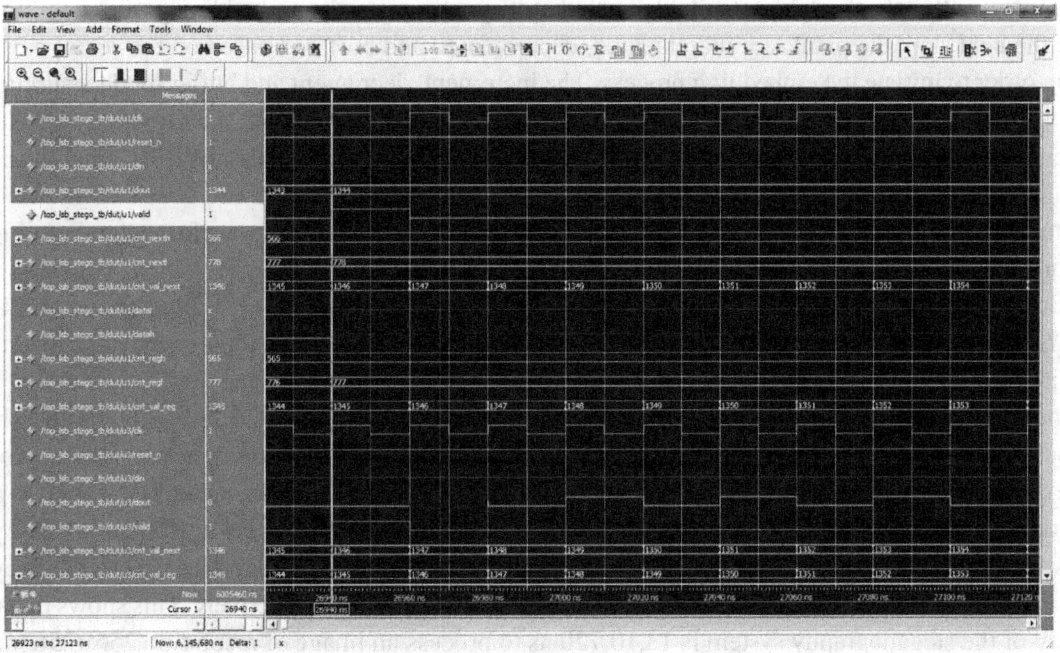

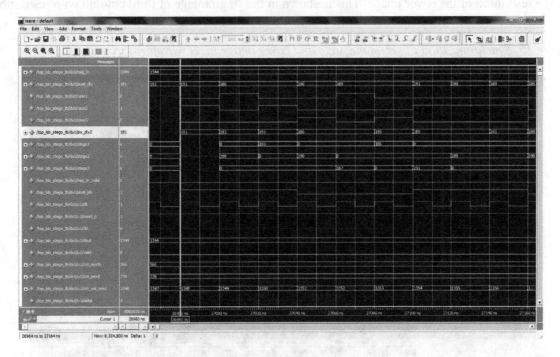

Figure 14. LSB output waveform generated from the cover image pixels

Figure 15 shows synchronization of secret message data and cover image pixel. Using pixel and message buffer, both data are delayed suitably such that it synchronizes the embedding process.

In the next step, the pixels are incremented, decremented and buffered based on the secret message data in order to initiate the embedding process. The increment, decrement and buffer module simulation waveforms are presented in Figure 16. This module works in accordance with the Equation 1 discussed earlier. For instance, if the LSB of the cover image pixel is '1' and secret message is '0', than the cover image pixel value '191' is decremented to '190' in the stego image pixel.

The first stego pixel is produced at 27,000 ns which is presented in the waveform shown in Figure 17 Further, stego pixels are produced at the rising edge of each clock cycle.

For an image size of 256×256 pixels, the last stego pixel is arrives at 13,37,720 ns shown in Figure 18. Therefore, complete embedding process ends at this timing instant. The total latency of the embedding module is 13,37,720 ns.

The timing diagram for illustration of steganography process is presented in Figure 19 and shows various stages involved in the steganography module. The input to the steganography embedding process is the cover image pixels and secret message in binary format. The input to the system is fed at 40 ns after asserting the clock signal. However, data enters into the system in the next rising edge of clock cycle. The system is designed such that it processes one pixel and generates one stego pixel in one clock cycle. First and last stego pixels are generated at 27,000 ns and 13,37,720 ns, respectively. This shows that total latency of the steganography system is 13,10,720 ns to process an image of resolution 256×256 pixels.

The simulation results of hardware implementation of steganography algorithm is presented in Figure 20. First column shows the original images of resolution 256×256 pixels. The second column shows the stego image produced by embedding the secret message. Embedding capacity is increased by increasing the resolution of the cover image. This is shown in the stego image of third column with resolution

Figure 15. Synchronization of secret message and cover image pixel waveform

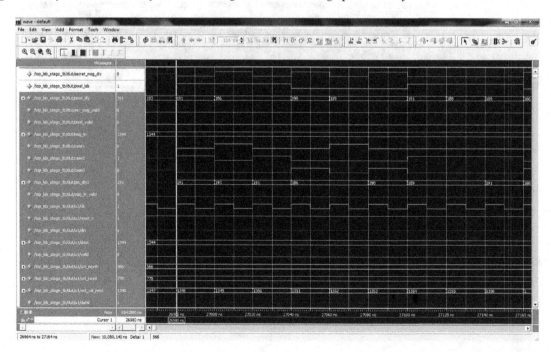

Figure 16. Waveform to increment, decrement and equal the pixel values in accordance with the secret message

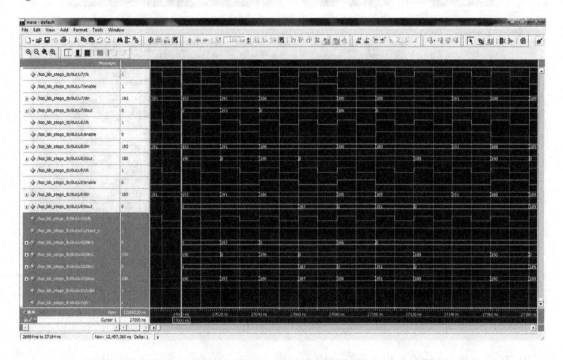

Figure 17. Stego pixels output module waveform

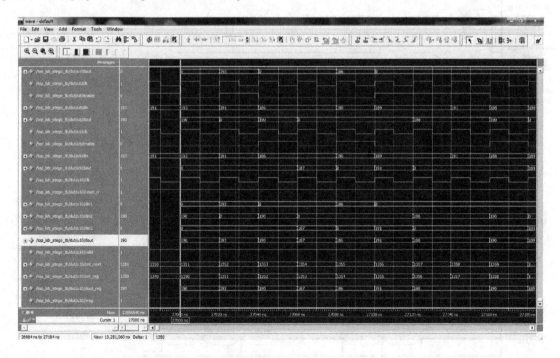

Figure 18. Last stego pixel output: end of embedding process at 13,37,720 ns

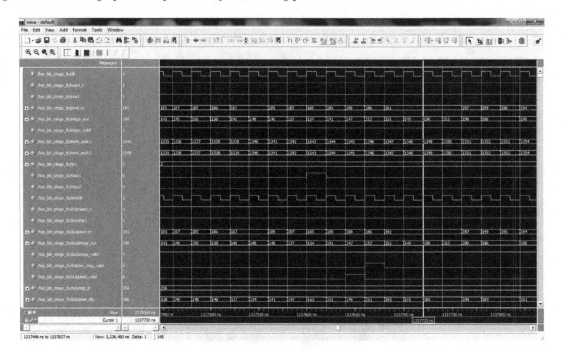

Figure 19. Timing diagram for illustrating the pipelining operation of steganography processor

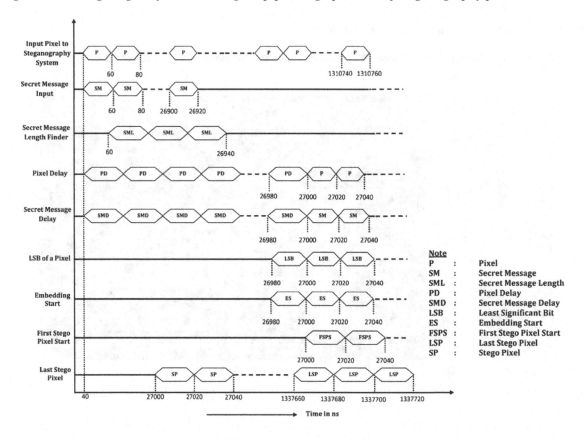

of 512×512 pixels. As is evident from the experimental results presented that the images are found to be of good quality. This is validated from the peak signal to noise ratio (PSNR) metric. Elaborated experiments are conducted for embedding secret information up to lower four nibbles of the cover image pixels. The quality of the image degrades, by embedding lower five bits of the cover image pixels. This shows the trade-off between embedding capacity and the image quality. To preserve image quality and to maintain embedding capacity, the idea presented in this chapter incorporates lower nibbles of the cover image pixels. With this trade-off, the method presented in this chapter is capable to embed maximum of 2,62,144 bits (32K Byte) of information in a 256×256 pixels cover image. However, the capacity is doubled for a cover image of size 512×512 pixels. Although, hardware simulation results are shown for monochrome images, frame work is repeated thrice for RGB channel. Below are hardware simulation results for the implementation of steganography system

As is evident from the experimental results presented that the system presented in this chapter is capable of embedding and extracting the secret message in real time. Although, the embedding algorithm presented in this work with more details, the extraction algorithm and its hardware implementation is a reverse process. Further, the algorithm is synthesized using Synplify tool (Version I-2014.03) and the results are shown in Figures 29 and 30, respectively. The design is synthesized and place and routed using XC2V500FG256-6 Xilinx FPGA device. From the synthesis report, it is found that the maximum frequency of operation is 144.3 MHz for an image size of 1920×1080 pixels. This shows that the work presented in this chapter is capable to work in real-time with reconstruction of an image at high resolution. Although, significant contributions have been made by various researchers for hardware architecture development and FPGA implementation, majority of those are either schematic based or block based implementation. The scheme proposed by Rajagopalan et al. (2014) has been implemented in FPGAs. However, the throughput achieved by this scheme is limited. Therefore, the method presented in this chapter is efficient in terms of area, speed and power consumption.

Figure 20. Original image of resolution 256 × 256 pixels

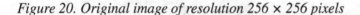

Figure 21. Stego image of resolution 256 × 256 pixels, PSNR: 78.03

Figure 22. Stego image of resolution 512 × 512 pixels, PSNR: 84.46

Figure 23. Original image of resolution 256 × 256 pixels

Figure 24. Stego image of resolution 256 × 256 pixels, PSNR: 78.10

Figure 25. Stego image of resolution 512 × 512 pixels, PSNR: 83.12

Figure 26. Original image of resolution 256 × 256 pixels

Figure 27. Stego image of resolution 256 × 256 pixels, PSNR: 77.84

Figure 28. Stego image of resolution 512 × 512 pixels, PSNR: 83.45

Table 4. Synthesis timing report from synplify tool

Performance Summary				
Worst Slack in Design: 3.070 ns				
Starting Clock	Requested Frequency (MHz)	Estimated Frequency (MHz)	Requested Period (ns)	Estimated Period (ns)
top_stego	100	144.3	10	6.930

Figure 29. RTL schematic view of the steganography system

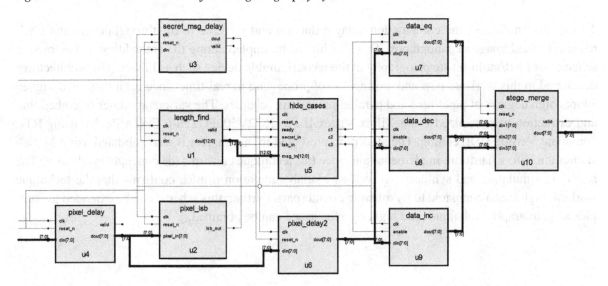

Figure 30. Detailed technology view of the steganography system

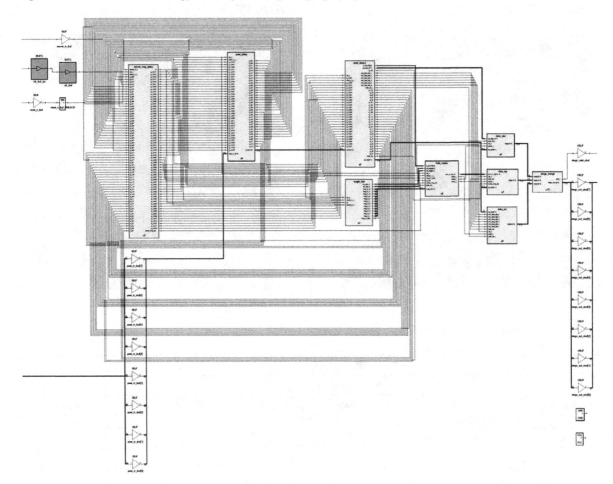

CONCLUSION

This chapter addresses the computation delay reduction and increase in the throughput for the LSB/ multi-bit based image steganography. This is achieved by implementing the embedding and extraction schemes of LSB/multi-bit steganography in the reconfigurable device such as FPGA. The architectures developed in this work are new and are capable of processing in real-time since each module designed adopts high degrees of pipelining and parallel processing schemes. The various modules of embedding and extraction were realized using Xilinx Virtex-II Pro XC2V500FG256-6 FPGA device using RTL compliant Verilog HDL coding. The algorithm presented in this chapter is also validated using Matlab simulation before hardware implementation since this technique confirms the concepts in advance. The hardware simulation and synthesis, as well as Matlab simulation results, confirms that the technique works at high speed compared to its software counterpart. Further, this scheme can be extended to complex steganography techniques and higher performance can be obtained.

REFERENCES

Amirtharajan, R., & Rayappan, J. B. B. (2012). An Intelligent Chaotic Embedding Approach to Enhance Stego-image Quality. *Information Sciences, 193*, 115–124. doi:10.1016/j.ins.2012.01.010

Chang, C. C., Lin, C. C., Tseng, C. S., & Tai, W. L. (2007). Reversible Hiding in DCT-based Compressed Images. *Information Sciences, 177*(13), 2768–2786. doi:10.1016/j.ins.2007.02.019

Crandall, R. (1998). *Some Notes on Steganography*. Posted on Steganography Mailing List.

Fan, L., Gao, T., & Cao, Y. (2013). Improving the Embedding Efficiency of Weight Matrix-based Steganography for Grayscale Images. *Computers & Electrical Engineering, 39*(3), 873–881. doi:10.1016/j.compeleceng.2012.06.014

Fan, L., Gao, T., & Chang, C. C. (2013, May). Mathematical Analysis of Extended Matrix Coding for Steganography. In *Sensor Network Security Technology and Privacy Communication System (SNS & PCS), 2013 International Conference on* (pp. 156-160). IEEE. doi:10.1109/SNS-PCS.2013.6553856

Fan, L., Gao, T., Yang, Q., & Cao, Y. (2011). An Extended Matrix Encoding Algorithm for Steganography of High Embedding Efficiency. *Computers & Electrical Engineering, 37*(6), 973–981. doi:10.1016/j.compeleceng.2011.08.006

Genovese, M., Bifulco, P., De Caro, D., Napoli, E., Petra, N., Romano, M., & Strollo, A. G. (2015). Hardware Implementation of a Spatio-temporal Average Filter for Real-time Denoising of Fluoroscopic Images. *Integration, the VLSI Journal, 49*, 114-124.

Gómez-Hernández, E., Feregrino-Uribe, C., & Cumplido, R. (2008, March). FPGA Hardware Architecture of the Steganographic Context Technique. In *Electronics, Communications and Computers, 2008. CONIELECOMP 2008, 18*th *International Conference on* (pp. 123-128). IEEE doi:10.1109/CONIELECOMP.2008.24

Hemalatha, R., Santhiyakumari, N., & Suresh, S. (2015, January). Implementation of Medical Image Segmentation using Virtex FPGA kit. In *Signal Processing and Communication Engineering Systems (SPACES), 2015 International Conference on* (pp. 358-362). IEEE. doi:10.1109/SPACES.2015.7058283

Hines, G. D., Rahman, Z. U., Jobson, D. J., & Woodell, G. A. (2004, July). DSP Implementation of the Retinex Image Enhancement Algorithm. In *Defense and Security* (pp. 13–24). International Society for Optics and Photonics.

Hussain, M., & Hussain, M. (2013). *A Survey of Image Steganography Techniques*. Academic Press.

Kipper, G. (2003). *Investigator's Guide to Steganography*. CRC Press. doi:10.1201/9780203504765

Mohd, B. J., Abed, S., Al-Hayajneh, T., & Alouneh, S. (2012, May). FPGA Hardware of the LSB Steganography Method. In *Computer, Information and Telecommunication Systems (CITS), 2012 International Conference on* (pp. 1-4). IEEE. doi:10.1109/CITS.2012.6220393

Rajagopalan, S., Amirtharajan, R., Upadhyay, H. N., & Rayappan, J. B. B. (2012). Survey and Analysis of Hardware Cryptographic and Steganographic Systems on FPGA. *Journal of Applied Sciences, 12*(3), 201–210. doi:10.3923/jas.2012.201.210

Rajagopalan, S., Prabhakar, P. J., Kumar, M. S., Nikhil, N. V. M., Upadhyay, H. N., Rayappan, J. B. B., & Amirtharajan, R. (2014). MSB Based Embedding with Integrity: An Adaptive RGB Stego on FPGA Platform. *Information Technology Journal*, *13*(12), 1945–1952. doi:10.3923/itj.2014.1945.1952

Shah, N. N., & Dalal, U. D. (2016). Hardware Efficient Double Diamond Search Block Matching Algorithm for Fast Video Motion Estimation. *Journal of Signal Processing Systems for Signal, Image, and Video Technology*, *82*(1), 115–135. doi:10.1007/s11265-015-0993-5

Tseng, Y. C., Chen, Y. Y., & Pan, H. K. (2002). A Secure Data Hiding Scheme for Binary Images. *Communications. IEEE Transactions on*, *50*(8), 1227–1231. doi:10.1109/TCOMM.2002.801488

Wang, K., Lu, Z. M., & Hu, Y. J. (2013). A High Capacity Lossless Data Hiding Scheme for JPEG Images. *Journal of Systems and Software*, *86*(7), 1965–1975. doi:10.1016/j.jss.2013.03.083

Xuan, G., Shi, Y. Q., Ni, Z., Chai, P., Cui, X., & Tong, X. (2007). Reversible Data Hiding for JPEG Images based on Histogram Pairs. In *Image Analysis and Recognition* (pp. 715–727). Springer Berlin Heidelberg. doi:10.1007/978-3-540-74260-9_64

Section 3
Segmentation, Classification and Registration based Image/ Video Processing

Video and Image segmentation/classification as well as registration have numerous techniques. Performance metrics are used to evaluate and benchmark such techniques. This section elaborates an overview of these techniques with an analysis of the evaluation metrics for Image and Video Segmentation methods. Moreover, a classification process is included to classify different types of crops, which can be applied for further video applications.

Chapter 8
Encoding Human Motion for Automated Activity Recognition in Surveillance Applications

Ammar Ladjailia
University of Souk Ahras, Algeria

Nouzha Harrati
University of Souk Ahras, Algeria

Imed Bouchrika
University of Souk Ahras, Algeria

Zohra Mahfouf
University of Souk Ahras, Algeria

ABSTRACT

As computing becomes ubiquitous in our modern society, automated recognition of human activities emerges as a crucial topic where it can be applied to many real-life human-centric scenarios such as smart automated surveillance, human computer interaction and automated refereeing. Although the perception of activities is spontaneous for the human visual system, it has proven to be extraordinarily difficult to duplicate this capability into computer vision systems for automated understanding of human behavior. Motion pictures provide even richer and reliable information for the perception of the different biological, social and psychological characteristics of the person such as emotions, actions and personality traits of the subject. In spite of the fact that there is a considerable body of work devoted to human action recognition, most of the methods are evaluated on datasets recorded in simplified settings. More recent research has shifted focus to natural activity recognition in unconstrained scenes with more complex settings.

INTRODUCTION

Much research within the computer vision community is dedicated towards the analysis of and understanding of human motion. The perception of human motion is one of the most important skills people possess, and our visual system provides particularly rich information in support of this skill. Yet, attempts and efforts to understand the human visual system or to devise an artificial solution for visual perception have proven to be a difficult task. Human motion analysis has received much attention from

DOI: 10.4018/978-1-5225-1022-2.ch008

researchers in the last two decades due to its potential use in a plethora of applications. This field of research focuses on the perception and recognition of human activities. As computing becomes ubiquitous in our modern society, the recognition of human activities emerges as a crucial topic where it can be applied to many real-life human-centric scenarios (Aggarwal & Ryoo, 2011). Furthermore, given the immense expansion of video data being recorded in everyday life from security surveillance cameras, movies production and internet video uploads, it becomes an essential need to automatically analyse and understand video content semantically. This is to ease the process of video indexing and fast retrieval of data when dealing with large multimedia content and big data. Hence, the importance of automated systems for human activity recognition is central to the success of such applications (Turaga, Chellappa, Subrahmanian, & Udrea, 2008). Further, due to the proliferating number of crimes and terror attacks as well as the vital need to provide safer environment, it becomes a necessary requirement to improve current state of surveillance systems via the use of computer vision methods to automate procedures of detecting suspicious human activities.

Human activity recognition aims to automatically infer the action or activity being performed by a person or group of people. For instance, recognizing whether someone is walking, raising hands or performing other types of activities. This usually involves the analysis and recognition of different motion patterns in order to produce a high-level semantic description for the human activities or interaction between people. This is vital to apprehend the human behavior and to determine whether their behavior is abnormal or normal via the use of automated methods (Ko, 2008). There have been considerable amount of work by the computer vision community dedicated to activity recognition with numerous approaches and methods being proposed to address different aspects and contexts of this area of research (Aggarwal & Ryoo, 2011; Poppe, 2010). Many of the early approaches have considered the use of video sequences recorded using a single camera with people being asked to perform basic actions in simplified settings and conditions. Various low-features have been proposed for encoding the human activity either at a temporal or spatial level such as edges, curvatures or complex features such as interest point descriptors. The detection of human motion is considered as a rudimentary component for constructing the activity descriptor in the majority of approaches either explicitly or implicitly for recovering other high level features. In fact, it is infeasible to detect human action from a still frame as even though achieving pose recovery can be possible from a single image, the perception of human activity can be challenging. Vishwakarma and Agrawal (2013) grouped the methods object detection through motion estimation into six conventional methods: background subtraction, statistical methods, temporal differencing and optical flow. Various recent surveys can be found in the literature on the representation of different features for human activity recognition (Poppe, 2010; Turaga et al., 2008; Vishwakarma & Agrawal, 2013). Interestingly, a new trend of research has emerged on activities recognition through the use of wearable sensors mounted to the human body (Lara & Labrador, 2013).

Applications for Activity Recognition

Research into automated recognition of human activities is fueled by the wide range of applications where human motion analysis can be deployed such as smart automated surveillance, behavioral biometrics, human computer interaction, animation and synthesis in addition to sport refereeing and analysis.

Smart Automated Surveillance

Traditionally, it is impossible for human operators to work simultaneously on different video screens in order to track and identify people of interest as well as analyze their behaviors across different places. Thus, it has become a vital requirement for scientists from the computer vision community to investigate visual-based alternatives to automate the process for human activity recognition over different views. Recently, various approaches were published in the literature to accomplish this task based on using basic features such as shape or color information. However, their practical deployment in real applications is very limited due to the complex nature of such problem (Bouchrika, 2008; Bouchrika & Nixon, 2006; Ko, 2008; Vishwakarma & Agrawal, 2013). In fact, the inability of human operators to monitor the increasingly growing numbers of CCTVs (Closed-Circuit television) installed in highly sensitive and populated areas such as government buildings, airports or shopping malls, has rendered the usability of such systems to be useless. According to the British Security Industry Association, the number of surveillance cameras deployed in the United Kingdom was estimated to be more than 5 million in 2015; this figure is expected to increase rapidly particularly after the terrorist attacks that a number of cities in Europe have witnessed. Despite the huge increase of monitoring systems, the question whether current surveillance systems work as a deterrent to crime is still questionable (Bouchrika, 2008). Security systems should not only be able to predict when a crime is about to happen but more importantly, by early recognition of suspicious individuals who may pose security threats, the system would be able to deter future crimes as it is a significant requirement to identify the perpetrator of a crime as soon as possible in order to prevent further offences and to allow justice to be administered. Furthermore, the use of smart visual surveillance technology has a wide spectrum of potential applications in addition to behavior analysis such as access control, crowd flux and congestion analysis (Ko, 2008; Vishwakarma & Agrawal, 2013).

Human Computer Interaction

Gestural interaction is becoming an integral part for newly systems from smart televisions to gaming consoles. The visual cues are the most important mode of non-verbal communication and their effective employment holds promising and innovative ways for people to interact with computers. This can even help to improve the accessibility and usability level for people with special needs and requirements. As featured in numerous science fiction movies where the actor can interact with computer systems via moving their hands and tapping their fingers in the air, it is now becoming a reality with the introduction of Microsoft Kinect and the cheap prices of depth sensing devices that sparked the rapid and abrupt advancement of gestural interaction from the advent of commercial products to a myriad of research projects (Ren, Meng, Yuan, & Zhang, 2011). Game players instead of using pads or joysticks, they can use their full body, hands and legs as an input method to control the game without wearing any special sensors or markers. Furthermore, many consumer electronic devices such as smart televisions have been developed with the capability to let users interact using hand gestures to swap between different channels or control the volume level. There are various development framework and programming toolkits being proposed to ease the process of gestural interaction using Kinect and other sensors (Deshayes, Mens, & Palanque, 2013; Suma et al., 2013). Moreover, there is a stream of research for creating interactive environments such as smart rooms that can react to various human gestures (Kühnel et al., 2011).

Video Indexing and Retrieval

With video sharing websites as Youtube facing relentlessly growth with gigabytes of multimedia content being uploaded every day, it becomes necessary to develop efficient ways for indexing and retrieving video data beyond the use of simple textual information and tags. This can be achieved through semantic attributes that can be extracted from the actual content of the video data. Content-Based video summarization has been gaining interest with advances in content-based image retrieval (Rui, Huang, & Chang, 1999). Most of the early methods have used simple sematic traits as colors and basic shapes for searching videos. Recent research efforts were geared towards object detection using various approaches remarkably the use of visual bag of words. This is commonly implemented as a histogram of the number of occurrences of specific visual patterns in a given image. The visual patterns are called words which are pre-constructed in a codebook using clustering techniques. In spite of their simplicity, bag of visual words were successfully applied to various challenging computer vision cases including recent studies to explore their applicability in automated human activity recognition. However, indexing human activities is still in its infancy due to the cumbersome challenges and complexities involved. In (Niebles, Wang, & Fei-Fei, 2008), the authors presented an approach for the non-supervised classification of human actions into different categories from video sequences. The basis of their method is the extraction of a collection of spatio-temporal words via the use of latent topic models.

MOTION FOR HUMAN ACTIVITY PERCEPTION

In spite of the fact that people can discern the state of the subject from a single static image to infer that they are doing, motion pictures provide even richer and reliable information for the perception of the different biological, social and psychological characteristics of the person (Blake & Shiffrar, 2007) such as emotions, actions and personality traits of the subject. Furthermore, this notion was also observed by Darwin (1872) in his book *"The Expression of Emotions in Man and Animals"* where it was stated: *"Actions speak louder than pictures when it comes to understanding what others are doing"*. The human visual system is very sensitive to motion as it tends to focus attention on moving objects. In contrast, static or motionless objects are not as straightforward to detect. Motion is a spatio-temporal event defined as the change of spatial location over time. Given some visual input, the visual perception of motion is regarded as the process by which the visual system acquires perceptual knowledge such as the speed and direction of the moving object (Derrington, Allen, & Delicato, 2004). Whilst this process is spontaneous for the human visual system, it has proven to be extraordinarily difficult to duplicate this capability into computer vision systems for automated understanding of human behavior.

Psychological studies carried out by the Swedish psychologist Johansson (1973), revealed that people are able to perceive human motion from Moving Lights Display (MLD). An MLD is a two-dimensional video of a collection of bright dots attached to the human body taken against a dark background where only the bright dots are visible in the scene. Different observers are asked to see the actors performing various activities. Based on these experiments observers can recognize different types of human motion such as walking, jumping, dancing and so on. Moreover, the observer can make a judgment about the gender of the performer (Kozlowski & Cutting, 1978), and even further identify the person if they are already familiar with their gait (Goddard, 1992). Cutting argued that the recognition is purely based on dynamic gait features as opposed to previous studies which were confounded by familiarity cues, size, shape or other non-gait sources of information. Although the different parts of the human body are not seen in the points and no links exist between the bright dots to show the skeleton structure of the human body, the observer can recover the full structure of the moving object. Thereby, the motion of the joints contains sufficient information for the perception of human motion (Bingham, Schmidt, & Rosenblum, 1995; Dittrich, 1993). There is a wealth of research which strives to document the capability of the human visual system to perceive the human motion from a small number of moving points as argued by early medical studies by Johansson, Cutting and Murray. Nevertheless, the underlying perceptual process is poorly understood and there is still a lack of research which explains the underlying principles for representing and retrieving the biological motion (Troje, Westhoff, & Lavrov, 2005). Two main theories have been put forward for the perception of human motion from the MLD: *structure-based* and *motion-based* (Cedras & Shah, 1995). The former theory claims that the initial step is recovering the 3D structure from the motion information observed from the MLDs, and then uses the recovered structure for the purpose of recognition. In the motion-based approach, recognition is based directly on the motion information without recovering the skeleton structure of the human body from the MLD; instead the motion information is extracted from a sequence of frames.

ACTION VS. ACTIVITY

In the computer vision literature, both terms "*Activity*" and "*Action*" are used interchangeably and contentiously but every term has its rough and gray definition (Poppe, 2010). An *action* is considered as a simple activity referring to simple pattern performed by a person during a short period of time lasting a few seconds. Examples of actions may include raising hands, bending, sitting and even walking. Poppe (2010) described additionally the term *action primitive* which refers to an atomic movement at the limb level. Vishwakarma and Agrawal (2013) has described the word *gesture* to refer to an elementary movement made by a part of the human body occurring in a very short span of time with low complexity such as waving a hand or stretching an arm. The term *action* can be considered similar to an extent to the term *gesture*. On the other hand, an *activity* is considered as a composite sequence of actions executed by either a single person or several people interacting with each other. Examples of activities are like leaving an unattended bag, shaking hands or assaulting a pedestrian. There is the term *interaction* which defines an activity or activities performed by two or more people spanning over longer times (Vishwakarma & Agrawal, 2013).

VISION-BASED SYSTEM FOR ACTION RECOGNITION

An automated vision-based system for human activity recognition through the use of motion features is designed to extract kinematic-based features without the need to use markers or special sensors to aid the extraction process. In fact, all that is required is an ordinary video camera linked to special vision-based software. Marker-less motion capture systems are suited for applications where mounting sensors or markers on the subject is not an option as the case of visual surveillance. Typically, the system consists of two main components:

1. Hardware platform dedicated for data acquisition. This can be a single CCTV camera or distributed network of cameras.
2. Software platform for data processing and recognition.

The architecture of the software side for human activity analysis is composed broadly of three main components:

1. Detection and tracking of the subject,
2. Feature extraction and
3. Classification stage.

Figure 1 shows the flow diagram for human action recognition outlining the different subsystems.

Subject Detection and Tracking

People detection is the first major milestone for automated system of human activity recognition. A walking subject is initially detected within a sequence of frames using background subtraction techniques to detect moving objects or via the use of other methods such as the Histogram of Oriented Gradients (HoG) (Bouchrika, Carter, Nixon, Morzinger, & Thallinger, 2010; Dalal & Triggs, 2005) which is capable of detecting people from still images at real-time. The HoG method requires no background subtraction and therefore avoiding the need of maintaining and updating a model for the background. Subsequently, intra-camera tracking is performed to establish the correspondence of the same person across consecutive frames. Tracking methods are supported by simple low-level features such as blob size, aspect-ratio, speed and color in addition to the use of prediction algorithms to estimate the parameters of moving objects in the following frames. This is based on motion models which describe how parameters change over time. The most popular predictive methods used for tracking is the Kalman filter (Welch & Bishop, 2001), the Condensation algorithm (Isard & Blake, 1998), and the mean shift tracker (Comaniciu, Ramesh, & Meer, 2000).

Figure 1. Overview system for marker-less human activity recognition

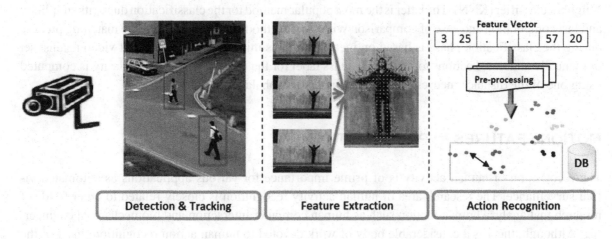

Feature Extraction and Representation

This is the most important stage for automated marker-less capture systems whether for human identification, activity classification or other imaging applications. This is because the crucial data required for the classification phase are derived at this stage. Feature extraction is the process of estimating a set of measurements either related to the configuration of the whole body or the configuration of the different body parts in a given scene and tracking them over a sequence of frames. The features should bear certain degree of the discriminability between the different clusters of human activities. Various types of features are employed such as the trajectories of the joints positions estimated via pose recovery of the different parts of the human body. Contour-based features are used in a number of recent studies via analyzing silhouettes data. Textural features are proved to offer promising results on the detection of similar human actions even for the case of single frames. Irani *et al.* (Blank, Gorelick, Shechtman, Irani, & Basri, 2005; Shechtman & Irani, 2007) have proposed a descriptor based on analyzing adjacent patches based on their internal correlation for comparing images where they showed its potency for action detection. However, the majorities of studies consider the use of motion-based features for understanding human activities (Shah & Jain, 2013) . Depending on how kinematic features are represented based on the spatial properties, features estimated at this level can be categorized into two major types:

- **Global Features:** Where the whole image or body region is considered meanwhile.
- **Local Features:** Refer to the characteristics which are extracted from smaller portions of the image.

Classification Phase

This is mainly a pattern recognition process which involves matching a test sequence with an unknown label against a group of labeled references considered as the gallery dataset. At this stage, a high-level description is produced from the features extracted during the previous phases to infer or confirm the subject identity. The classification process is normally preceded by pre-processing stages such as data normalization, feature selection and dimensionality reduction of the feature space through the use of statistical methods. A variety of pattern recognition methods are employed in vision-based systems for human activity recognition including Neural Networks, Support Vector Machines (SVM) and K-Nearest Neighbor classifier (KNN). The latter is the most popular method for the classification due to its simplicity and fast computation and ease of comparison with most methods in the literature. The matching process during the classification phase is based on measuring the similarity between the test video against set of manually annotated actions to predict the class label for the unseen data. The similarity is computed using one of the distance metrics such the Euclidian or Mahalanobis distance.

MOTION FEATURES REPRESENTATION

The recognition of human activity is of prime importance for various applications as automated visual surveillance. The research area of human activity recognition is closely related to other fields of research that analyze human motion such as human computer interaction and biomechanical engineering. Although, there is a considerable body of work devoted to human action recognition, most of the methods are evaluated on datasets recorded in simplified settings. More recent research has shifted focus to natural activity recognition in unconstrained scenes with more complex settings (Oshin, Gilbert, & Bowden, 2014). Poppe (2010) and Vishwakarma (2013) surveyed the recent methods, research studies and datasets devoted to this area of research. Existing methods can be broadly classified into two major categories in terms of image representation which are either global or local representation. There are recent studies that consider the use of a hybrid model by fusing both types of features as it was suggested to be more suitable for encoding human actions. In another study, Weinland *et al* have categorized three major classes of features for human action representation which are: body models, image models and sparse features (Weinland, Ronfard, & Boyer, 2011). The last two categories refer to the global and local features respectively discussed in most surveys. The body models aims to recover the spatial structure of the different parts of the human body via fitting a prior model. From another perspective, the temporal dimension is taken into account explicitly for most image representations in addition to the spatial information meanwhile other methods extract image features on a frame by frame basis. In this research work, three major categories for the various approaches devoted to markerless human activity recognition are considered and discussed in this section.

Pose-Based Approaches

For action recognition using pose-based representation, the parts of the human body are first recovered or reconstructed through the use of specific models. Although model-based approaches tend to be complex requiring high computational cost, these approaches are the most popular for human motion analysis due to their advantages (Yam & Nixon, 2009). The model can be either a 2 or 3-dimension structural model, motion model or a combined model. The structural model describes the topology of the human body parts as head, torso, hip, knee and ankle by measurements such as the length, width and positions. This model can be made up of primitive shapes based on matching against low-level features as edges. The stick and volumetric models are the most commonly used structural-based methods. Akita (1984) proposed a model consisting of six segments comprising of two arms, two legs, the torso and the head. Guo *et al* (Guo, Xu, & Tsuji, 1994) represented the human body structure by a stick figure model which had ten articulated sticks connected with six joints. Rohr (1994) proposed a volumetric model for the analysis of human motion using 14 elliptical cylinders to model the human body. Karaulova *et al.* (Karaulova, Hall, & Marshall, 2000) used the stick figure to build a hierarchical model of human dynamics represented using Hidden Markov Models (HMMs). Gavrila *et al.* (Gavrila & Davis, 1995) described a 3D model for pose recovery based on conducting a search of synthesized images against real images using the chamfer distance for different views. The main merit of using 3D models is the viewpoint invariance provided the pose estimation is done accurately (Weinland, Özuysal, & Fua, 2010).

Global-Based Approaches

For the global representations which are called occasionally holistic methods, the region of interest (ROI) of a person is encoded as a whole. In most cases, the labeling or detection of body parts are not required. Instead, the features are computed densely on a grid bounded by region of interest. The subject is usually derived from an image through applying background subtraction. The processing of global representations is based on low-level information taken from silhouettes, edges or optical flow (Poppe, 2010). However, these methods are susceptible to noise, occlusions and variations in camera viewpoint. Many research studies argued that silhouette data provides strong cues for activity recognition with the benefit of being insensitive to texture, contrast and color changes (Weinland et al., 2011). However, silhouette-based methods depend on the accuracy of background segmentation which cannot be guaranteed in outdoor scenes. Recent studies argued that noisy silhouettes can be employed for activity recognition through the use of better matching techniques including the chamfer distance, phase correlation or shape context descriptor derived from silhouette data (Ogale, Karapurkar, & Aloimonos, 2007; Oikonomopoulos, Patras, & Pantic, 2005). Another important type of features used for global representation is optical flow extracted from consecutive frames to represent the motion whilst the subject performs an activity.

Wang (Wang, Huang, & Tan, 2007) applied the *R* transform on the extracted silhouettes reporting that the obtained representation is translation and scale invariant. The main benefit of the *R* transform is its low computational cost as well as it geometric invariance. A set of HMMs are employed for training the extracted features in order to recognize activities. Yamato *et al* quantized silhouette images into super pixels such that each pixel indicates the ratio of black to white pixels within the considered smaller region (Yamato, Ohya, & Ishii, 1992). Weinland (Weinland & Boyer, 2008) described a compact and efficient representation which is based on matching a set of discriminative static landmark pose models. The method does not depend on or take into account the temporal ordering of sequences. In their work, silhouette models are matched against edge data using the Chamfer distance and therefore eliminating the need for background segmentation. For the use of optical flow, Polana and Nelson computed the temporal texture to recognize events based on their motion. For human activity recognition, features are based on the optical flow magnitude contained within non-overlapping cells of a regular grid (Nelson & Polana, 1992; Polana & Nelson, 1994). In a different study, Ali and Shah derived a set of kinematic-based features from the optical flow such as divergence, velocity, symmetric and anti-symmetric flow fields. Multiple instance learning method is used together with Principal Component Analysis to determine the kinematic modes (Ali & Shah, 2010).

Local-Based Approaches

For activity recognition using local representations, a collection of independent patches within an image are analyzed to generate a discriminative feature vector for the observed activity. Local representations do not require accurate localization or background subtraction and enjoy the benefits of being to some extent invariant to appearance transformation, background clutter and partial occlusion (Poppe, 2010). Local patches are described by local grid-based descriptors that would summarize locally the observation within grid cells for the case of still frames. In contrast to the global representation, the local features are not linked or related to specific body parts or spatial positions of an image. Actions or activities are encoded based on the statistics of the sparse features. The main benefit of using local features is the un-necessity for people detection or the localization of the different body parts (Weinland et al., 2011). Space-time interest point descriptors which are analogous to classical 2D interest points as SURF and SIFT, have become the most popular type of local features being used for action recognition (Laptev, 2005).

For the use of motion-oriented features for human activity recognition, Yeffet (Yeffet & Wolf, 2009) proposed a local trinary pattern descriptor for encoding human motion from a sequence of frames. The trinary number is generated from a matching process of patches of a given frame against adjacent patches residing on both the previous and next frames respectively. The matching process is based on the self-similarity descriptor for textures (Shechtman & Irani, 2007). The encoding of action is done in the same way as the local binary operator to describe the displacement of patches between adjacent frames. A histogram-based feature vector is constructed from the concatenation resulting from the image divided into a grid. As an extension of their work, Kliper-Gross (Kliper-Gross, Gurovich, Hassner, & Wolf, 2012) employed the same approach of the local trinary motion pattern renamed as Motion Interchange Pattern (MIP) for the automated recognition of human activities. Kliper-Gross presented a suppression mechanism in order to decouple static edges from edges related to motion. Further, in order to account for camera movement, motion compensation procedure is integrated within the actual local motion description based on affine transformation. For the classification stage, bag of visual features are used together with support vector machines. Oshin (Oshin et al., 2014) presented the Relative Motion descriptor for activity recognition in unconstrained scenarios using motion induced cues only. The descriptor is based on the relative distribution of spatio-temporal interest points my measuring the response strength of such points within localized regions.

The optical flow is used also for human action recognition via local representation of features. Chaudhry (Chaudhry, Ravichandran, Hager, & Vidal, 2009) argued that the recent use of complex histogram-based descriptors can fail at some point as they live on a non-Euclidean space. A Histogram of oriented optical flow (HOOF) is proposed with the merit of scale or motion direction invariance. The HOOF features are derived at every frame without the need for prior segmentation or background subtraction. The Binet-Chauchy kernels are extended to allow the matching of non-linear histograms of time series. The method was evaluated on the Weizmann human action dataset reporting a high classification rate of 95.66%. Ikizler (Ikizler, Cinbis, & Duygulu, 2008) combined the use of boundaries of a human figure fitted via small line segments together with motion information estimated via optical flow. The Hough transform is applied to detect line segments. The compact representation presented in their work was tested in different challenging conditions with high accuracy for action recognition. Feature selection is applied to compact the original feature space from 108 dimensions into a smaller space of 30 features. Martinez (Martínez, Manzanera, & Romero, 2012) computed optical flow to approximate the velocity for every pixel. The obtained flow vectors are accumulated into a per-frame histogram weighted by the norm whilst the motion orientations are quantized into 32 main directions. A histogram-based descriptor of 192 bins is obtained for every action. Results conducted on the Weizmann dataset shows the method can achieve an average accuracy of 95% using the support vector machine classifier.

In a different study by (Ladjailia, Bouchrika, Merouani, & Harrati, 2015b), the authors proposed an approach to encode a sequence of frames into a feature vector describing the performed action by a person. The method does not depend on background subtraction for the derivation of motion features. This is because it is computationally expensive and complex to deploy background subtraction for real-time surveillance applications due the process of updating the background model which is influenced by a number of factors such as background clutter, weather conditions and other outdoor environmental effects. Inspired by the work of Kliper-Gross et al., (2012) for proposing the Motion Interchange Pattern for action recognition together with the fact that local descriptors are known for their effectiveness and robustness for encoding texture for recognition purposes including biometrics, the local descriptor which captures the motion of the local structure based on estimating optical flow. Provided that there is a motion of a small patch at frame *t* to the next frame *t+1*, there is a high probability that a similar patch would be induced within the neighboring region of the original patch position at the previous frame. The proposed descriptor is based on constructing a feature that reflects the patch displacement from frame to frame based.

Because of the common increase of image self-similarity regions, the block matching using simple similarity operators can fail in distinguishing to between similarity caused by motion and similar static textures. In addition, the matching can be difficult as moving patches may have their appearances changed due to the non-rigid nature of the human motion. The optical flow is instead harnessed to better estimate the motion information from video sequences. Optical flow is one of the most active research areas in computer vision due to their central role in various fields of applications such as autonomous vehicle or robot navigation, visual surveillance and fluid flow analysis. The main basis of optical flow is to observe the displacement of intensity patterns (Fortun, Bouthemy, & Kervrann, 2015). This pattern is a result of the apparent motion of objects, surfaces, and edges in a visual scene caused by the relative movement between an observer and the scene (Burton & Radford, 1978). In other words, optical flow can arise either from the relative motion of the object or camera. For a given image I, the constraint for optical flow states that the gray intensity value of a moving pixel $I(x,y,t)$ at time t stays constant over time as expressed as:

$$I\left(x,y,t\right) - I\left(x+V_x, y+V_y, t+1\right) = 0 \tag{1}$$

To solve Equation 1 which has two unknowns, constraints are required to be added to ease finding a solution. There are several solutions proposed in the literature. Differential methods are the most used method. The method of Lucas, Kanade, et al., (1981) is considered for estimating the optical flow vector. The method is based on the principle that relative motion of brightness content between two successive images is small and approximately constant within a local neighborhood of a given point p. Therefore, the optical flow equation is assumed to hold for all points within the smaller neighborhood region centered at p. The lucas-kanade method solves the inherent ambiguity of the optical flow equation via combining information for several close pixels.

Based on a triplet of frames denoted as *previous*, *current* and *next*, a descriptor number d is constructed for every pixel for the current image through computing two optical flow images for v_{prev}: (*previous, current*) and v_{next}: (*current, next*). Thresholding is applied such that it is based on the magnitude of the velocity flow considering only values greater than *tau=0.5*. Based on the location of the angular values within the polar coordinate system which is equally divided into 8 numbered sections of 45 degrees from 1 to 8, the optical flow vector is converted into a number reflecting the order within the eighth circular portions. This is denoted using the function *AngIndex* as expressed in Equation 2. The zero indexes refer that there is no motion where the magnitude of the optical flow is less than the threshold *tau* . Both of the two digits resulting at every pixel from the next and previous frames are concatenated together to generate a number of base 9 which is converted to a decimal number.

$$d = AngIndex\left(v_{prev}\right) + AngIndex\left(v_{next}\right) * 9 \tag{2}$$

The number d serves as a descriptor for the motion at a pixel level. Experimentally, it is observed that a simple action can be fully contained within only 15 frames based on video recorded at a frame rate of 25. Therefore, the encoding process is performed for every pixel for the seven different triplets of consecutive frames taken from a video. The motion orientation histogram for a triplet is computed as shown in Equation 3. *b* is a Boolean function returning 1 for true cases and 0 for false conditions

$$H_i = \sum_{x,y} b\Big(d\big(x,y,t\big) == i\Big)$$ (3)

Figure 2 outlines the procedure to estimate the histogram of motion-based features using optical flow. Various features that could potentially describe better the motion are generated based on simple fusion operations including summation and statistical operators being applied on the set of motion orientation histograms for the triplets of frames. Equation 4 shows the obtained feature vector by concatenation of different histograms. H^t refers the histogram obtained at t^{th} triplet of frames. STD is an abbreviation for the standard deviation. The resulting action vector consists of features describing purely local motion features of the human body without any information describing neither the global structure of the activity nor the anthropometric measurements of the human body.

$$F = \left[H^1 ... H^7 \quad Mean\big(H^{1...7}\big) \quad STD\big(H^{1...7}\big) \quad \sum_{t=1}^{7} H^t \right]$$ (4)

The feature selection process is considered in this research to derive the most discriminative features and suppress the redundant and irrelevant components which may degrade the classification rate. Because it is infeasible to conduct a brute force search procedure for all possible combinations of subsets to derive the optimal feature subset due to the high dimensionality of the raw feature vector. Alternatively, the Adaptive Sequential Forward Floating Selection (ASFFS) search algorithm (Somol, Pudil, Novovi\vcová, & Pacl\ik, 1999) is harnessed to reduce the number of features. The feature subset selection procedure is purely based on an evaluation function that assesses the discriminativeness of each component or set of features in order to derive the optimal subset of features for the classification process. Validation-based evaluation criterion is described to pick up the subset of features that would minimize the classification

Figure 2. Construction of histogram using motion-based features via optical flow

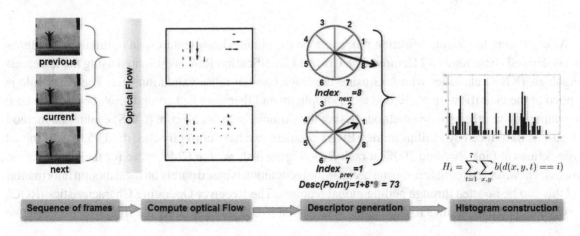

errors and ensure larger inter-class separability between the different classes. As opposed to the voting paradigm employed by the *KNN* classifier, the evaluation function utilizes coefficients *w* that signify the significance of the most nearest neighbors of the same class. The probability score for a candidate s_c to belong to a cluster *c* is expressed in the following Equation 5 as:

$$f\left(s_c\right) = \frac{\sum_{i=1}^{N_c-1} z_i w_i}{\sum_{i=1}^{N_c-1} w_i} \tag{5}$$

Where N_c is the number of instances within cluster *c*, and the coefficient w_i for the ith nearest instance is inversely related to proximity as given:

$$w_i = \left(N_c - i\right)^2 \tag{6}$$

The value of z_i is defined as:

$$z_i = \begin{cases} 1 & if \ nearest\left(sc, i\right) \in c \\ 0 & otherwise \end{cases} \tag{7}$$

Such that the *nearest(s_c,i)* function gives the ith nearest instance to the instance s_c. The Euclidean distance metric is used to deduce the nearest neighbours from the same class. The significance for a subset of features is based on the validation-based metric which is computed using the leave-one-out cross-validation rule. The human action signature is made as the subset of features *S* among the feature space *F* attaining the maximum value which is the average sum of *f* computed across the *N* instances *x* as expressed the following equation:

$$Signature = \arg\max\left(\frac{\sum_{x=1}^{N} f_S\left(x\right)}{N}\right) \tag{8}$$

After running the feature selection procedure on the obtained raw features, an optimal action signature is derived containing 648 features. The Correct Classification Rate is estimated using the K-nearest neighbour (KNN) classifier with *k=3* using the leave-one-out cross-validation rule. The KNN rule is applied at the classification phase due to its simplicity and therefore fast computation besides the ease of comparison to other existing methods. Using the Cumulative Match Score (CMS) evaluation method which was introduced by Phillips in the FERET protocol, we have correctly classified 95.02% of the 20 basic actions at rank *R=1* and 100% at rank *R=9*. Figure 3 shows the CMS curve for the classification process. The achieved results promising because the recognition is based purely on local motion information and this can be boosted through adding global features. The Receiver Operating Characteristics (ROC) curves are plotted in Figure 3 to express the verification results for estimating the similarity between two instances across all pairs. In the verification process, the instances from database are verified to check if they belong to the claimed class labels based on computing the Euclidean distance. The thresholding function described in Feature Selection section is used to express whether the two pairs belong to the claimed class. In order to plot the False Acceptance Rate (FAR) versus the False Rejection Rate (FRR), different score thresholds are used. Using the human action signature derived from dynamics, the system achieved equal error rate of 1.89% is obtained (Ladjailia, Bouchrika, Merouani, & Harrati, 2015a)

DATASETS FOR HUMAN ACTIVITIES

There are several datasets made publicly available to the research community to validate their methods for automated activity recognition and provide a common ground for researchers to compare their results on the same dataset. Most of the early datasets are constructed with a single camera containing a dozen of simple actions for a limited number of people. Recording is usually done in controlled environment with simple settings. There are recent emerging datasets based manual annotations of video clips being taken from movies and videos uploaded to online services.

KTH Dataset

The KTH dataset is constructed by the KTH Royal Institute of Technology in Stockholm, Sweden (Schüldt, Laptev, & Caputo, 2004). The dataset consists of 2,391 video sequences containing six types of human actions including: running, walking, jogging, boxing, hand clapping and hand waving. The

Figure 3. Classification results for human activity recognition: (a) cumulative match score plot. (b) receiver operating characteristic curve

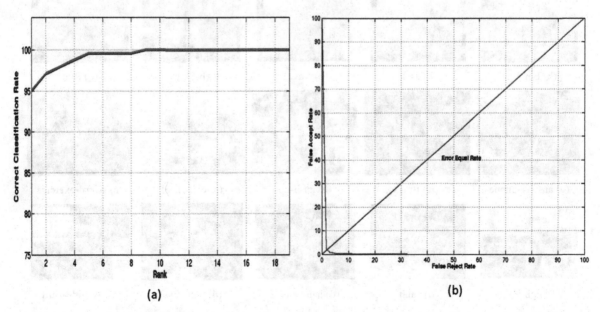

(a)   (b)

actions are performed by 25 people with three different outdoor scenarios and an indoor session. For the outdoor sessions, there are variations in terms of illumination, scale and clothing appearance. All video sequences were recorded over homogeneous backgrounds with a static camera with a frame rate of 25 frames per second. The videos are resized downward to the spatial resolution of 160×120 pixels with an average duration of four seconds. As for benchmarking, the authors suggested dividing the dataset with respect to the individuals into a training set containing 8 people, a validation set with subjects and a test dataset consisting of 9 persons. The dataset is made online publicly available for download as AVI video files Figure 4 shows examples taken from the KTH dataset.

Weizmann Dataset

The Weizmann dataset (Blank et al., 2005) contains 90 video sequences with low-resolution of *180x144* recorded at frame rate of 50 frames per second in de-interlaced mode. There are nine different people, each performing 10 natural activities. The performed actions include: walk, run, skip, jumping-jack (or shortly "jack"), jump forward on two legs (or "jump"), jump in place on two legs (or "pjump"), gallop

Figure 4. Human Activity Datasets: (a) KTH (Schüldt et al. 2004), (b) Weizmann (Blank et al. 2005), (c) HMDB51 (Kuehne, 2001), (d) Hollywood2 (Marszalek et al. 2009)

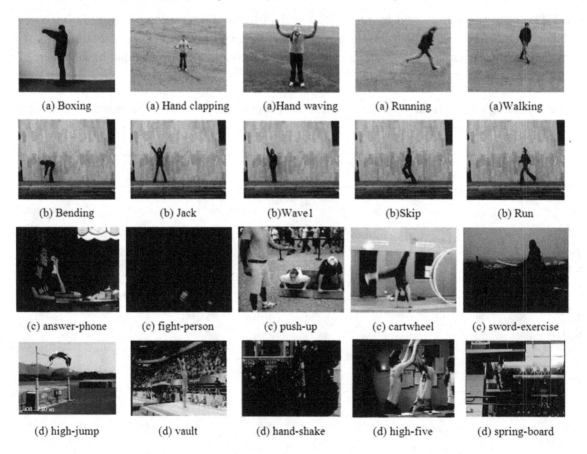

(a) Boxing	(a) Hand clapping	(a)Hand waving	(a) Running	(a)Walking
(b) Bending	(b) Jack	(b)Wave1	(b)Skip	(b) Run
(c) answer-phone	(c) fight-person	(c) push-up	(c) cartwheel	(c) sword-exercise
(d) high-jump	(d) vault	(d) hand-shake	(d) high-five	(d) spring-board

sideways (or "side"), wave two hands (or "wave2"), wave one hand (or "wave1") and bend. Silhouettes from the video sequences are provided with the dataset generated via subtracting the median background from each of the sequences. Examples from the Weizmann dataset are shown in Figure 4. In (Ladjailia et al., 2015b), the authors manually collected a dataset containing 241 video sequences for 19 different basic actions by decomposing an activity into primitive actions. Each video consists of 15 frames which are all checked to better describe the complete action.

Hollywood Dataset

There are two versions of the Hollywood dataset (Marszalek, Laptev, & Schmid, 2009). The first release covers only 8 basic actions with a limited number of video clips. The second version of the Hollywood dataset contains 12 different classes of human actions and 10 classes of scenes distributed over 3,669 clips with a total duration of approximately 20.1 hours of video footage. The dataset is setup with the aim to provide a comprehensive benchmark for human activity recognition in realistic and challenging environments. The dataset is constructed by taking video clips from 69 movies through the aid of automated movie script processing to retrieve scene descriptions. The list of actions contained in this dataset include: Answer Phone, Drive Car, Eat, Fight Person, Get Out Car, Hand Shake, Hug Person, Kiss, Run, Sit Down, Sit Up and Stand Up. The videos contained in the dataset are subjected to various factors as occlusions, camera movements and dynamic backgrounds which would make it more challenging. The database is split into a training and test subsets such that the two subsets do not share samples from the same film.

HMDB51 Dataset

The Human Motion DataBase (HMDB51) (Kuehne, Jhuang, Garrote, Poggio, & Serre, 2011) contains 51 different human action categories such that every activity class comes with at least 101 video clips with a total of 6,766 videos. The video are extracted from a wide variety of sources including Youtube. com. The authors claim that the HMDB51 dataset is the largest and most realistic database devoted for human activity recognition. Each video clip is manually annotated and validated by at least two people to ensure the consistency. Information meta tags are provided to allow better and precise selection of testing data and training for flexible evaluation of the performance of the proposed approach. The tags for each video describe the camera view-point, the presence of camera movement, the video quality, and the number of people in the scene. The original videos taken to extract the activity clips vary in size and frame rate. Therefore, In order to ensure consistency across the dataset, the heights of all clips are resized to 240 pixels. The width is rescaled accordingly to maintain the aspect ratio constant. The frame rate is resampled to 30 frames per second for all video clips.

CHALLENGES AND DIFFICULTIES

Despite the recent outstanding advancements in computer vision and pattern recognition technologies, the automated marker-less extraction and recognition of human activities are proven to be a challenging task. Although, the problem can be stated in simple terms, given a sequence of frames with one or more people performing a given activity, can an automated system recognize the activity being performed?

The solution is difficult to devise or implement. The difficulties stem from a large number of factors that can be related to one three following classes:

- **Person:** Most of the existing methods proposed for human activity recognition rely on sensors or special markers mounted on the subject (Lara & Labrador, 2013). For a marker-less approach, the articulated nature of human body which encompasses a wide range of possible motion transformations in addition to self-occlusion and appearance variability, exacerbate further complexity on the task of visual feature extraction for the process of human activity recognition (Moeslund, Hilton, & Krüger, 2006). Even through, there is ample research about pedestrian detection for real-time applications with reported higher accuracy, the localization of people is still hard to achieve in cluttered environments with the desired performance (Poppe, 2010; Tang, Andriluka, & Schiele, 2014). Furthermore, there is a substantial variation in terms of the appearance and the time needed for performing an action by different people. The variation is determined and influenced by various factors such as age, emotional state and fatigue which can severely change the way we perform actions.

- **Acquisition Environment:** Challenges related to the acquisition environment may include background clutter, illumination, camera movement and viewpoint as well as occlusion by other objects in the scene. Dynamic background adds further complexity for foreground segmentation and extracting motion or kinematic features related to people. Further, the challenges become even harder when using a moving camera. For the change of camera viewpoint or position, the same action can be represented and understood differently when changing the viewpoint or even the distance from the camera (Weinland et al., 2010). Low resolution and poor video quality due the temporal and spatial down-sampling are common in current surveillance technologies which exacerbate further obstacles (Rahman, See, & Ho, 2015). Even though recent research studies argued about the possibility of recognizing human actions from a number of limited frames (Schindler & Van Gool, 2008), it is still a difficult process to achieve an acceptable classification rate for cases of low-frame rates or frames being dropped.

- **Activity Understanding:** An activity can be performed at various ways by different people depending on the context (Zhu, Nayak, & Roy-Chowdhury, 2013) or even culture of the performer. For instance, human gestures or actions to express joy and happiness can take different ways and forms. Inversely, the same activity performed by different people can have different semantic meanings. Furthermore, activities can interleave within each other and performed in parallel rather than a sequential fashion. For instance a person can use their computer whilst eating at the same time or answering the phone. Hence, the system needs to infer between primary and secondary activities in the scene.

CONCLUSION

The perception of human motion is one of the most important skills people possess, and our visual system provides particularly rich information in support of this skill. Yet, attempts and efforts to understand the human visual system or to devise an artificial solution for visual perception have proven to be a difficult task. Human motion analysis has received much attention from researchers in the last two decades due to its potential use in a plethora of applications. This field of research focuses on the perception and recognition of human activities. The recognition of human activity is of prime importance for various applications as automated visual surveillance. The research area of human activity recognition is closely related to other fields of research that analyze human motion such as human computer interaction and biomechanical engineering. Although, there is a considerable body of work devoted to human action recognition, most of the methods are evaluated on datasets recorded in simplified settings. More recent research has shifted focus to natural activity recognition in unconstrained scenes with more complex settings. Various types of features are considered for the representation of human actions that can be grouped in three major categories: Pose-based, global and local methods. There are several datasets made publicly available to the research community to validate their methods for automated activity recognition and provide a common ground for researchers to compare their results on the same dataset.

REFERENCES

Aggarwal, J. K., & Ryoo, M. S. (2011). Human activity analysis: A review. *ACM Computing Surveys*, *43*(3), 16. doi:10.1145/1922649.1922653

Akita, K. (1984). Image Sequence Analysis of Real World Human Motion. *Pattern Recognition*, *17*(1), 73–83. doi:10.1016/0031-3203(84)90036-0

Ali, S., & Shah, M. (2010). Human action recognition in videos using kinematic features and multiple instance learning. *Pattern Analysis and Machine Intelligence. IEEE Transactions on*, *32*(2), 288–303.

Bingham, G. P., Schmidt, R. C., & Rosenblum, L. D. (1995). Dynamics and the Orientation of Kinematic Forms in Visual Event Recognition. *Journal of Experimental Psychology. Human Perception and Performance*, *21*(6), 1473–1493. doi:10.1037/0096-1523.21.6.1473 PMID:7490589

Blake, R., & Shiffrar, M. (2007). Perception of human motion. *Annual Review of Psychology*, *58*(1), 47–73. doi:10.1146/annurev.psych.57.102904.190152 PMID:16903802

Blank, M., Gorelick, L., Shechtman, E., Irani, M., & Basri, R. (2005). Actions as space-time shapes. In *Computer Vision, 2005. ICCV 2005. Tenth IEEE International Conference on* (Vol. 2, pp. 1395-1402).

Bouchrika, I. (2008). *Gait Analysis and Recognition for Automated Visual Surveillance*. University of Southampton.

Bouchrika, I., Carter, J. N., Nixon, M. S., Morzinger, R., & Thallinger, G. (2010). Using gait features for improving walking people detection.*20th International Conference on Pattern Recognition (ICPR)* (pp. 3097-3100). doi:10.1109/ICPR.2010.758

Bouchrika, I., & Nixon, M. S. (2006). Markerless Feature Extraction for Gait Analysis.*IEEE SMC Chapter Conference on Advanced in Cybernetic Systems*.

Burton, A., & Radford, J. (1978). *Thinking in perspective: critical essays in the study of thought processes*. Methuen.

Cedras, C., & Shah, M. (1995). Motion-based Recognition: A survey. *Image and Vision Computing*, *13*(2), 129–155. doi:10.1016/0262-8856(95)93154-K

Chaudhry, R., Ravichandran, A., Hager, G., & Vidal, R. (2009). *Histograms of oriented optical flow and binet-cauchy kernels on nonlinear dynamical systems for the recognition of human actions.* Paper presented at the Computer Vision and Pattern Recognition, 2009. CVPR 2009. IEEE Conference on. doi:10.1109/CVPR.2009.5206821

Comaniciu, D., Ramesh, V., & Meer, P. (2000). Real-time Tracking of Non-Rigid Objects using Mean Shift.*Proceedings. IEEE Conference on Computer Vision and Pattern Recognition*, 2. doi:10.1109/CVPR.2000.854761

Dalal, N., & Triggs, B. (2005). Histograms of oriented gradients for human detection. *Computer Vision and Pattern Recognition, 2005. CVPR 2005. IEEE Computer Society Conference on.* doi:10.1109/CVPR.2005.177

Derrington, A. M., Allen, H. A., & Delicato, L. S. (2004). Visual mechanisms of motion analysis and motion perception.*Annual Review of Psychology, 55*(1), 181–205. doi:10.1146/annurev.psych.55.090902.141903 PMID:14744214

Deshayes, R., Mens, T., & Palanque, P. (2013). *A generic framework for executable gestural interaction models.* Paper presented at the Visual Languages and Human-Centric Computing (VL/HCC), 2013 IEEE Symposium on. doi:10.1109/VLHCC.2013.6645240

Dittrich, W. H. (1993). Action Categories and the Perception of Biological Motion. *Perception, 22*(1), 15–22. doi:10.1068/p220015 PMID:8474831

Fortun, D., Bouthemy, P., & Kervrann, C. (2015). Optical flow modeling and computation: A survey. *Computer Vision and Image Understanding, 134*, 1–21. doi:10.1016/j.cviu.2015.02.008

Gavrila, D., & Davis, L. (1995). *Towards 3-d model-based tracking and recognition of human movement: a multi-view approach.* International workshop on automatic face-and gesture-recognition.

Goddard, N. H. (1992). *The Perception of Articulated Motion: Recognizing Moving Light Displays.* University of Rochester.

Guo, Y., Xu, G., & Tsuji, S. (1994). Understanding Human Motion Patterns. *Pattern Recognition, Conference B: Computer Vision & Image Processing., Proceedings of the 12th IAPR International. Conference on, 2.*

Ikizler, N., Cinbis, R. G., & Duygulu, P. (2008). *Human action recognition with line and flow histograms.* Paper presented at the Pattern Recognition, 2008. ICPR 2008. 19th International Conference on. doi:10.1109/ICPR.2008.4761434

Isard, M. C., & Blake, A. C. (1998). CONDENSATION: Conditional Density Propagation for Visual Tracking. *International Journal of Computer Vision, 29*(1), 5–28. doi:10.1023/A:1008078328650

Johansson, G. (1973). Visual Perception of Biological Motion and a Model for its Analysis. *Perception & Psychophysics, 14*(2), 201–211. doi:10.3758/BF03212378

Karaulova, I. A., Hall, P. M., & Marshall, A. D. (2000). A Hierarchical Model of Dynamics for Tracking People with a Single Video Camera. In *Proceedings of the 11th British Machine Vision Conference, 1*, 352-361. doi:10.5244/C.14.36

Kliper-Gross, O., Gurovich, Y., Hassner, T., & Wolf, L. (2012). Motion interchange patterns for action recognition in unconstrained videos. *European Conference on Computer Vision*, (pp. 256-269). doi:10.1007/978-3-642-33783-3_19

Ko, T. (2008). *A survey on behavior analysis in video surveillance for homeland security applications.* Paper presented at the Applied Imagery Pattern Recognition Workshop, 2008. AIPR'08. 37th IEEE. doi:10.1109/AIPR.2008.4906450

Kozlowski, L. T., & Cutting, J. E. (1978). Recognizing the Gender of Walkers from Point-Lights Mounted on Ankles: Some Second Thoughts. *Perception & Psychophysics*, *23*(5), 459. doi:10.3758/BF03204150

Kuehne, H., Jhuang, H., Garrote, E., Poggio, T., & Serre, T. (2011). *HMDB: a large video database for human motion recognition.* Paper presented at the Computer Vision (ICCV), 2011 IEEE International Conference on. doi:10.1109/ICCV.2011.6126543

Kühnel, C., Westermann, T., Hemmert, F., Kratz, S., Müller, A., & Möller, S. (2011). I'm home: Defining and evaluating a gesture set for smart-home control. *International Journal of Human-Computer Studies*, *69*(11), 693–704. doi:10.1016/j.ijhcs.2011.04.005

Ladjailia, A., Bouchrika, I., Merouani, H. F., & Harrati, N. (2015a). Automated Detection of Similar Human Actions using Motion Descriptors. *16th international conference on Sciences and Techniques of Automatic control and computer engineering (STA).* IEEE.

Ladjailia, A., Bouchrika, I., Merouani, H. F., & Harrati, N. (2015b). On the Use of Local Motion Information for Human Action Recognition via Feature Selection.*4th IEEE International Conference on Electrical Engineering (ICEE).* doi:10.1109/INTEE.2015.7416792

Laptev, I. (2005). On space-time interest points. *International Journal of Computer Vision*, *64*(2-3), 107–123. doi:10.1007/s11263-005-1838-7

Lara, O. D., & Labrador, M. A. (2013). A survey on human activity recognition using wearable sensors. *IEEE Communications Surveys and Tutorials*, *15*(3), 1192–1209. doi:10.1109/SURV.2012.110112.00192

Lucas, B. D., & Kanade, T. et al. (1981). An iterative image registration technique with an application to stereo vision. *IJCAI*, *81*, 674–679.

Marszalek, M., Laptev, I., & Schmid, C. (2009). *Actions in context.* Paper presented at the Computer Vision and Pattern Recognition, 2009. CVPR 2009. IEEE Conference on. doi:10.1109/CVPR.2009.5206557

Martínez, F., Manzanera, A., & Romero, E. (2012). *A motion descriptor based on statistics of optical flow orientations for action classification in video-surveillance. In Multimedia and Signal Processing* (pp. 267–274). Springer.

Moeslund, T. B., Hilton, A., & Krüger, V. (2006). A survey of advances in vision-based human motion capture and analysis. *Computer Vision and Image Understanding*, *104*(2), 90–126. doi:10.1016/j.cviu.2006.08.002

Nelson, R. C., & Polana, R. (1992). Qualitative recognition of motion using temporal texture. *CVGIP. Image Understanding*, *56*(1), 78–89. doi:10.1016/1049-9660(92)90087-J

Niebles, J. C., Wang, H., & Fei-Fei, L. (2008). Unsupervised learning of human action categories using spatial-temporal words. *International Journal of Computer Vision, 79*(3), 299–318. doi:10.1007/s11263-007-0122-4

Ogale, A. S., Karapurkar, A., & Aloimonos, Y. (2007). *View-invariant modeling and recognition of human actions using grammars. In Dynamical vision* (pp. 115–126). Springer.

Oikonomopoulos, A., Patras, I., & Pantic, M. (2005). Spatiotemporal salient points for visual recognition of human actions. *Systems, Man, and Cybernetics, Part B: Cybernetics. IEEE Transactions on, 36*(3), 710–719.

Oshin, O., Gilbert, A., & Bowden, R. (2014). Capturing relative motion and finding modes for action recognition in the wild. *Computer Vision and Image Understanding, 125,* 155–171. doi:10.1016/j.cviu.2014.04.005

Polana, R., & Nelson, R. (1994). *Low level recognition of human motion (or how to get your man without finding his body parts).* Paper presented at the Motion of Non-Rigid and Articulated Objects. doi:10.1109/MNRAO.1994.346251

Poppe, R. (2010). A survey on vision-based human action recognition. *Image and Vision Computing, 28*(6), 976–990. doi:10.1016/j.imavis.2009.11.014

Rahman, S., See, J., & Ho, C. C. (2015). *Action Recognition in Low Quality Videos by Jointly Using Shape, Motion and Texture Features.* Paper presented at the IEEE Int. Conf. on Signal and Image Processing Applications. doi:10.1109/ICSIPA.2015.7412168

Ren, Z., Meng, J., Yuan, J., & Zhang, Z. (2011). *Robust hand gesture recognition with kinect sensor.* Paper presented at the 19th ACM international conference on Multimedia. doi:10.1145/2072298.2072443

Rohr, K. (1994). Towards Model-Based Recognition of Human Movements in Image Sequences. *CVGIP. Image Understanding, 59*(1), 94–115. doi:10.1006/ciun.1994.1006

Rui, Y., Huang, T. S., & Chang, S.-F. (1999). Image retrieval: Current techniques, promising directions, and open issues. *Journal of Visual Communication and Image Representation, 10*(1), 39–62. doi:10.1006/jvci.1999.0413

Schindler, K., & Van Gool, L. (2008). *Action snippets: How many frames does human action recognition require?* Paper presented at the Computer Vision and Pattern Recognition, 2008. CVPR 2008. IEEE Conference on. doi:10.1109/CVPR.2008.4587730

Schüldt, C., Laptev, I., & Caputo, B. (2004). *Recognizing human actions: a local SVM approach.* Paper presented at the Pattern Recognition. doi:10.1109/ICPR.2004.1334462

Shah, M., & Jain, R. (2013). *Motion-based recognition* (Vol. 9). Springer Science & Business Media.

Shechtman, E., & Irani, M. (2007). *Matching local self-similarities across images and videos.* Paper presented at the Computer Vision and Pattern Recognition. doi:10.1109/CVPR.2007.383198

Somol, P., Pudil, P., Novovičová, J., & Paclík, P. (1999). Adaptive floating search methods in feature selection. *Pattern Recognition Letters, 20*(11), 1157–1163. doi:10.1016/S0167-8655(99)00083-5

Suma, E. A., Krum, D. M., Lange, B., Koenig, S., Rizzo, A., & Bolas, M. (2013). Adapting user interfaces for gestural interaction with the flexible action and articulated skeleton toolkit. *Computers & Graphics*, *37*(3), 193–201. doi:10.1016/j.cag.2012.11.004

Tang, S., Andriluka, M., & Schiele, B. (2014). Detection and tracking of occluded people. *International Journal of Computer Vision*, *110*(1), 58–69. doi:10.1007/s11263-013-0664-6

Troje, N. F., Westhoff, C., & Lavrov, M. (2005). Person Identification from Biological Motion: Effects of Structural and Kinematic Cues. *Perception & Psychophysics*, *67*(4), 667–675. doi:10.3758/BF03193523 PMID:16134460

Turaga, P., Chellappa, R., Subrahmanian, V. S., & Udrea, O. (2008). Machine recognition of human activities: A survey. *Circuits and Systems for Video Technology. IEEE Transactions on*, *18*(11), 1473–1488.

Vishwakarma, S., & Agrawal, A. (2013). A survey on activity recognition and behavior understanding in video surveillance. *The Visual Computer*, *29*(10), 983–1009. doi:10.1007/s00371-012-0752-6

Wang, Y., Huang, K., & Tan, T. (2007). Human activity recognition based on r transform. *Computer Vision and Pattern Recognition, 2007. CVPR'07. IEEE Conference on* (pp. 1-8).

Weinland, D., & Boyer, E. (2008). Action recognition using exemplar-based embedding. *Computer Vision and Pattern Recognition, 2008. CVPR 2008. IEEE Conference on* (pp. 1-7).

Weinland, D., Özuysal, M., & Fua, P. (2010). *Making action recognition robust to occlusions and viewpoint changes. In Computer Vision–ECCV 2010* (pp. 635–648). Springer.

Weinland, D., Ronfard, R., & Boyer, E. (2011). A survey of vision-based methods for action representation, segmentation and recognition. *Computer Vision and Image Understanding*, *115*(2), 224–241. doi:10.1016/j.cviu.2010.10.002

Welch, G., & Bishop, G. (2001). An Introduction to the Kalman Filter. *ACM SIGGRAPH 2001 Course Notes*.

Yam, C.-Y., & Nixon, M. (2009). Gait Recognition, Model-Based. In Encyclopedia of Biometrics, (pp. 633-639). Academic Press.

Yamato, J., Ohya, J., & Ishii, K. (1992). Recognizing human action in time-sequential images using hidden markov model. *Proceedings CVPR*, *92*, 1992.

Yeffet, L., & Wolf, L. (2009). Local trinary patterns for human action recognition. *Computer Vision, 2009 IEEE 12th International Conference on* (pp. 492-497).

Zhu, Y., Nayak, N. M., & Roy-Chowdhury, A. K. (2013). Context-aware activity recognition and anomaly detection in video. *Selected Topics in Signal Processing. IEEE Journal of*, *7*(1), 91–101.

KEY TERMS AND DEFINITIONS

Action: Is considered as a simple activity referring to simple pattern performed by a person during a short period of time lasting a few seconds. Examples of actions may include raising hands, bending, sitting and even walking.

Activity: Is defined as a composite sequence of actions executed by either a single person or several people interacting with each other. Examples of activities are like leaving an unattended bag, shaking hands or assaulting a pedestrian.

Feature Extraction: Is the process of estimating a set of measurements either related to the configuration of the whole body or the configuration of the different body parts in a given scene and tracking them over a sequence of frames.

Global Feature: This is the visual characteristics taken from an image in holistic fashion such that the region of interest of a person is encoded as a whole. In most cases, the features are computed densely on a grid bounded by region of interest.

Human Activity Recognition: Is the process to automatically infer the action or activity being performed by a person or group of people via the use of computer vision methods. This may involve the analysis and recognition of different motion patterns in order to produce a high-level semantic description for the human activities.

Local Feature: Is a type of low-level cues which are extracted from smaller portions of the image with no connection made their spatial locations within the human body.

Motion: Is a spatio-temporal event defined as the change of spatial location over time. Given some visual input, the visual perception of motion is regarded as the process by which the visual system acquires perceptual knowledge such as the speed and direction of the moving object.

Chapter 9
Object–Based Surveillance Video Synopsis Using Genetic Algorithm

Shefali Gandhi
Dharmsinh Desai University, India

Tushar V. Ratanpara
Dharmsinh Desai Univerisity, India

ABSTRACT

Video synopsis provides representation of the long surveillance video, while preserving the essential activities of the original video. The activity in the original video is covered into a shorter period by simultaneously displaying multiple activities, which originally occurred at different time segments. As activities are to be displayed in different time segments than original video, the process begins with extracting moving objects. Temporal median algorithm is used to model background and foreground objects are detected using background subtraction method. Each moving object is represented as a space-time activity tube in the video. The concept of genetic algorithm is used for optimized temporal shifting of activity tubes. The temporal arrangement of tubes which results in minimum collision and maintains chronological order of events is considered as the best solution. The time-lapse background video is generated next, which is used as background for the synopsis video. Finally, the activity tubes are stitched on the time-lapse background video using Poisson image editing.

INTRODUCTION

The ability of videos to represent dynamic activities makes them more powerful compared to still images. Because of increasing demands of video surveillance and the arrival of inexpensive network cameras, there is an explosive growth of surveillance videos, which are used by governments and other organizations for intelligence gathering, prevention of crime, the safety of a person, group or organization, or investigating crime. According to a recent survey carried out by The British Security Industry Authority (BSIA), 2013, there are up to 5.9 million closed-circuit television cameras in the UK only, that is, one

DOI: 10.4018/978-1-5225-1022-2.ch009

for every 11 people. (Berret, 2013). With this easily deployed and remotely managed, network cameras that operate 24 hours a day, amount of captured video is also growing. It becomes necessary to view the whole video to find the activity of interest. As browsing and retrieval is of such large videos is very time consuming and tedious task, most of the captured videos are never watched or examined. Continuous monitoring of such videos by human operators is also impractical due to high cost and reliability issues (e.g., operator fatigue). So, it is becoming more and more important to develop automated tools to abstract these infinite videos for easy browsing and retrieval such that the human involvement in the process can be reduced. Video synopsis is a step taken in the direction of sorting through video for creating its summary and it is especially advantageous for surveillance videos.

Video synopsis is an approach used to create short video summary of a long input video, while preserving the essential activities of the original video. Most of the video abstraction techniques discard inactive frames. Video synopsis represents video in a temporally compact way by simultaneously displaying the multiple activities, which are occurred at different times in original video, thus reducing temporal as well as spatial redundancy in the video. The activities are shifted from their original time intervals to other time interval when no activity is taking place at that spatial region. It provides a condensed representation of a video sequence, which significantly reduces the computational complexity of many video content analysis and video retrieval applications, like, object detection and segmentation. The abstracted video can be used for video indexing. It can also be used to provide the user with efficient links for accessing activities in the videos. It is an effective tool for browsing and indexing long surveillance videos (Rav-Acha, Pritch, & Peleg, 2006; Pritch, Rav-Acha, & Peleg, 2008). For example, it can be used as a powerful tool by security agencies or police to browse a large number of surveillance videos quickly, and thus speeding up analysis of any case and saving cost of human resources (Yogameena, & Priya, 2015). It can also be employed for home security, in which video synopsis of videos captured by home monitoring cameras are sent to cellular wireless networks and then users can view these video synopsis by smart devices, such as smart phones and tablets, for home remote monitoring, elder or child care, etc. Figure 1 shows frames of an input video consisting of a walking person and, after a period of inactivity; a moving car comes in it. A condensed video synopsis of this video is also shown, which is produced by playing the movements of person and car simultaneously.

Figure 1. Video synopsis

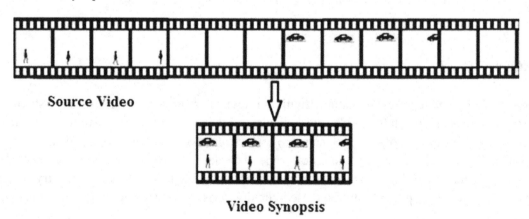

The rest of this chapter is organized as follows.

- Section 2 outlines some related works on video synopsis.
- Section 3 describes the detailed proposed approach of video synopsis system,
- Section 4 presents experimental results and
- Section 5 concludes this chapter.

RELATED WORK

In past decades, many significant researches have been done in the field of video abstraction. Basically, two different types of video abstracts are there: still image abstract, also called a static storyboard and moving image abstracts, also called a moving storyboard or video skims. Still image abstract is a small set of salient images generated from source video. But this representation loses dynamics of video. Moving image abstract is a set of image sequences along with the corresponding audio. Thus, moving image abstracts are video clips only but shorter in length compared to source video (Li, Zhang, & Tretter, 2001; Oh, Wen, Hwang, & Lee, 2005). Fast forwarding and video skimming can effectively summarize the long video by omitting some of the frames. However, the skipped frames may contain some important information. And to overcome this problem, adaptive fast forwarding technique has been developed (Pertovic, Jojic, & Huang, 2005). It omits frames from the time period of low interest or the time period having lower activity, while keeping frames from time period of interest or period having more activity. But it results in lower compaction rate. Information based adaptive fast forward can also be used for surveillance video, where the method adapts the playback velocity of the source video to temporal information density (Höferlin, Höferlin, Weiskopf, & Heidemann, 2011; Truong, & Venkatesh, 2007).

However, in all aforementioned approaches summary is achieved by dropping frames without considering the concept of events, which may cause loss of important events. In space-time video montage, both spatial and temporal information distribution of a video are analyzed simultaneously and informative space-time parts of it are extracted and packed together (Kang, Matsushita, Tang, & Chen, 2006). It produces much compact and highly informative video but it is computationally expensive (Padhiyar & Joshi, 2015).

Video synopsis is a technique that addresses all above issues by dropping all inactive parts of the source video and preserving all events information to generate highly compact video. The approach was first presented by Pritch, Rav-Acha, and Peleg (2008), which includes two major phases. In first phase, endless stream of video is converted into objects and activities and stored into the database and in second phase, called a response phase, video synopsis is generated as a response to the user's query. Two methods to generate video synopsis are explained, region based and object based. The region based method represents energy function as 3D MRF, where each node represents a pixel in the 3D volume of the output video. This makes it computationally complex. The object based approach reduces complexity. It identifies moving objects from the video and performs the optimization on the detected objects. It is much faster and enables the use of object based constraints. Temporal median is used to model the background and background cut algorithm is used to detect foreground. Energy function considers activity cost, collision cost and temporal consistency cost, which is minimized using simulated annealing. The time-lapse background is generated by interpolation of temporal histogram and activity histograms, on which the selected objects are stitched using Poisson Image Editing. Creating video synopsis of an endless video stream is done in two major phases. First phase is an online phase, which runs in parallel to video capture and detects and tracks moving objects and stores them into the object queue. And response phase starts when a user gives a query to the system, which specifies the Time Period of Interest in the input video and the length of the synopsis video. All objects in the period of interest are selected and packed into the synopsis range in the way that minimizes the energy cost. After computing all the cost values, the optimal temporal arrangement is generated. If there are K objects in the period of interest and T time steps, there are TK possible arrangements.

Yildiz, Ozgur, and Akgul (2008) have derived a highly efficient method to generate video synopsis using dynamic programming. It finds the space-time surfaces in the video volume which have less motion information. Video volume is projected onto a plane orthogonal to one of its axes to find a minimum cost path. The minimum cost path is back-projected down to the video volume and as a result it gives the space time surfaces which can be discarded. Applying this process iteratively generates the summarized video. This method requires small memory and makes energy minimization less complex, making it more appropriate for real time applications. The pipe-line style arrangement of buffering and processing units is used for this. The buffered frames are processed periodically by processing unit and only the synopsis video is stored. But this method forces the paths to be always connected. A method of dividing the projection image into bands is a solution to difficult videos. Dividing the projection image into bands lets the minimum cost path to be composed of individual paths in each band.

Pritch, Ratovitch, Hendel, and Peleg (2009) have proposed a new methodology to generate clustered summaries that simultaneously shows multiple objects doing similar activity. A novel method called online principal background selection (OPBS), to generate time-lapse background for video synopsis is presented by Feng, Liao, Yuan, and Li (2010). Liu and Yang (2010) developed an interactive method that used both content and context based information to generate dynamic video synopsis. The main issue in video synopsis is of occlusion. Nie and Xiao (2013) developed a novel global video synopsis approach which solves this problem by expanding background and thus providing more space for moving objects. So, the temporal as well as spatial shifting of objects becomes possible. Lu, Wang and Pan (2013) have introduced an approach that combines Gaussian Mixture Model and texture based method to detect foreground. Thus, solving the problem of continuous activity(tube) from a single moving object is being separated into a few small pieces. Hao, Cao, and Li (2013) have used the Graph Cut based GrabCut segmentation algorithm for the moving object matting in single camera surveillance video synopsis, which gives more accurate results. A novel approach was developed by Nie, Sun, Li, Xiao and Ma (2014) based on movements of part of object to generate synopsis. Hsia and Chiang (2013) have proposed a fast and efficient video retrieval technique using low-complexity range tree algorithm instead of linear search, which improves the effectiveness of the search of the objects that match the requested conditions from the video synopsis. Huang, Chung, Yang, Chen, and Huang (2014) have formulated the problem of synopsis video generation as a maximum a posterior probability (MAP) estimation problem. In practice, to monitor a large area, usually several surveillance cameras are combined as a network. However, monitoring activity of a moving object across a wide area covered by several cameras is very tough. Zhu, and Liao (2015) have introduced a novel technique to generate video synopsis of the surveillance videos captured by multiple cameras. It is very difficult to analyze a large surveillance video for any abnormal activity. Yogameena and Priya (2015) have presented an approach for analyzing the behaviour and find abnormalities in the surveillance videos captured in the public environment like railway station, airport etc. Chou, Lin, Chiang, Chen, and Lee (2015) proposed a novel coherent event-based video synopsis system that condenses the video contents and provides synopsized video with clear views for human by simultaneously displaying multiple events with similar trajectories. Jeeshna and Kuttymalu (2015) have proposed object movement based video synopsis method which focuses on the movement of a single video object and removes the redundancies which are there in the object movement.

PROPOSED APPROACH

The chapter describes the system to generate video synopsis, which is temporally compressed representation of long input video. Firstly, the space-time activity tubes are extracted by modelling background and detecting moving objects from the source video. Then the activity tubes are rearranged on time axis using genetic algorithm (GA). Next, the time-lapse background is generated, on which the moving objects are stitched to produce synopsis video. The detailed description of abstract model methodology for each operation is described in following sections. The abstract model of the proposed approach for generating video synopsis is shown in Figure 2.

Generally video synopsis is useful to generate short summary of the surveillance videos. Therefore, a surveillance video is taken as an input. The length of video is 1 hour, 4 minutes and 52 second. Its frame rate is 25 fps and total number of frames is 97309. If the frame rate is reduced up to certain value, it doesn't affect the working of system. However, if the frame rate is very low, it causes the difficulties

Figure 2.

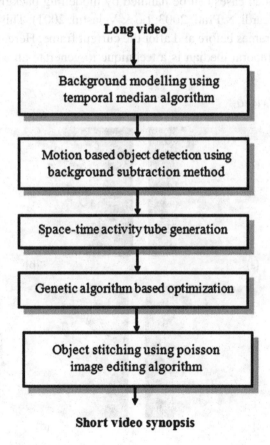

tracking the moving objects. In the videos with very low frame rate, the continuity of the moving object is often very poor. The position of object in current frame appears very far away from its position in the previous frame, which causes it to be detected as two separate objects. Such problems in tracking objects occurs when the distance between the object detected in current frame and object detected in previous frame is more than the size of object itself. Some of the frames from source video are shown in Figure 3.

Background Modelling Using Temporal Median Algorithm

The first step before extraction of moving foreground objects is background estimation. The background image must represent the scene with no moving objects and must be able to obtain accurate detection of moving objects (Piccardi, 2004). In order to detect moving objects accurately, the background must be updated regularly. Typically, the static image (when there is no motion in the image) (Shoushtarian & Ghasem-aghaee, 2003), or the first frame of the video is considered as the background image. If video clip is not so long, the background doesn't get changed much over time. In such scenario, using first frame or temporal median over the entire video can be used as background frame. But in long surveillance videos, because of change in lighting (day-night transition, change in direction of sun light or other source of illumination, shadows, and clouds), change in background objects etc., background changes over time. The temporal averaging or median over the entire video clip is also not sufficient to produce background that adapts the changes occurring in background with time. Another issue with this method is that it will not consider the moving objects that become part of background later or part of the background that starts moving. Such cases can be handled by modelling background at regular interval of time (Cucchiara, Cranna, Piccardi, & Prati, 2003; Lo & Velastin, 2001). This can be done by computing temporal median of several frames before and after the current frame. Here, for the simplicity a median over 30 seconds is used. Temporal median is a technique to generate an adequate background image

Figure 3. Frames from input video

(a)

(b)

(c)

(d)

which reflects the changes in the background. Thus, to handle sudden changes in the scene, background is modelled by computing the temporal median over a few frames before and after the current frame. The temporal median filter also provides good accuracy with high frame rate and less memory requirements.

Steps to model the background using temporal median are as follows:

Step 1: From the input video, N/2 frames before the current frame k and N/2 frames after the current frame are extracted, where N is the total number of frames considered for calculating total median.

Step 2: The frames are processed pixel by pixel. The temporal median of (x, y) pixel of the k-th frame, I_{med}(x, y, k), is obtained by (Hung, Pan, & Hsieh, 2014),

$$I_{med}\left(x, y, k\right) =$$
$$med(I\left(x, y, k - \left(N/2-1\right)\right), I\left(x, y, k - \left(N/2-2\right)\right), \ldots, I\left(x, y, k\right), I\left(x,y,k+1\right), \ldots \ldots$$
$$\ldots, I\left(x,y, k+N/2\right))$$

$$(1)$$

where I(x, y, k−N/2), . . ., I(x, y, k+N/2), denote pixel values located at (x,y) over the N frames and med() is the median operation.

Step 3: The calculated median is set as the value of background at that particular pixel.

$$B\left(x,y,k\right) = I_{med}\left(x,y,k\right)$$

$$(2)$$

Step 4: Repeat step (2) and (3) for all pixels in the frame.

For example, if N is set to 900 then N/2 is 450. Then the Equation 1 for modelling background for 1000th frame becomes,

$$I_{med}\left(x,y,1000\right) =$$
$$med(I\left(x, y, 551\right), I\left(x, y, 552\right), \ldots, I\left(x, y, 1000\right), I\left(x, y, 1001\right), \ldots \ldots,$$
$$I\left(x, y,1449\right), I\left(x, y, 1450\right)$$

and Equation 2 becomes,

$$B\left(x, y, 1000\right) = I_{med}\left(x,y,1000\right)$$

Figure 4 shows two background frames with varying illumination and background objects. There is a visible difference in illumination in figure 4(a) and 4(b). Moreover, the parked car will be considered as a part of background in figure 4(a) but not in 4(b).

Figure 4. Background frames modelled using temporal median algorithm

(a)   (b)

Motion Based Object Detection Using Background Subtraction Method

The next step towards generating space-time activity tube after modeling backgrounds is to extract moving objects from the video. Background subtraction method is used to detect the foreground objects. The algorithm uses either spatial or temporal information in the frames to detect moving objects detection. The intensity of pixels is most commonly used method. The moving regions of current frame are detected by subtracting the current frame from the corresponding reference background frame generated in previous stage. This subtraction is done pixel-by-pixel.

Following algorithm is used to extract the foreground objects:

Step 1: Each frame of the video is one by one compared with the corresponding background frame generated using temporal median algorithm, in the previous stage.

Step 2: If the difference is greater than some threshold *Th*, the pixel is considered as a part of foreground. For the experiments, the threshold value is set to 30.

Formula used to compute the difference at pixel (x, y) in t[th] frame is,

$$foreground\_mask = \left| I\left(x,\ y,\ t\right) - B\left(x,\ y,\ t\right) \right| > Th \tag{3}$$

where I(x, y, t) is the pixel in current input frame and B(x, y, t) is corresponding pixel in the background frame.

Step 3: The difference is returned in form of binary mask, which has the same size as the input frame. In the mask, pixels with a value 0 correspond to the background, and pixels with a value 1 correspond to the foreground.

Step 4: Morphological operations like, close, open and erode are performed further on the resulting binary mask to eliminate noise from the mask and to fill the holes in the remaining blobs.

Step 5: Using blob analysis the groups of connected pixels are detected, which are likely to correspond to moving objects.

Figure 5 shows the frames with detected moving objects.

Figure 5. Moving objects detected using background subtraction

(a) (b)

(c) (d)

All the detected objects are not considered. Objects with size more than 50 are taken in consideration.

Space-Time Activity Tubes Generation

Objects detected in previous stage are defined and viewed as tubes in space-time volume. After the detection of moving objects, each object should be tracked along its path, in order to create space-time activity tubes. The Particle filter, Kalman filter, SIFT features based and Mean-shiftare the commonly used abject tracking methods. The Kalman filter is employed in this approach. It estimates motion of each detected object. It is used to predict the location of moving objects in each frame. It is recursive, the new estimate of the object's position is computed from the previous estimate and the current input data. The kalman filter is computationally more efficient and it is not required to store all the previously observed data as well (Cuevas, Zaldivar, & Rojas, 2005; Zhu, WenLiu, & Yuan, 2014). It also determines the likelihood of each moving object to be considered as continuation of already existing tube or as a new tube. As moving objects are represented as activity tubes in the space-time volume, the words "objects" and "tubes" can be used interchangeably (Yilmaz, Javed, & Shah, 2006). The basic space-time representation of several tubes in the original video, are shown in Figure 6. Here, X- axis denotes location of moving object and on Y-axis time is denoted. And path of each object is shown as a curve in x verses t graph.

Let B be the set of all activity tubes in the original video, b_i be the ith tube and each tube $b \in B$, appears during the finite time segment t_b in the original video.

Figure 6. Shows an example of extracted tube of a walking man flattened over the corresponding background

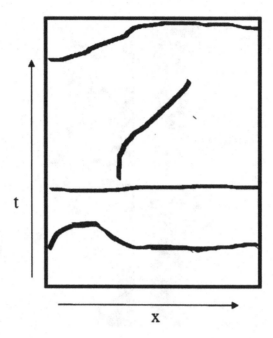

The time period t_b can be defined as,

$$t_b = \begin{bmatrix} st_b & et_b \end{bmatrix}$$

(5)

where st_b is the frame number from which object appears in the video and et_b is the last frame number up to which object remains visible in the source video. Total 264 tubes are detected from the source video of 1 hour, 4 minutes and 52 seconds.

Figure 7. Extracted tube flattened over background

In the generated synopsis video, each tube b is temporally shifted from its original time period t_b in the source video to the new time segment t'_b, which can be defined as,

$$t'_b = \begin{bmatrix} st'_b & et'_b \end{bmatrix} \tag{6}$$

where st'_b and et'_b indicate the start time and end time of activity tube in synopsis video, respectively. For example, in the source video, time segment of the tube shown in Figure 7 is 15591 ~ 15961, so as per Equation 5,

$$t_b = \begin{bmatrix} 15591 & 15961 \end{bmatrix}$$

If it is shifted to a new time segment that starts from frame number 100 then according to Equation 6,

$$t'_b = \begin{bmatrix} 100 & 471 \end{bmatrix}$$

Each tube b is represented by following characteristic function (Pritch et al., 2008),

$$C_b\left(x, \ y, \ t\right) = \begin{cases} \|\ I(x,y,t) - B(x,y,t)\ \| & \text{if } t \in t_b \\ 0 & \text{otherwise} \end{cases} \tag{7}$$

where B(x, y, t) is a pixel in the background image and I(x, y, t) is the related pixel in the input image. And t_b is the time interval for which this specific object appears in the video. Here, the tube b is defined by a characteristic function C_b(x, y, t) indicating whether the pixel(x, y) in frame t is detected as a part of foreground for tube b. If it is detected as foreground then the difference, ‖I (x, y, t) – B (x, y, t)‖, will be positive and will be added in C_b(x, y, t) otherwise 0. Collection of all such pixels in all the frames within st_b ~ et_b, makes a tube b.

Genetic Algorithm Based Optimization

The meta-heuristic optimization algorithms are classified into two types, population based and trajectory based. In trajectory based algorithms, only one solution is considered at a time. Hill-climbing and Simulated Annealing(SA) are important examples of trajectory based algorithms. The population based algorithms, like Particle Swarm Optimization(PSO) and GA, use multiple solutions in one iteration. GA is a search heuristic that is an effective way for solving optimization problems. It doesn't guarantee the best solution but offers reasonably optimal solution in the short time (Sahu & Singh, 2014). GA is based on genetic operators like recombination and mutation. The recombination is the main genetic operator. It ensures that beneficial aspects of current solutions are preserved and undesirable solutions are eliminated. GA eliminates the weak candidates from the process of reproduction, which increases the probability of algorithm to converge towards optimized solutions in few generations. The mutation and crossover operators will help GA to jump the discontinuity in the search space and lead to better exploration. The other advantages of GA like, ease of implementation, intuitiveness, ability to solve nonlinear optimization problems effectively and computation speed, makes the algorithm more popular. In video synopsis, the idea of GA is used to deal with the optimization problem of temporal shifting of tubes. It uses techniques based on natural evolution like, survival of fittest individuals, selection, recombination and mutation, to solve the optimization problems (Bajpai & Kumar, 2010). It avoids local optimal solutions. The steps to realize a basic GA are listed below.

Step 1: Use the starting frame number of each tube in synopsis video to represent the problem domain as a chromosome. Decide basic parameters of GA like, size of population, crossover probability Pc and mutation probability Pm.

Step 2: Generate an initial population from the desired duration of video synopsis.

Step 3: Calculate the fitness of each individual in the current iteration using fitness function to measure its performance in the problem domain, where the fitness function consists of collision cost and temporal consistency cost.

Step 4: Select the fittest pairs of chromosomes from the current population for mating. Generate offspring of the next population by applying crossover and mutation genetic operators on the selected fittest pairs of parent chromosomes.

Step 5: Replace the initial (parent) population with the new population.

Step 6: Go to Step 3, and repeat the process until the current iteration count exceeds the maximum iterations.

The above mentioned steps to get optimal solution of temporal shifting of tubes using GA are explained in detail below.

Encoding and Initial Population Generation

Let x be the permutation of tubes in chronological order in the population. Denote X as the set of all the populations and population at i^{th} iteration is denoted as X(i). So, X(0) becomes the initial population. The M number of individual chromosomes are generated to form the initial population X(0). The required size of resulting video synopsis is used for it. The population size M, is set to 12 for the experimental purpose. The value of crossover probability Pc is set 0.5 and mutation probability is set to 0.06. Let T be the maximum iteration number.

Here, the optimization is needed to decide the chronological order of tubes in the synopsis video. So, Starting frame number of a tube in synopsis video is used to encode of an individual in population. Each individual is encoded in such a way that i^{th} element in the individual indicates the starting frame number of tube i in the synopsis video and number of elements in each individual is same as the number of detected tubes in the source video. Let N be the total number of tubes. The starting frame of ith tube (moving object) in synopsis video is denoted as S_i ($1 \le i \le N$).

The jth individual, R_j in iteration is represented as,

$$R_j = S_1 \quad S_2 \quad S_3 \quad S_4 \ldots\ldots\ldots\ldots S_{N-1} \quad S_N \tag{8}$$

For example, if there are 7 activity tubes, R_j can be encoded as follows,

$R_j = 4\ 1\ 20\ 60\ 35\ 5$ 25

which means that the first tube will start from frame number 4, second tube will start from frame number 1, third from frame number 20 and so on, in the video synopsis. The duration of the video synopsis is a user defined parameter. For the experiments, it has been determined from total duration of all the extracted tubes.

Fitness Computation

In GA based optimization, to evaluate the goodness of an individual, its fitness is computed. And based on the fitness, its probability to be inherited in next iteration is calculated. In this step, fitness of each individual in the current population will be calculated.

The goal is to create synopsis video with maximum activities covered from the original video and minimum collisions and overlaps between the objects. In order to let the tubes reasonably shifted in a video synopsis, the following problems are needed to be considered:

1.   The important objects and activities should be preserved as much as possible.
2.   The length of a video synopsis should be comparatively shorter than the original video.
3.   Objects collision should be avoided as far as possible to provide users better visual experience of video synopsis.
4.   It the chronological order of activities in both original video and video synopsis should be consistent.

From the above mentioned problems, first is solved by foreground detection and motion tracking in the previous stages. Most of the surveillance videos contain redundant spatial and temporal information, which are removed by rearranging tubes on time scale. Therefore, the compression ratio is guaranteed. The main problems that should be solved by temporal shifting of tubes are related to object collision and chronological order.

The optimal new positions for the object tubes are computed by minimizing the following costs:

1.   Collision Cost (E_c)
2.   Temporal Consistency Cost (E_t)

1. Collision Cost E_c

It encodes collision cost of all pairs of extracted activity tubes. If two shifted tubes b_i and b_j collides in video synopsis, the collision cost for overlapping parts of the two shifted tubes b_i and b_j is defined as sum of the products of pixel activities of the two objects tubes (Pritch et al., 2008).

$$E_c\left(b_i,\ b_j\right) = \sum_{(x,y,t)\in t'_{bi}\ \cap\ t'_{bj}} C_{bi}\left(x,\ y,\ t\right)\mathbf{C}_{bj}\left(x,\ y,\ t\right) \tag{9}$$

Collision cost is set to zero, if two shifted tubes do not collide with each other. The multiplication of characteristic function values is considered because if two tubes t'_{bi} and t'_{bj} both have active pixel(positive value of characteristic function) in a frame t, at same location (x, y) then only the multiplication will be positive, otherwise it will be zero. So, collision cost increases with the number of colliding pixels of a pair of tubes.

2. Temporal Consistency Cost E_t:

The temporal consistency cost E_t allows to preserve the chronological order of events. Thus, the events occurring after other events in the source video should not appear before them in the synopsis video. Preserving chronological order is more important in the cases where tubes are having a strong interaction. The cost of reversing the chronological order of two events in the video synopsis is the sum of their activities. When chronological order is maintained the value of E_t is zero.

Steps to compute temporal consistency cost, E_t are:

1. Let st_{bi} and st_{bj} be the first frames of two tubes b_i and b_j respectively.
2. Compute the value of a variable,

$$x = \left[st_{bi} - st_{bj} \right] \cdot \left[st'_{bi} - st'_{bj} \right] \tag{10}$$

3. Determine whether the chronological order of tubes b_i and b_j is reversed by the function,

$$\tau(x) = \begin{cases} 0 & if \ x \geq 0 \\ 1 & otherwise \end{cases} \tag{11}$$

4. Compute the temporal consistency cost by,

$$E_t(b'_i, b'_j) = \tau(x) \left[\sum_{(x,y,t) \in b_i} \mathbf{C}_{bi}(x, y, t) + \sum_{(x,y,t) \in b_j} \mathbf{C}_{bj}(x, y, t) \right] \tag{12}$$

The fitness of an individual chromosome is computed from collision cost and temporal consistency cost. The fitness $f(x_i)$ of the individual x_i can be defined as:

$$f(x_i) = \sum_{b_i, b_j \in B} (\alpha E_t(b'_i, b'_j) + \beta E_c(b'_i, b'_j)) \tag{13}$$

where, α (weight of temporal consistency cost) and β (weight of collision cost), are user defined weights. Reducing the value of β, generates a dense synopsis video, where more objects may overlap and more activity is covered. Its value can be increased to generate a sparse synopsis of the video, where overlapping of objects is reduced but it covers less activity. And the goal is to minimize the total collision cost and temporal consistency cost for all pairs of object tubes. The Figure 8 below shows three activity tubes from different time segments of the original video.

The time segments of these tubes b1, b2 and b3, in the source video are:

Figure 8. Various space-time activity tubes (a) tube b1 (b) tube b2 (c) tube b3

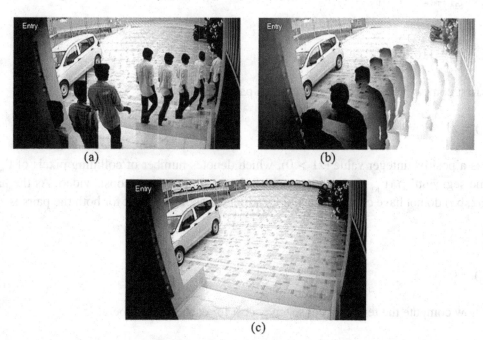

(a)   (b)

(c)

$t_{b1} = [st_{b1}\ et_{b1}] = [15530\quad 15630]$

$t_{b2} = [st_{b2}\ et_{b2}] = [16030\quad 16145]$

$t_{b3} = [st_{b3}\ et_{b3}] = [900\quad 989]$

So, here N = 3 and let the value of M be 2, which means there are 3 tubes in the video and 2 individuals in the initial population.

Suppose individuals of initial population are,

$R_1 = 15\ 20\quad 10$

$R_2 = 17\ 25\quad 20$

Let us evaluate the fitness of each individual one by one.

Considering R_1 first, time segments of the tubes in the synopsis video will be,

$t'_{b1} = [st'_{b1}\ et'_{b1}] = [15\quad 116]$

$t'_{b2} = [st'_{b2} \ et'_{b2}] = [20 \quad 126]$

$t'_{b3} = [st'_{b3} \ et'_{b3}] = [10 \quad 100]$

Let us first compute the collision cost for each pair of tubes.

$E_c(t'_{b1}, t'_{b2}) = z1$

where $z1$ is a positive integer value ($z1 > 0$), which denotes number of colliding pixels of t'_{b1} and t'_{b2} during time segment $t'_{b1} \cap t'_{b2}$ (from frame number 20 to 116) in the synopsis video. As the pairs (b1, b3) and (b2, b3) do not have common spatial locations, the collision cost for both the pairs is zero.

$E_c(t'_{b1}, t'_{b3}) = 0$

$E_c(t'_{b2}, t'_{b3}) = 0$

Let us now compute the temporal consistency cost for each pair of tubes.

$E_t(t'_{b1}, t'_{b2}) = 0$

$E_t(t'_{b1}, t'_{b3}) = 0$

$E_t(t'_{b2}, t'_{b3}) = 0$

The temporal consistency cost for all pairs of tubes is zero, because chronological order in the original video and synopsis video has been kept same for all the pairs.

According to Equation 14, fitness value of individual R_1 will be,

$$f(x_1) = \alpha(0+0+0) + \beta(z1+0+0) = \beta z1 \tag{14}$$

Now, in case of R_2, time segments of the tubes in the synopsis video will be,

$t'_{b1} = [st'_{b1} \ et'_{b1}] = [17 \quad 118]$

$t'_{b2} = [st'_{b2} \ et'_{b2}] = [25 \quad 131]$

$t'_{b3} = [st'_{b3} \ et'_{b3}] = [20 \quad 110]$

Let us first compute the collision cost for each pair of tubes.

$E_c(t'_{b1}, t'_{b2}) = z2$

where z2 is a positive integer value (z2>0), which denotes number of colliding pixels of t'_{b1} and t'_{b2} during time segment $t'_{b1} \cap t'_{b2}$ (from frame number 25 to 118) in the synopsis video.

$$E_c(t'_{b1}, t'_{b3}) = 0$$

$$E_c(t'_{b2}, t'_{b3}) = 0$$

As the pairs (b1, b3) and (b2, b3) do not have common spatial locations.

Let us now calculate the temporal consistency cost for each pair of tubes.

$$E_t(t'_{b1}, t'_{b2}) = 0$$

$$E_t(t'_{b1}, t'_{b3}) = z3$$

$$E_t(t'_{b2}, t'_{b3}) = 0$$

Here, temporal consistency cost for the pairs (b1, b2) and (b1, b3) is zero as the chronological order is not changed for these pairs. But for the pair (b2, b3), it is reversed. So, according to Equation 11, value of $\tau(x)$ will be 1. And as per Equation 12, the value of $E_t(t'_{b2}, t'_{b3})$ is z3, where z3 is some integer value greater than zero.

According to Equation 15, fitness value of individual R_2 will be:

$$f(x_2) = \alpha(0+z3+0) + \beta(z2+0+0) = \alpha z3 + \beta z2 \tag{15}$$

Selection of Fittest Individuals

The selection operation allows the fittest individuals to survive. The selection is done from the set of parent individuals considering their fitness $f(x_i)$. And roulette strategy will be used for it. These selected individuals will be inherited to the next generation.

Steps of selection using roulette strategy are as follows:

Step 1: Find the probability of ith individual in the current population using,

$$P(x_i) = \frac{f(x_i)}{\sum_{j=1}^{M} f(xj)} \tag{16}$$

Step 2: Find the expected count,

$$EC(x_i) = N * P(x_i) \tag{17}$$

Select the individuals with minimum value of expected count.

Recombination (Crossover)

Recombination will be used to exchange a part of genes of an individual pair according to the crossover probability, which forms a new pair of individuals. In this case, two crossover schemes, single-point crossover and two-point crossover are used to enhance the effect of it.

Steps:

1. Select M/2 fittest individuals from the previous iteration.
2. Get two integer values a1 and a2, where $1 \leq a1 \leq a2 \leq N$.
3. Randomly select a number d from the set { 0, 1, 2} If d = 0, then exchange the genes from 0 to a1 in the pair of individuals. If d = 1, then exchange the genes from a1 to a2 in the pair of individuals. If d = 2, then exchange the genes from a2 to N in the pair of individuals.
4. Repeat the steps until new population contains M intermediate individuals, which are called offsprings.

Figure 9 shows the example of single-point crossover and two-point crossover.

Mutation

If only the crossover operator will be used to produce offspring, one potential problem that may arise is that if all the chromosomes in the initial population have the same value at a particular position then all future offspring will have this same value at this position. Mutate the M intermediate individuals generated after crossover according to the mutation probability P_m and form M individual candidates. The gene locus mutation strategy is used for it. According to gene locus mutation, a position $k(1 \leq k \leq N)$ is selected randomly in the code string Rj and then change the gene in this position.

Convergence

A stopping criterion must be there to define the condition to converge the optimization of energy function. When the current iteration count t exceeds maximum iteration number T, i.e. $t > T$, stop the evolution. And the individuals with maximum fitness (minimum cost) in all the populations X(0) to X(T) is the optimal solution. Otherwise set $t = t + 1$ and jump to step (3) for calculating fitness of newly created offspring.

Object Stitching Using Poison Image Editing

In this stage, moving objects from different time stamps in the source video are stitched to the new timestamps in the synopsis video. For stitching object on the frames, Poisson Image Editing method is used. The mathematical tool, on which poisson image editing is based, is the Poisson partial differential equation with Dirichlet boundary conditions (Perez, Gangnet, & Blake, 2003; Farbman, Hoffer, Lipman, Cohen-Or, & Lischinski, 2009). It gives the Laplacian of an unknown function on the region of interest, along with the unknown function values on the boundary of the region. The poisson image editing al-

Figure 9. Example of (a) single-point crossover (b) two-point crossover

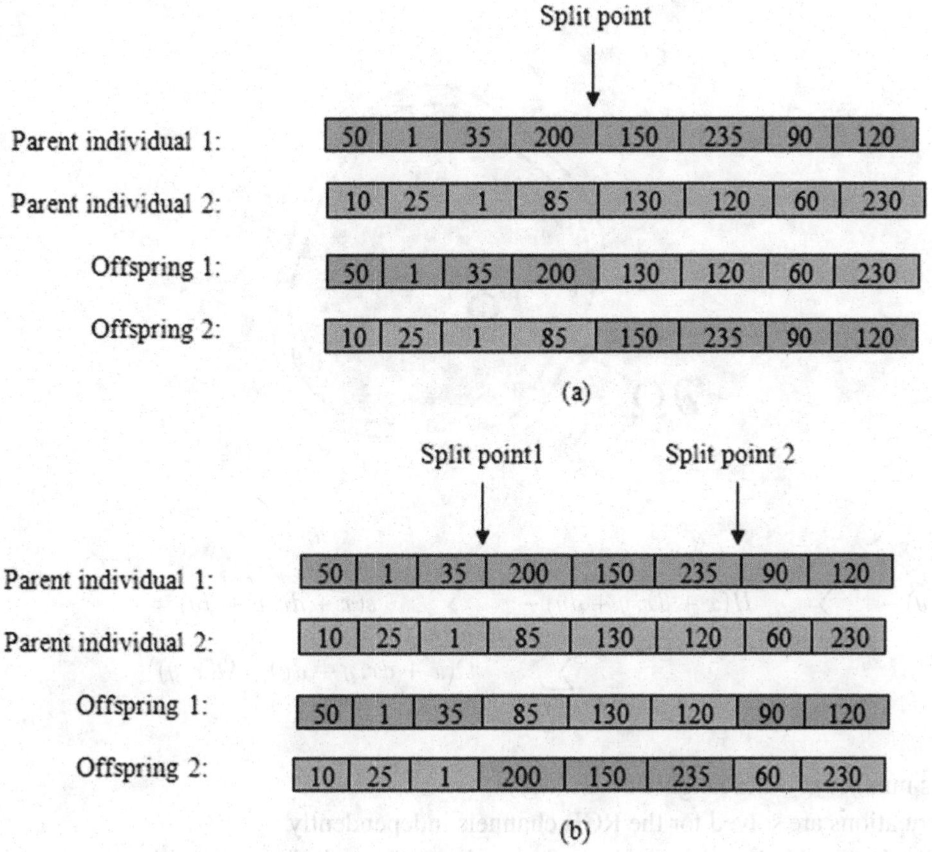

lows for the tweaking of absolute information of object to be stitched (colors), but preserves the relative information (image gradient) of it as much as possible after it is pasted.

Let image definition domain S, be the closed subset of R^2 and Ω be the closed subset of S with boundary $\partial \Omega$. Let H be the improved version of Ω that blends well with S. Figure 10 shows these notations. The pixels of $\partial \Omega$ should be exactly same as the pixels of S on that boundary.

$$H\left(x,y\right) = S\left(x,y\right) \qquad \forall\, (x,y) \in \partial\Omega \qquad (19)$$

The gradient of the pixels on interior of H should be equal to the gradient of the pixels on the interior of Ω. Gradient of an image at a pixel (x, y) can be defined as,

$$|\nabla\Omega\left(x,y\right)| = 4\Omega(x,y) - \Omega(x-1,y) - \Omega(x+1,y) - \Omega(x,y-1) - \Omega(x,y+1) \qquad (20)$$

If one of the neighbour pixels is on boundary, its value will be fixed. And if it is out of the selected bound, then it will be discarded. So, the difference equation for a pixel (x, y) becomes,

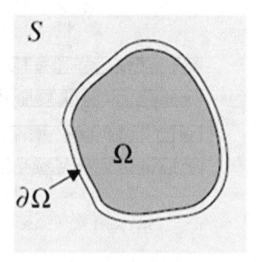

$$| N | H(x,y) - \sum_{(dx,dy)+(x,y)\in \mathbb{C}} H(x+dx,y+dy) - \sum_{(dx,dy)+(x,y)\in \partial \Omega} s(x+dy,y+dy) = \\ \sum_{(dx,dy)+(x,y)\in \Omega \cup \partial \Omega} \Omega(x+dx,y+dy) - \Omega(x,y)) \quad (21)$$

where, N is number of valid neighbours of (x,y).

These equations are solved for the RGB channels independently.

Following Figure 11 shows three frames belonging to three different activities in original video and a frame where foreground objects from these frames are stitched on background frame, using poison image editing. Figure 11(a) shows a frame with walking person, 11(b) shows a frame with moving bus and a frame in 11(c) shows another walking person. Figure 11(e) shows a frame where all the three foreground objects are stitched on the background frame shown in 11(d).

After shifting all the moving objects to the new time segment, the resulting video synopsis represents the activities of the source video in very short time duration. is be considerably short in length. The approach presented here extracts the moving objects as space-time activity tubes from the long input surveillance video and rearranges those objects at different time segments using the idea of GA, to generate the much condensed video synopsis.

EXPERIMENTAL RESULTS

The proposed algorithm has been implemented using Matlab 2013a and executed on intel(R) core i5-4210U 1.70GHz processor with 8GB RAM. The proposed work has been evaluated using three video sequences, labelled as VS1, VS2 and VS3. Video sequence VS1 is from ViSOR repository and VS2 and VS3 are captured from a real outdoor surveillance camera monitoring the entry area of a school. The VS1 contains 3795 frames at size of 384 X 288, VS2 contains 16500 frames at the size of 510 X 250 and VS3 has 97309 frames of 280 X 148 size. From the input video moving objects are extracted and

Figure 11. Foreground objects stitched on background frame

tracked to generate space-time activity tubes. The surveillance videos generally consists many frames without any activity, which are automatically skipped in this phase. The processing of this phase for the video of 1 hour took less than one hour. Table 1 shows number of objects detected after this phase along with ground truth for each video sequence. The objects of very small size are not detected as foreground objects. Most of the surveillance companies provide hardware implementation for detecting and tracking moving objects, which can be used instead.

In the next phase GA is employed to generate the optimal temporal arrangements by minimizing the cost for the rearranging objects in video synopsis. This is done by computing value of collision cost and temporal consistency cost for each pair of tubes. The minimum cost obtained after all the iterations increases as the synopsis duration decreases and vice versa. This relationship between cost and synopsis duration is shown in Table 2 where, total duration indicates addition of age(#frames) of all extracted activity tubes. As depicted in Table 2, the cost value increases as the synopsis duration decreases. The cost computation took 70 seconds for the video sequence with 264 objects.

Table 1. Total number of objects detected

Video sequence	Length (#frames)	Number of Objects Detected	Ground Truth	% of Object Detection
VS1	3795	19	22	86.36%
VS2	16500	38	41	92.68%
VS3	97309	264	283	93.28%

Table 2. Relationship between cost and synopsis duration

Video Sequence	Minimum Cost		
	TotalDuration/2	TotalDuration/5	TotalDuration/10
VS1	480	985	1471
VS2	1756	3659	7506
VS3	119760	215504	373416

In the final phase, all the objects are stitched on the sequence of background images to generate condensed synopsis video. Table 3 depicts the length of the video sequences and length of the corresponding video synopsis. Figure 12 shows an example of video synopsis results of VS1 which depicts how the activities are condensed and displayed in short period of time. It also explains that if a synopsis duration is reduced, more condensed output videos are generated with more number of colliding objects. Figure 13 shows the frames from synopsis video of VS3.

In the synopsis video, with each activity tube a time stamp is shown which indicates the time at which that activity is occurred in the original video. This enables the viewers to compare the chronological order of multiple activities being displayed simultaneously in the video synopsis.

CONCLUSION

An approach to produce object based video synopsis of a long surveillance video under the idea of GA is presented in this chapter. The temporal median algorithm is used to model the background images from the long input video and due to its computation simplicity; foreground extraction is employed using background subtraction method. The average accuracy of object detection is 90.78% as per the results shown in Table 1. Extracted foreground objects are tracked using kalman filter and represented in form

Table 3. Video synopsis results

Video Sequence	Original Video Length (#frames)	Synopsis Length (hh:mm:ss)	Synopsis Length (#frames)	Synopsis Length (hh:mm:ss)
VS1	3795	00:06:19	723	00:00:52
VS2	16500	00:11:00	955	00:01:03
VS3	97309	01:04:52	8048	00:05:36

Figure 12. Video synopsis of VS1. (a) A typical frame from the input video (379 seconds). (b) A frame with condensed activity from synopsis video (62 seconds). (c) & (d) Frames from a more condensed synopsis video (51 seconds).

(a)   (b)

(c)   (d)

Figure 13. Video synopsis of VS3. (a) A frame from original video sequence. (b) A frame from condensed synopsis video.

(a)   (b)

of space-time activity tubes. In order to obtain the condensed synopsis video, the optimum temporal shifting of these activity tubes is required to be done. The idea of GA is employed for the optimization of temporal shifting, which considers collision cost & temporal consistency cost as the measure to compute fitness of individuals and decides the new time segments for each moving object. The moving objects are stitched on the background images using poisson image editing. Only the static backgrounds are considered for experiments in this approach, but time-lapse background video can be generated selecting more frames from active time periods and a few frames from inactive region as well. The approach produces a synopsis video from the original long video and thus making the browsing and retrieval of such long video easy.

Though the proposed approach produces much more condensed video, there are several features that can be added to it. The system can be made interactive, where user can specify the time period of interest as well as length of synopsis video. The system can accept the query like "show the synopsis of past 24 hours video footage in 10 minutes" and generate the synopsis video accordingly. The same approach can be extended to design online video synopsis system, which can deal with the endless video by extracting and storing moving object details in real time and constructs the synopsis video at the time of user query.

The presented approach generates much compressed video synopsis but it has few a limitations listed below:

- If the original video is already having dense activities all the time, the used system cannot generate much compressed synopsis of such videos. For example surveillance videos captured at busy malls or busy airports.
- The given system is not interactive, where user can provide query like "generate 5 min synopsis of the surveillance video of past 3 hours".
- The kalman filter is employed to track the object in the stated system, which causes flickering effect if the object being tracked is divided into multiple parts due to obstacles or if two objects are interacting.

REFERENCES

Berret, D. (2013, July 10). Online surveillance camera for every 11 people in Britain, says CCTV survey. *The Telegraph.*

Chou, C. L., Lin, C. H., Chiang, T. H., Chen, H. T., & Lee, S. Y. (2015, June). Coherent event-based surveillance video synopsis using trajectory clustering. In *Multimedia & Expo Workshops (ICMEW), 2015 IEEE International Conference on*, (pp. 1-6).

Cucchiara, R., Grana, C., Piccardi, M., & Prati, A. (2003). Detecting moving objects, ghosts, and shadows in video streams. *Pattern Analysis and Machine Intelligence. IEEE Transactions on, 25*(10), 1337–1342.

Cuevas, E. V., Zaldivar, D., & Rojas, R. (2005). *Kalman filter for vision tracking*. Academic Press.

Farbman, Z., Hoffer, G., Lipman, Y., Cohen-Or, D., & Lischinski, D. (2009, July). Coordinates for instant image cloning. *ACM Transactions on Graphics, 28*(3), 67. doi:10.1145/1531326.1531373

Feng, S., Liao, S., Yuan, Z., & Li, S. Z. (2010, August). Online principal background selection for video synopsis. In *Pattern Recognition (ICPR), 2010 20th International Conference on*, (pp. 17-20). doi:10.1109/ICPR.2010.13

Hao, L., Cao, J., & Li, C. (2013, June). Research of GrabCut algorithm for single camera video synopsis. In *Intelligent Control and Information Processing (ICICIP), 2013 Fourth International Conference on*, (pp. 632-637). doi:10.1109/ICICIP.2013.6568151

Höferlin, B., Höferlin, M., Weiskopf, D., & Heidemann, G. (2011). Information-based adaptive fast-forward for visual surveillance. *Multimedia Tools and Applications*, *55*(1), 127–150. doi:10.1007/s11042-010-0606-z

Hsia, C. H., Chiang, J. S., Hsieh, C. F., & Hu, L. C. (2013, November). A complexity reduction method for video synopsis system. In *Intelligent Signal Processing and Communications Systems (ISPACS), 2013 International Symposium on*, (pp. 163-168). doi:10.1109/ISPACS.2013.6704540

Huang, C. R., Chung, P. C. J., Yang, D. K., Chen, H. C., & Huang, G. J. (2014). Maximum a Posteriori Probability Estimation for Online Surveillance Video Synopsis. *Circuits and Systems for Video Technology. IEEE Transactions on*, *24*(8), 1417–1429.

Hung, M. H., Pan, J. S., & Hsieh, C. H. (2014). A fast algorithm of temporal median filter for background subtraction. *Journal of Information Hiding and Multimedia Signal Processing*, *5*(1), 33–40.

Jeeshna, P. V., & Kuttymalu, V. K. (2015). A Technique for Object Movement Based Video Synopsis. IEEE Transaction on International Journal of Engineering and Advanced Technology, 20(9), 1303-1315.

Kang, H. W., Matsushita, Y., Tang, X., & Chen, X. Q. (2006, June). Space-time video montage. In computer vision and pattern recognition, 2006 IEEE computer society conference on, (vol. 2, pp. 1331-1338).

Li, Y., Zhang, T., & Tretter, D. (2001). *An overview of video abstraction techniques*. Technical Report HPL-2001-191. HP Laboratory.

Liu, A., & Yang, Z. (2010, October). An interactive method for dynamic video synopsis generation. In *Computer Application and System Modeling (ICCASM), 2010 International Conference on*.

Lo, B. P. L., & Velastin, S. A. (2001). Automatic congestion detection system for underground platforms. In *Intelligent Multimedia, Video and Speech Processing, 2001.Proceedings of 2001 International Symposium on*, (pp. 158-161). doi:10.1109/ISIMP.2001.925356

Lu, M., Wang, Y., & Pan, G. (2013, May). Generating fluent tubes in video synopsis. In *Acoustics, Speech and Signal Processing (ICASSP), 2013 IEEE International Conference on*, (pp. 2292-2296). doi:10.1109/ICASSP.2013.6638063

Nie, Y., Sun, H., Li, P., Xiao, C., & Ma, K. L. (2014). Object Movements Synopsis via Part Assembling and Stitching. *Visualization and Computer Graphics. IEEE Transactions on*, *20*(9), 1303–1315.

Nie, Y., Xiao, C., Sun, H., & Li, P. (2013). Compact video synopsis via global spatiotemporal optimization. *Visualization and Computer Graphics. IEEE Transactions on*, *19*(10), 1664–1676.

Oh, J., Wen, Q., Hwang, S., & Lee, J. (2004). Video abstraction. *Video data management and information retrieval*, 321-346.

Padhiyar, S., & Joshi, D. (2015). *A Survey. International Journal of Engineering Sciences & Research Technology*.

Pérez, P., Gangnet, M., & Blake, A. (2003, July). Poisson image editing. *ACM Transactions on Graphics*, *22*(3), 313–318. doi:10.1145/882262.882269

Petrovic, N., Jojic, N., & Huang, T. S. (2005). Adaptive video fast forward. *Multimedia Tools and Applications, 26*(3), 327–344. doi:10.1007/s11042-005-0895-9

Piccardi, M. (2004, October). Background subtraction techniques: a review. In Systems, man and cybernetics, 2004 IEEE international conference on, (vol. 4, pp. 3099-3104). doi:10.1109/ICSMC.2004.1400815

Pritch, Y., Ratovitch, S., Hende, A., & Peleg, S. (2009, September). Clustered synopsis of surveillance video. In *Advanced Video and Signal Based Surveillance, 2009. AVSS'09. Sixth IEEE International Conference on*, (pp. 195-200). doi:10.1109/AVSS.2009.53

Pritch, Y., Rav-Acha, A., & Peleg, S. (2008). Nonchronological video synopsis and indexing. *Pattern Analysis and Machine Intelligence. IEEE Transactions on, 30*(11), 1971–1984.

Rav-Acha, A., Pritch, Y., & Peleg, S. (2006, June). Making a long video short: Dynamic video synopsis. In *Computer Vision and Pattern Recognition, 2006 IEEE Computer Society Conference on*, (vol. 1, pp. 435-441).

Shoushtarian, B. (2003). A practical approach to real-time dynamic background generation based on a temporal median filter. *Journal of Sciences, 14*(4), 351–362.

Tanaka, M., Kamio, R., & Okutomi, M. (2012, November). Seamless image cloning by a closed form solution of a modified poisson problem. In SIGGRAPH Asia 2012 Posters (p. 15). ACM. doi:10.1145/2407156.2407173

Truong, B. T., & Venkatesh, S. (2007). Video abstraction: A systematic review and classification. *ACM Transactions on Multimedia Computing, Communications, and Applications, 3*(1). doi:10.1145/1198302.1198305

Yildiz, A., Ozgur, A., & Akgul, Y. S. (2008, October). Fast non-linear video synopsis. In *Computer and Information Sciences, 2008. ISCIS'08. 23rd International Symposium on*, (pp. 1-6). doi:10.1109/ISCIS.2008.4717951

Yilmaz, A., Javed, O., & Shah, M. (2006). Object tracking: A survey. *ACM Computing Surveys, 38*(4). doi:10.1145/1177352.1177355

Yogameena, B., & Priya, K. S. (2015, January). Synoptic video based human crowd behavior analysis for forensic video surveillance. In *Advances in Pattern Recognition (ICAPR), 2015 Eighth International Conference on*, (pp. 1-6). doi:10.1109/ICAPR.2015.7050662

Zhu, J., Liao, S., & Li, S. Z. (2015). *Multi-Camera Joint Video Synopsis*. Academic Press.

KEY TERMS AND DEFINITIONS

Activity Tubes: In video synopsis, each detected object is tracked and represented as an activity tube in the space-time domain, which are then temporally rearranged in order to generate condensed synopsis video.

Background Subtraction: In the field of image and video processing, background subtraction is a technique employed to extract the foreground objects. It is also known as Foreground Detection.

Collision Cost: It is a cost imposed when the two shifted objects in synopsis video collides with each other.

Crossover Probability: It defines how often the crossover will be performed. The crossover probability of 100% indicates that all offsprings are generated by crossover. And if it is 0%, all the offsprings of new generation are exact copies of chromosomes from previous population.

Genetic Algorithm (GA): GA is an adaptive heuristic based search algorithm which is used to solve unconstrained and constrained optimization problems. It simulates process of biological evolution and natural selection.

Mutation Probability: It depicts how frequently the parts of chromosome will be mutated. If offspring is taken after crossover without any change, the mutation probability is 0%. If mutation probability is set to 100%, all the bits of a chromosome are changed. Mutation prevents GA from being trapped into local maxima.

Simulated Annealing: It is a probabilistic approach for solving both the unconstrained and bound-constrained problems of optimization and finding the global optimization of a given function in a large search space.

Temporal Consistency Cost: It is a cost associated with chronological order of the event in video synopsis. This cost is imposed when the temporal order of two activities in the synopsis is altered compared to original video.

Video Abstraction: Video abstraction is a technique which allows to browse huge collection of video data sets quickly and to access and index the content efficiently.

Video Synopsis: Video synopsis is an approach used to build a short summarized video from a long video by simultaneously displaying multiple activities even though they took place at different time stamps in the original video.

Chapter 10

Technical Evaluation, Development, and Implementation of a Remote Monitoring System for a Golf Cart

Claudio Urrea
Universidad de Santiago de Chile, Chile

Víctor Uren
Universidad de Santiago de Chile, Chile

ABSTRACT

A technical evaluation of the sensing, communication and software system for the development and implementation of a remote monitoring system for an electric golf cart is presented. According to the vehicle's characteristics and the user's needs, the technical and economic aspects are combined in the best possible way, thereby implementing its monitoring at a distance. The monitoring system is used in two important stages: teleoperation and the vehicle complete autonomy. This allows the acquisition of video images on the vehicle, which are sent wirelessly to the monitoring station, where they are presented through a user-friendly interface. With the purpose of complementing the information sent to the remote user of the vehicle, several important teleoperation variables of a land vehicle, such as voltage level, current and speed are sensed.

INTRODUCTION

Monitoring is, and has been, the mean by which reference is made to the supervision necessary for the execution of a certain action plan. It also allows the detection of possible interferences that might appear during an action and their subsequent correction. Monitoring, task that in the past was supervised exclusively by the human eye, has evolved over time as technology advances. These advances are mainly

DOI: 10.4018/978-1-5225-1022-2.ch010

reflected in the evolution of electronic devices, whose beginning dates back to the early 20th century. Monitoring and the sciences devoted to the creation of electronic devices that permit its implementation have made significant advance as the industrial world grows bigger. The industrial area continuously demands the use of new technologies, improvements and innovation in monitoring, since they are essential for the new devices to operate in a market that calls for advanced automation in the industries.

Over the last few years, important progress in wireless monitoring and telecommunications has been made. At the beginning of the 20th century, Guglielmo Marconi revolutionized telecommunications by creating the wireless telegraph and, thus, being the first man that sent signals wirelessly ("Guglielmo Marconi," n. d.).

The fact of establishing monitoring from a remote place without needing cables becomes very useful, since in this way it is not necessary to be present in a process to know its state. Thus, decision-making in certain events is faster and more comfortable, as information is within reach, and physical integrity is not exposed during the execution of dangerous tasks.

RELATED WORK

Wireless monitoring of vehicles is an area that has developed only in the last few years and is now covered by some companies that offer remote monitoring service for vehicles, mostly using Global Positioning System (GPS) technology, i.e., by means of a device installed in the vehicle it is possible to know its location through the interaction of the device with satellites. In some companies that offer that service, the information on the vehicle's position is carried together with other signals coming from different sensors that monitor the various events desired by the client, using mainly the Global System for Mobile Communications (GSM) and General Packet Radio Service (GPRS) technologies (Glasgow et al., 2004; Chunga & Oh, 2006; Li et al., 2013).

Wireless monitoring in vehicles has also been and is a research subject at educational centers, such as, for example, the Instituto de Investigación en Ingeniería de Aragón (I3A) of the Universidad de Zaragoza, created in 2002 by a decree of the Government of Aragón, Spain, where there is an applied research group called "VEHIVIAL" (Castejón et al., 2006). These researchers have developed teleoperated vehicles, showing that adequate monitoring is essential. On the other hand, in the same institution the Grupo de Tecnologías de las Comunicaciones (GTC) works at the automobile industry with the purpose of conceiving vehicles that can be viewed as mobile terminals that require telematic information and services in an interactive manner (Satyanarayana, & Mazaruddin, 2013; Kantharia et al. 2014).

The same as in other countries, in Chile there are developments in this field. In Santiago, a group of students of the Universidad de Chile developed the first Chilean solar vehicle, called "EOLIAN," which is equipped with a monitoring system. After the data capturing made by the vehicle, they are sent via radiofrequency to a computer that displays them on screen by means of an interface created with the LabVIEW software.

In terms of the latest technologies applied to remote monitoring of vehicles, the Formula 1 is one of the main representatives due to the high competitiveness between the different constructors and the great availability of resources. Tire treatment is one of the most critical factors in racing cars performance. Specifically, temperature is the principal factor that provides information on the tires workload. Normally, it is necessary to wait until the car returns to the pits to register this information. Nevertheless, Danese et al. (2008) propose and develop an innovative system of acquisition of information, which delivers temperature through infrared cameras located on the vehicle. Thanks to this technology, the engineers receive real-time information on the tires performance of the single-seaters that are competing, with the purpose of advising the driver and keeping him informed on what is happening in his vehicle as well as on the actions to be taken. The system may be adapted to acquire traditional images, which are useful to study deformation in several parts of the vehicle.

Another field in which remote monitoring has great importance is the military, where work is continuously going on in innovations with the purpose of being ahead with respect to their potential adversaries, especially when dealing with pilotless vehicles that are capable of getting important data that are mostly classified as secret. This type of remote systems has also been implemented in the mining industry, such as the SGA3722 vehicle made in the Shougang Heavy Vehicle Factory. The system uses sensors distributed in the vehicles to collect information on pressure, speed, mileage, fuel levels and data in the ARM (Advances RISK Machines). The wireless network transmits data in real time from the terminal to the monitoring station by Wi-Fi (Wireless Fidelity). In conclusion, the system is very useful to enhance the efficiency of the transportation in mining and security vehicles (Wu, Liu, & Yuan, 2010).

GENERALITIES OF THE MONITORING SYSTEM

Description of the Golf Cart

The golf cart, purchased by the Departamento de Ingeniería Eléctrica of the Universidad de Santiago de Chile (DIE-USACH) corresponds to the electrical vehicle division from the MOTORMAN Company, whose devices are obtained from an Asian supplier. MOTORMAN is a company dedicated to the mechanization of tasks of different economic sectors, by the sale and rental of their machinery. Figure 1 depicts the vehicle:

The vehicle has the technical characteristics presented in Table 1 (http://www.motorman.cl/).

Figure 1. Golf cart purchased by the DIE-USACH

Table 1. Characteristics of the golf cart

Model	Voyager
Passenger capacity	2
Motor	4 Kw
Batteries	Trojan 48 Volt (6 Volt/8 Units)
Controller	Curtis
Size	2350·1180·1750 mm
Maximum speed	30 Km/Hr
Equipment	• Headlamps and rear lights • Folding windshield • Back-up alarms • Alloy wheels

General Description of the Monitoring System

The designed and implemented monitoring system can be divided schematically into three blocks:

1. The first, sensing, contains the sensing elements on board the vehicle;
2. The second, communication channel, includes the elements in charge of transmitting and receiving the data that will be sent; and
3. The third consists basically of a visual interface implemented by means of a PC that allows displaying the data transmitted to the remote user.

Figure 2 shows this schematically.

A brief description of the blocks involved in the monitoring is presented below.

Figure 2. Referential monitoring scheme
Source: Mainville Electric Motor CO, n. d.

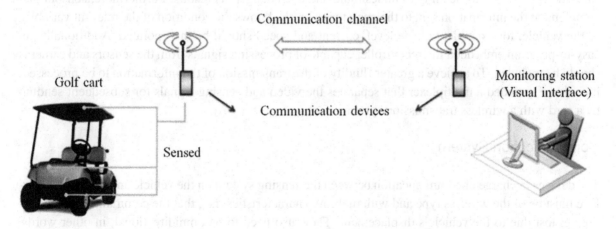

223

Sensing

The sensing system includes several video cameras on board the electrical golf-cart as well as sensors that allow complementing the information sent to the remote operator. In addition, elements necessary to establish the processing order of the sensed information and the control of other relevant variables that will be presented further on are included.

Transmission and Reception

The transmission and reception system is composed of all the equipment in charge of transmitting and receiving video images and data obtained in the sensing process (in which the communication channel is air). The protocol used for the sending of information is determined by technical and economic specifications.

Visual Interface

Visual interface is a graphic program that permits the person in charge of monitoring to see the images captured by the video cameras on board the golf cart and the data collected by the sensors. The presentation of video and data should be attractive and friendly so that the operator of the visual interface interacts with it easily.

Monitoring System Requirements

Sensing System

In general, the monitoring implemented in the DIE-USACH electric golf cart must be sufficiently complete to allow the remote operator to know the condition of the different variables involved in the vehicle's displacement. Special attention is given to the number of video cameras on board the vehicle, because it should be the one required to achieve, at a distance, adequate and safe driving. The video cameras must operate in real time, because any significant delays in acquiring and sending their video signals would cause delayed reactions, leading to undesirable and even dangerous results. Furthermore, sensors that complement the information sent to the operator so he/she knows the condition of the relevant variables of the vehicle, for example, voltage level, current and speed should be incorporated. Additionally, an easy-to-program embedded microcontroller capable of processing signals from the sensors and cameras should be included. To achieve a greater fluidity of the transmission of the information to be processed, it could be included a multiplexer that separates the video and sensing signals for subsequent sending by a card with a wireless transmission system.

Communication System

The devices in charge of communication between the sensing system on the vehicle and the visual interface must be of the wireless type and with mobility characteristics, i.e., that the communication should not get lost due to the vehicle's displacement. They also need to be omnidirectional, in other words, that may be oriented or used in any direction or sense. Moreover, it also must be sufficiently robust to support the adverse conditions of the settings through which the vehicle moves. Furthermore, for the monitoring system to have an adequate travel autonomy, it must have an information sending/receiving range and a bandwidth defined according to the desired autonomy.

Visual Interface

A graphic user interface must be designed and implemented so that the operator can know the various captured video images as well as the signals coming from the sensors. It must contain subscreens showing the different views offered by the cameras installed on the vehicle, subframes that show the sensed variables and the buttons that allow interacting with them. The information presented on the screen must be updated in real time. This interface must be as graphic, simple and intuitive as possible, so that any operator can use it without the need to have previous training on its use.

TECHNICAL EVALUATION OF THE MONITORING SYSTEM

Sensing System

Technical Considerations for Camera Selection

In the market, one can find a wide range of video cameras. The choice of the most adequate one for the monitoring system depends on the technical characteristics that they have, the most important of which are the following:

- **Size**: Their length should not exceed about 20 to 25 cm, since to equip the golf-cart with larger cameras would be invasive and unnecessary.
- **Weight**: Altogether, the equipment implemented in the golf cart should not have excessive weight; ideally, each camera should not weigh more than 300 g. This in order not to impair the golf cart performance and to command it easily.
- **Angle of Vision**: This angle must be as large as possible, providing the remote driver a wide field of view and allowing him to know what is happening in the neighborhood of the golf cart ("Angulo_de_vision," n. d.).
- **Resolution**: Ideally, the resolution of each camera should not be less than 640 x 480 pixels. This resolution allows the acquisition of clear visual information from the environment in which the cart moves so its movement is optimal.
- **Reach**: The reach of each camera should be such that, looking at the far horizon, braking can be done without great inconvenience if there is an obstacle. For instance, if the reach of a camera set in the cart were 20 meters, and supposing that the cart moved at maximum speed, 30 km/h, the driver would have 2.4 seconds to react and put the foot on the brake in case an obstacle appeared at that distance. However, this time is only a reference, since the braking is not instant.
- **Power Supply**: Video cameras powered by rechargeable batteries are used. It is easier to include continuous current power supplies —through batteries— than other power generation systems in a vehicle in motion.

- **Connection Method**: Since sending the images from the vehicle to the remote operator must be wireless, it is necessary to incorporate equipment such as routers, for example, that can carry out those kinds of tasks.
- **Frames Per Second**: The number of frames per second must be such that the human eye cannot see irregularities in the image. The higher this number the better the quality of the perceived image, since the human eye is capable of detecting from 30 to 40 frames per second, depending on the person ("Frame_rate," n. d.).
- **Real Time Work**: It is extremely important for the cameras not to have any delay in processing the images. This situation might lead to delayed reactions in the remote operator, thus, causing accidents.

The above mentioned technical characteristics, together with other considerations, economic factors among them, will determine the choice of the cameras. In the market there are different types of cameras, among which those shown in Figure 3 stand out.

Analogic Cameras

The main characteristic of analogic cameras is that their video signal is analogic, that is, it can be directly connected to a monitor, video recorder, etc. The sensor that captures the image is CCD-type (Charge-Coupled Device) and its resolution is measured in TVL (Television lines). Currently, analogic cameras are widely used in surveillance systems known as CCTV (Closed-circuit television). In this regard, the commercial market mainly provides outdoor cameras whose dimensions are too large for locating them in the golf-cart (their length is 25 cm approximately). In spite of this, there is a number of cameras of smaller dimensions, but with a considerable weight due to their case-type. An example of this type of camera is the Lummix color CCD Sony 1/3 540 TVL 3.6mm IR 20 M White, whose length does not exceed 10cm. As may be seen in Figure 4, this camera was made with a protection system to withstand the conditions of outdoor environments. The reason for selecting this model was its small size, its 1/3effective range (i.e. a range of 60° in horizontal vision, which is larger than the average of 45°) and its 4540 TVL.

Figure 3. Camera options for the implementation

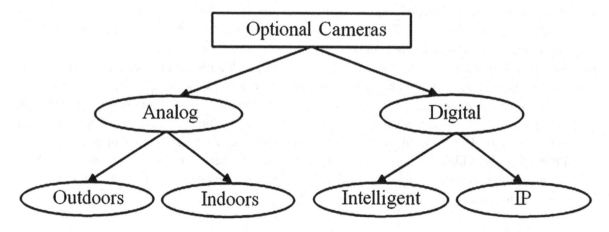

Indoor cameras have smaller sizes that their outdoor counterparts. In particular, the Samsung D/N SCB-2000 indoor camera has the following characteristics: 5.8 x 5.2 x 12.1 cm, 1/3 effective range Super HAD color CCD, minimum illumination of 0.05 Lux and 0.0001 Lux (using Sens-up), a high resolution of 600-TVL and DUAL power of 24V AC and 12V DC, among others (see Figure 5) ("Samsung Techwin CO," 2010).

It must be noted that analogic cameras, both indoor and outdoor, need devices that allows them to send information signals via Ethernet —which is the medium of communication routers use in order to send their signal through a net. These devices are called DVR (Digital Video Recorder). In closed-circuit television, images from analogic cameras are sent to a DVR, to which are connected by a coaxial cable. The DVR stores the captured images in the hard disks within it. An DVR broadcasts live the images received from the cameras by means of a monitor connected to it. The user can interact with the images through an interface offered by the DVR. Figure 6 presents a DVR device.

Digital Cameras

Digital cameras are characterized by treating and storing digital images. Their resolution is measured in pixels, i.e. the more the pixels, the better the resolution. Digital cameras represent an option for the monitoring of the golf cart. There are two types: intelligent and IP (Internet Protocol) digital cameras.

Figure 4. Lummix color CCD Sony 1/3540 TVL 3.6mm IR 20 M white
Source: Gobar LTDA, n. d.

Figure 5. Samsung D/N SCB-2000
Source: Artilec, n. d.

Figure 6. DVR 4 channels lummix full D1 dual core H.264

Intelligent Cameras

These cameras possess an integrated system that is oriented to artificial vision.

Intelligent cameras have, besides electronic components that permit the capture of images, a processor that allows them to treat images without an external CPU. In general, they permit Ethernet or serial port connections, digital inputs and outputs, control lines and different levels of illumination ("Camara_Inteligente," n. d.).

A camera that stands out among digital models is the CMUcam3, which is the third version of a sequence of intelligent cameras created by the University of Carnegie Mellon[1]. This model is composed of a camera in digital format and an open-source development system, which allows the recognition of shapes, color detection and following of objects in motion, etc. Figure 7 depicts a CMUcam3 digital camera.

The CMUcam3 is an attractive option for the monitoring of the golf cart. Since it is capable of recognizing shapes and colors, such camera can be implemented to detect traffic signs with colors like traffic lights. Nevertheless, its 352 x 288-pixel resolution does not provide clear images, so its use is more convenient for the vehicle automation than for capturing the images to be presented to the operator.

There are other types of intelligent cameras like the CMUcam3, which are easily found in websites oriented to robotics, such as http://www.robotshop.com/. In addition, it should be underlined that these cameras have been developed mainly for the treatment of images. Consequently, they are not the best option for a monitoring whose principal objective is solely seen the medium in which the vehicle moves clearly.

IP Cameras

IP cameras are surveillance video cameras whose distinctive feature is the capacity of sending the video signals (sometimes also audio signals) directly to an ADSL router to visualize images live through a local area network (LAN) or through any equipment connected to the Internet (WAN or Wide Area

Figure 7. CMUcam3
Source: Carnegie Mellon University, 2007

Figure 8. Camera IP TrendnetProView TV-IP501P PoE
Source: Trendnet, n. d.

Network). This feature permits the user to be located in any place in the world. In this regard, a camera model whose characteristics make it adequate for the golf cart is the IPTrendnetProView TV-IP501P PoE, which may be observed in Figure 8 ("Cámaras IP," n. d.).

This camera transmits high-quality video in real time through the network and in up to 30 frames per seconds. In addition, its vision angle is 45 horizontal degrees and 35 vertical degrees. Furthermore, it may be supplied from an electrical outlet or by PoE (Powerover Ethernet), has a digital zoom of 4x, a small microphone, captures images with a 640 x 480 pixel resolution, and its image sensor is a ¼" color CMOS. Lastly, its dimensions are 19.5 x 9 x 5 cm and its weigh is about 133 grams (its base weighs 121 grams) ("Trendnet," n. d.).

Therefore, the TrendnetProView TV-IP501P has characteristics that make it eligible for being implemented in the golf cart. These are good resolution, light weight, real-time transmission, a nice outline and zoom. In addition, as it is IP-type, the camera can be connected directly to the router in charge of transmitting the signals from the cart to the network, without needing intermediate devices that might somehow affect the speed of transmission.

Remarks

Taking into account the characteristics of each camera above described, it is concluded that the best option for implementing a monitoring system, i.e. having the technical characteristics detailed above, are IP video cameras, which are connected directly to a router that sends the different signals from the golf cart. The adequate size, weight and cost of these kinds of cameras also validate this election. Another camera type eligible for the golf cart is the BIPcam model, since it offers good features for the vehicle automation, e.g. the detection of the lines of the route.

Communication System

To carry out the monitoring and remote distance control wirelessly, it is necessary to know what kind of communication system or platforms that meet their requirements are in the commercial market.

The communication system must be stable, have great reach, low energy consumption, low noise level, and elements that take up as little space as possible. Described below are some communication systems.

Wi-Fi Communication System

Wi-Fi (which means Wireless Fidelity) is the name of the certification granted by the Wi-FiAlliance, formerly WECA (Wireless Ethernet Compatibility Alliance). This organization is made up of several companies interested in promoting a common standard for the internet connections. It is also the creator of all the Wi-Fi standards since the IEEE (Institute of Electrical and Electronics Engineers) 802.11. This was the first standard and it was created in 1997. The now obsolete 802.11 supported a maximum speed of 2 Mbps. Currently, it has been replaced by the 802.11a, 802.11b, 802.11g and 802.11n standards, which are known as physical standards.

WiMax Wireless Communication System

Expanding ADSL services to rural or geographically inaccessible areas has not always been economically feasible, because only the wiring implies high costs. However, there is a continuous demand for access to the network in these zones, as in the majority of the world population. Such reasons encouraged the development of a wireless standard that reaches a greater number of users and promotes the introduction of new and better telecommunication services. This standard is WiMax (World Wide Interoperability for Microwave Access) and its most important characteristics are the following:

- Higher productivity within more distant ranges (up to 50 Km).
- Better bits/second/HZ rates in long distances.
- Scalable system.
- Easy channel addition: maximizes the capacity of the cells.
- Flexible bandwidths that allow the use of licensed and unlicensed spectrums.

ZigBee Wireless Communication System

ZigBee is a Wireless technology of short reach and low-consumption, whose protocol is IEEE 802.15.4. It originates from the former Alliance HomeRF and it was defined as a wireless solution of low capacity for home applications, such as security and automation.

This technology is not intended to reach high speeds, since it only has a rate of 20 to 250 Kbps in a range of 10 to 75 meters, but to attain sensors whose transceivers have a very low consumption. In fact, some devices with two AA batteries maintain their electrical energy supply for up to 2 years without the need of changing batteries. This is possible since such devices spend most of the time in a dormant state. In general, this technology is used in home automation, industrial automation, interactive toys and medicine.

Comparison of Wi-Fi, WiMAX and ZigBee

These three technologies possess features applicable to communication media, according to the requirements stated in this applied research study. In this relation, Table 2 contrasts the characteristics of the advantages and disadvantages of the Wi-Fi, WiMAX and ZigBee communication systems.

It is seen that ZigBee has short reach and low speed, so it is not the best option for sending data and video from the golf cart to the user; WiMAX, on the other hand, has great reach and high speed, in ad-

Table 2. Comparison of the different wireless communication technologies

Characteristics	Technology		
	Wi-Fi	**WiMax**	**ZigBee**
Reach	20 – 250 m	40 – 70 km	13 – 154 m
Speed	11 – 300 Mbit/s	124 Mbit/s	28 – 200 kbit/s
Frequency	2.4 – 5 GHz	2.4 – 3.5 – 5.8 GHz	0.86 – 0.91 – 2.4 GHz
Bandwidth	1.2 Mbit/s	0.64 – 1.024 Mbit/s	0.250 Mbit/s
Power	100 mW	100 mW	1 mW
Applications	Internet (portable computers, PDA)	Wideband long-reach service	Domotics, industrial automation, remote recognition, etc.
Advantages	Speed	Speed and reach	Low energy consumption
Disadvantages	Low reach	High price	Low reach and speed

dition to good bandwidth, making it an excellent option, but its high cost makes its purchase prohibitive; Wi-Fi has good speed (depending on the number of users), good bandwidth, but low reach, but that reach is considered from one antenna to the next, which means that if there are several antennas connected to a network in order to increase the coverage, the communication reach between the user and the golf cart increases considerably, and it would only be necessary for the user and the vehicle to be connected to the same network as the antennas. The Universidad de Santiago de Chile already has a network of this kind through which it provides internet to its employees and students, and its use would provide the vehicle with a good action range to move and remain communicated with the remote operator. If this is feasible, it implies purchasing a wireless router that satisfies the technical specifications of the USACH network and of the golf cart. A quite convenient option is to use a model WRT54GL Linksys wireless router.

Software for the Creation of Interfaces

To create a Graphic User Interface (GUI) it is necessary to make a survey of specialized applications in this field. Designing these interfaces requires a software that works at the object level and facilitates the use of windows, buttons and other kinds of elements that allow interaction with the operator in a simple and clear way. The most widely known and representative are Java, MatLab, and Visual Basic.

Java

The Java platform is based on the language oriented to the objects of the same name (very similar to C++). This platform permits the creation of GUI thanks to its AWT (Abstract Window Toolkit). It possesses a large set of user interface components, a robust event management model, graphic and image tools that include shape, color and font, layout handlers to work with flexible windows that not depend on a specific size or resolution and, finally, types of data transfer to copy and paste by using the clipboard of the platform in which an application is executed.

MATLAB

The MATLAB software offers a programming environment called GUIDE. This environment is very similar to Visual Basic and Visual C++ and permits easily designing and executing customized simulation programs. In addition, it has all the basic characteristics of the other visual programs, such as Visual Basic o Visual C++.

A GUIDE function comprises two files: one ".m" (executable) and other ".fig"(the graphic part). These two parts are joined through callback subroutines. Once the files are recorded from the broadcast console, the program may be executed in the MATLAB command window by simply writing the file's name. The just created ".m" file has a predetermined structure: first a header and then the code corresponding to the subroutines.

Visual Basic

The Visual Basic software aims at programing objects, including properties and methods.

The compiler of Visual Basic generates a code that requires the use of the Dynamic Link Libraries to operate. These DLL provide the functions implemented in the language and contain routines in executable code, which are charged on demand. In addition, it contains a great number of DLLs that enhance the access to the majority of the operating system functions and the integration with other applications.

In the Integrated Development Environment (IDE) of Visual Basic, a running program may be executed, i.e. the program is pseudo-compiled rapidly and, then, executed. In addition, IDE also allows the creation of an executable ".exe" of the specific program in order to use it outside the Visual Basic environment, as well as the generation of an installer module that contains the executable program and the DLLs necessary for it. Thanks to this module, the generated application is distributed and can be installed in any equip with a supported operating system.

IMPLEMENTATION OF THE MONITORING SYSTEM

Configuration and Connection of the Elements

The first device implemented is the Linksys WRT54GL router, whose function is to transmit and receive the data on board the golf cart. Its configuration in the repeater mode allows the router to repeat the signal emitted by some other nearby device or router with the purpose of allowing that signal to get to places further away from the original source. After the configuration of the router, the corresponding elements are connected to its output ports, as shown in Figure 9.

As to the remaining devices, i.e., voltage and current sensor, GPS, and servomotors, they must be connected directly to the Arduino Mega 2560 board ("Trojan Battery Company," 2015).

Figure 10 shows, in a general way, how all the elements that compose the designed and implemented system are connected.

Figure 9. Elements connected to the router via Ethernet
Source: Wireless Marketing Company Limited, n. d.

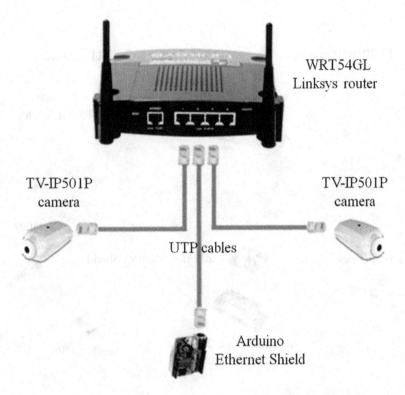

Implementation on Board the Golf Cart

The implemented video cameras are shown in Figure 11. Both cameras are supported by Pan/Tilt systems, designed and implemented with servomotors, which allow the cameras to cover a field of view greater than that covered by their fixed focus. A short distance away from the front video camera (Figure 11a) the GPS of the golf cart is seen.

In the case of the voltage and current sensor, it is connected between the eight batteries of the golf cart wired in series. Figure 12 shows the golf cart batteries together with the implemented autopilot sensor, shown in a circle, and its close-up.

To command the elements corresponding to the driving and signaling of the golf cart, a control system is implemented by means of an Arduino Mega 2560 board. This controls the lights, the horn, and the turn signals, which are located on the steering wheel column.

Figure 13 shows the location of all the elements implemented in the golf cart and its description is presented in Table 3.

Operator Interface

For monitoring the images and the golf cart's variables of interest, a very easy to use operator interface was designed, and it was implemented with HTML code using a Mozilla Firefox browser. An access code must be used to enter this interface. After accessing it, the interface displays the images captured

Figure 10. Connection scheme of the elements

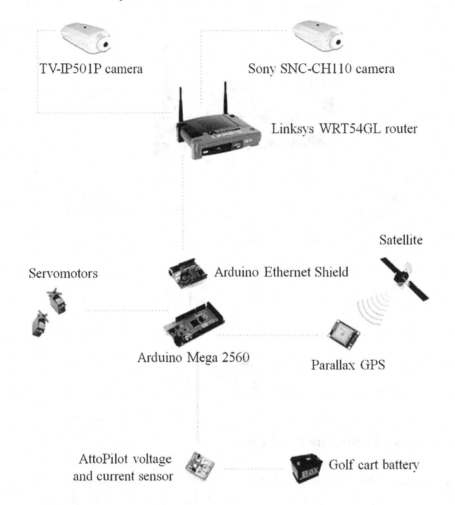

Figure 11. Cameras implemented on board the golf cart vehicle. a) Front camera. b) Rear camera

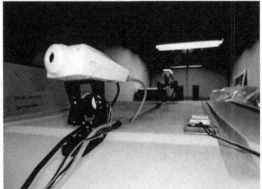

Figure 12. Voltage and current sensor implemented on board the golf cart

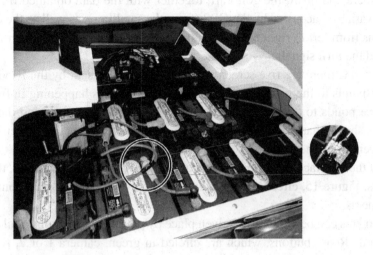

Figure 13. Location of the implemented elements
SOURCE: Beaudaniels-illustration, n. d.

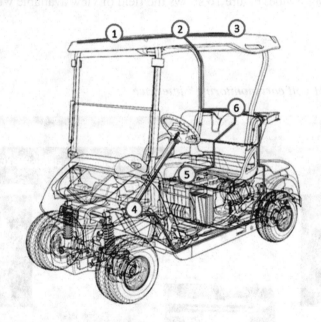

Table 3. Comparison of the different wireless communication technologies

Tag	Description
1	Front camera with standard GS-3630BB servomotor Pan/Tilt
2	GPS on mini protoboard
3	Rear camera with small servomotor Pan/Tilt
4	Relay box
5	Autopilot voltage and current sensor
6	Arduino Mega 2560, router, power source, and complements

by the IP video cameras on board the golf cart, together with the data obtained from the voltage and current sensor, the variables acquired by the GPS (from the satellites), as well as the commands needed to move the cameras from left to right or up and down, and send the signals for turning on and off the lights, the horn, and the turn signals.

As seen in Figure 14, there are five screens to show the images sent by the IP video cameras. The center screen corresponds to the main camera, which shows what is happening in front of the vehicle. The top screen corresponds to the camera that shows what is happening behind the vehicle. The two screens on the sides show what is happening on the sides of the vehicle. The screen on the bottom left shows what is happening on the vehicle.

At the bottom of the interface are the buttons meant to control the video cameras, the lights, the horn, and the turn signals. Figure 15, circled in green and red, shows the buttons that control the movement of the IP video cameras.

The IP video cameras can be pointed at a given place by pressing the buttons circled in red. Together with the "Front" and "Rear" buttons, which are circled in green, camera 1 or 2, respectively, can be chosen, and this choice appears as a number in the middle of the movement buttons.

It should be stated that, thanks to the Pan/Tilt systems, the viewing angle of the operator in the monitoring interface is very wide. Figure 16 shows the field of view available when scanning with any of the IP video cameras.

Figure 14. "DIE-USACH golf cart monitoring" interface

Figure 15. Buttons for moving the IP video cameras

Figure 16. a) Horizontal field of view. b) Vertical field of view

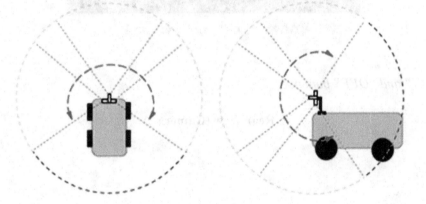

It is seen that by operating the servomotors of the IP video camera the user has a complete view of what is happening in front and on the sides of the golf cart, covering an angle of approximately 225 degrees over the horizontal plane. Similarly, the servomotor system allows scanning up and down, covering a Field of view equal to the previous one, but vertically. Also, a zooming system makes it possible to close-up on the captured images by pressing the "x1," "x2," "x3" and "x4" buttons which are below the screens that show the images from the IP video cameras.

For every camera an "ON" and an "OFF" button is included which allow turning on or off the sound around the golf cart detected by the microphones included in the cameras.

The command buttons area (Figure 19), shows the first three buttons, "Left," "Off," "Right," which operate the vehicle's turn signals. The following button, "Horn," activates the vehicle's horn; and the last two, "On" and "Off," turn the vehicle's lights on and off.

On the bottom right part of the interface there is a screen that constantly gives relevant information on the golf cart, such as the geographic coordinates, the voltage and current of the battery that feeds the cart, the course (in degrees), the speed at which it moves (in kilometers per hour) and the height above sea level (in meters):

On the bottom left part of the interface there is a button that says "Google Maps". When this button is pressed, on the top left of the screen a window opens with the Google Maps page. The purpose of including the Google Maps option in the monitoring interface is for the user, if he so desires, to enter the geographic coordinates of the vehicle's location and in that way know in what place in the planet it is found.

Figure 17. Zoom buttons

Figure 18. "ON" and "OFF" buttons

Figure 19. Buttons "Left," "Off," "Right," "Horn," "On" and "Off"

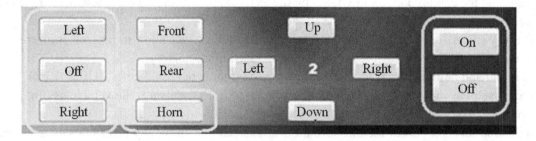

Operation and Startup Tests

After all the implementations aboard the vehicle had been made, the general parameters of the whole system were adjusted, ensuring their proper operation. From various operating tests carried out, satisfactory results were obtained controlling remotely the general lights, the turn signals, the horn and

Figure 20. Sensed variables in the golf cart

Lat/Long	:	-33.55339,-70.61756
Altitude (meters)	:	600.70
Course (degrees)	:	90.77
Speed (km/h)	:	0.00
Voltage (volts)	:	49.00
Current (amps)	:	0.00

the servomotors, all this with practically immediate results. The images from the IP video cameras and the data from the voltage and current sensor were also received remotely. However, there were short information delivery delays from the IP video cameras of less than 0.4 seconds. With respect to the router's operating tests, they were even more satisfactory, because the information delivery delays were less than 0.2 seconds.

Finally, and after getting all these satisfactory results, the golf cart was monitored wirelessly while it was moving, keeping, for safety reasons, an adequate distance from human beings. Figure 21 shows a picture of the golf cart moving, and a notebook from which the different elements were monitored and controlled satisfactorily.

FUTURE RESEARCH DIRECTIONS

This project implies the possible development of further work. On the one hand, the system implemented on the golf cart might be improved, making it more robust in general. More sensors could be added to create a network in which they communicate with one another and are commanded by a central control-

Figure 21. Tests of the general system

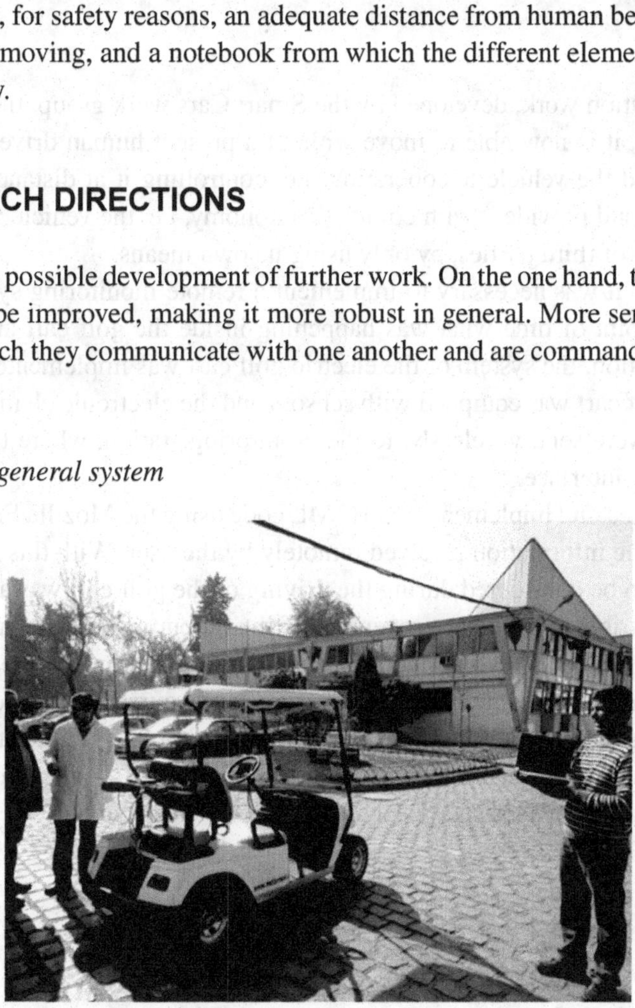

ler. Thanks to this measure, a completely automated golf cart vehicle would be obtained. This vehicle could also be equipped with "artificial intelligence" to deal with situations that are more complex. Furthermore, more video cameras could be implemented in the golf cart to obtain an even more elaborated vision of the mobile surroundings.

On the other hand, if more artificial vision elements were included in the vehicle, objects, distances and signals present in the dynamic environments across which the mobile moves could be recognized in a completely automatized way.

Regarding the interface developed in HTML, a new Web interface that offers more interconnectivity services might be designed based on it. Consequently, a new interface could be available in the Internet and be used practically from any location in the world, allowing one to know the state of the diverse variables related to the golf cart, for example, the images captured by the video cameras and the general environment.

Finally, the versatility of the developed system has the advantage that it might be extrapolated to multiple applications. Thus, through the improvement and introduction of other features, such system could be implemented in environments such as the mining or transport industries in general.

CONCLUSION

By means of this innovation work, developed by the Smart Cars work group, the electric golf cart was given autonomy, that is, it is now able to move without a present human driver. The project had two stages: the first achieved the vehicle teleoperation, i.e. controlling it at distance in the way a human driver does, and the second provide it with complete autonomy, i.e. the vehicle was capable of moving without the intervention of third parties, by only using its own means.

To achieve this goal, it was necessary to implement a remote monitoring system in the vehicle, in order to know at any point of time what was happening inside the golf cart and in its surroundings. After a technical evaluation, the system of the electric golf cart was implemented satisfactorily. In this implementation, the golf cart was equipped with sensors and the electronic elements needed for acquiring video images that were sent, wirelessly, to the monitoring station, where they were presented by means of a user-friendly interface.

The interface, designed and implemented in HTML code using the Mozilla Firefox browser, allowed easy understanding of the information received remotely by the user. With this interface, access to the important information to be considered during the driving of the golf cart was achieved.

Thus, this work contributed to the improvement of the autonomy and security systems associated with vehicles in general, since a great number of accidents occurring in this area are caused by human error, e.g. due to the tiredness generated after excessive driving hours. These type of monitored vehicles will help to reduce the latent risks always present in mining and in tasks related to driving. This may be translated into an increase in the security, comfort and quality of life of people in general.

All the implemented system's tests gave excellent performance results.

ACKNOWLEDGMENT

This work has been supported by Proyectos Basales and the Vicerrectoría de Investigación, Desarrollo e Innovación (VRIDEI), Universidad de Santiago de Chile, Chile.

REFERENCES

Angulo de vision. (n.d.). In *Wikipedia*. Retrieved from https://es.wikipedia.org/wiki/%C3%81ngulo_de_visi%C3%B3n

Artilec. (n.d.). *Artilec Seguridad Electrónica*. Retrieved June 07, 2012, from http://www.artilec.cl/site/productos.php?r=01&s=5

Cámara inteligente. (n.d.). In *Wikipedia*. Retrieved from https://es.wikipedia.org/wiki/C%C3%A1mara_in-inteligente

CámarasIP. (n.d.). *CámarasIP*. Retrieved June 07, 2012, from http://www.camarasip.cl/es/

Carnegie Mellon University. (2007). *Embedded vision processor, CMUcam3 datasheet* [Data file]. Retrieved from www.cmucam.org/.../48/CMUcam3_datasheet.pdf

Castejón, L., Miravete, A., & Cuartero, J. (2006). Composite bus rollover simulation and testing. *International Journal of Heavy Vehicle Systems*, *13*(4), 281–297. doi:10.1504/IJHVS.2006.010584

Chat, I. R. C. (n.d.). In *Wikipedia*. Retrieved from http://es.wikipedia.org/wiki/C%C3%A1mara_inteligente

Chung, W. Y., & Oh, S. J. (2006). Remote monitoring system with wireless sensors module for room environment. *Sensors and Actuators. B, Chemical*, *113*(1), 64–70. doi:10.1016/j.snb.2005.02.023

Danese, G., Giachero, M., Leporati, F., Nazzicari, N., & Nobis, M. (2008). *An embedded acquisition system for remote monitoring of tire status in F1 race cars through thermal images*. Paper presented at 11thEuromicro Conference on Digital System Design, Parma, Italy. doi:10.1109/DSD.2008.55

Daniels, B., & Daniels, A. (n.d.). *Beaudaniels-illustration*. Retrieved from http://www.beaudaniels-illustration.com/technical-drawing-site-2/Cutaway.html

Electrónica. (n.d.). In *Encyclopedia Libre Wikipedia*. Retrieved from http://es.wikipedia.org/wiki/Electr%C3%B3nica

Frame rate#Visible frame rate. (n.d.). In *Wikipedia*. Retrieved from https://en.wikipedia.org/wiki/Frame_rate#Visible_frame_rate

Glasgow, H. B., Burkholder, J. M., Reed, R. E., Lewitus, A. J., & Kleinman, J. E. (2004). Real-time remote monitoring of water quality: A review of current applications, and advancements in sensor, telemetry, and computing technologies. *Journal of Experimental Marine Biology and Ecology*, *300*(1–2), 409–448. doi:10.1016/j.jembe.2004.02.022

Gobar, L. T. D. A. (n.d.). *Gobar, Seguridad - Electronica Electricidad*. Retrieved June 07, 2012, from http://www.gobar.cl/?ver=detalle&id=1142

Guglielmo Marconi. (n.d.). In *Encyclopedia Asifunciona online*. Retrieved from http://www.asifunciona. com/biografias/marconi/marconi.htm

Kantharia, P., Patel, T., & Thakker, M. (2014). Design of sensor fault detection and remote monitoring system for temperature measurement. *International Journal of Current Engineering and Technology*, *4*(2), 504–508.

Li, S., Tian, J., Yang, Z., & Qiao, F. (2013). Research and implement of remote vehicle monitoring and early-warning system based on GPS/GPRS. In *Proceedings of SPIE 8768- International Conference on Graphic and Image Processing (ICGIP 2012)*. Singapore: SPIE. doi:10.1117/12.2010751

Mainville Electric Motor Co. (n.d.). *Motores de carro de golf*. Retrieved from http://www.manvillemotor. com/golf_cart_motors_spanish.htm

Monitorización. (n.d.). In *Encyclopedia Libre Wikipedia*. Retrieved June 09, 2011, from http://es.wikipedia. org/wiki/Monitorizaci%C3%B3n

Motorman, S. A. (n. d.). *Motorman*. Retrieved April 01, 2016, from http://www.motorman.cl/

Rodríguez, S. (n.d.). *Masadelante.com, servicios y recursos para tener éxito en internet*. Retrieved from http://www.masadelante.com/faqs/resolucion

Samsung Techwin Co. Ltd. (2010). *High resolution day & night camera SCB-2000, specifications* [Data file]. Retrieved from http://www.hanwhatechwinamerica.com/SAMSUNG/upload/Product_Specifica-tions/SCB-2000_Datasheet.pdf

Satyanarayana, G. V., & Mazaruddin, S. D. (2013). *Wireless sensor based remote monitoring system for agriculture using ZigBee and GPS*. Paper presented at Conference on Advances in Communication and Control Systems 2013 (CAC2S 2013), India.

Trendnet. (2009). *ProView poe internet camera tv-ip501p, specifications* [Data file]. Retrieved from http://downloads.trendnet.com/tv-ip501p/datasheet/en_spec_tv-ip501p(v1.0r)-091009.pdf

Trendnet. (n.d.). *Trendnet*. Retrieved April 01, 2016, from https://www.trendnet.com/langsp/products/ proddetail?prod=165_TV-IP501P

Trojan Battery Company. (2015). *T-605 datasheet* [Data file]. Retrieved from http://www.trojanbattery. com/pdf/datasheets/T605_Trojan_Data_Sheets.pdf

Wireless Marketing Company Limited. (2011). *Linksys WRT54GL*. Retrieved from http://www.talad-wireless.com/product_info.php?products_id=265

Wu, H., Liu, L., & Yuan, X. (2010). *Remote monitoring system of mine vehicle based on wireless sensor network*. Paper presented at 2010 International Conference on Intelligent Computation Technology and Automation, Changsha, China. doi:10.1109/ICICTA.2010.209

ADDITIONAL READING

Madiña, Y., & Leonardo, A. (2010). *Diseño, construcción y programación de un robot móvil, controlado mediante visión artificial.* Santiago, Chile: Universidad de Santiago de Chile.

Mosqueira, R. J. (2010). *Detección de vehículos y pistas en carreteras utilizando visión artificial.* Santiago, Chile: Universidad de Santiago de Chile.

Osorio, L. (n. d.). Diseño e implementación del control remoto de un móvil con retroalimentación visual internet y JAVA. *Universidad de Santiago de Chile, Santiago, Chile.*

Siegwart, R., Nourbakhsh, I., & Scaramuzza, D. (2004). *Introduction to autonomous mobile robots* (2nd ed.). London, England: The MIT Press.

Vargas, M. A. (2008). *Sistema de monitoreo y control remoto para una central micro-hidráulica.* Santiago, Chile: Universidad de Chile.

KEY TERMS AND DEFINITIONS

Frames-Per-Second: Speed at which a device shows images called frames.

Interface: Virtual environment useful for linking devices or systems.

Resolution: Indicated the quantity of details that can be observed in an image. The value is expressed in pixels.

Single-Seaters: Vehicle or airplane for one person.

Teleoperation: Technical Word used to indicate the remote control of a machine at distance.

User-Friendly: Easy to use or understand, referred to a machine or system.

Wireless: Type of communication to transfer information between two points that are not connected by an electrical conductor.

ENDNOTE

[1]   Carnegie Mellon University (CMU). Located in Pittsburgh (Pennsylvania), it is one of the most renowned higher research centers on informatics and robotics in the United Stated.

Chapter 11

Intelligent Traffic Monitoring System through Auto and Manual Controlling using PC and Android Application

Paromita Roy
*Bengal College of Engineering and Technology,
India*

Amira S. Ashour
Tanta University, Egypt

Nilanjan Dey
Techno India College of Technology, India

Nivedita Patra
*Bengal College of Engineering and Technology,
India*

Satya Priya Biswas
*Bengal College of Engineering and Technology,
India*

Amartya Mukherjee
IEM Kolkata, India

ABSTRACT

Traffic congestion in cities is a major problem mainly in developing countries. In order to counter this, many models of traffic systems have been proposed by different scholars. Different ways have been proposed to make the traffic system smarter, reliable and robust. A model is proposed to develop an Intelligent Traffic Monitoring System (ITMS) which uses infrared proximity sensors and a centrally placed microcontroller and uses vehicular length along a lane to implement auto controlling of the traffic. The model also provides mean to control the traffic manually through a PC software and an Android application.

INTRODUCTION

Road network is the lifeline in all cities. Nowadays, traffic jam becomes a daily life problem in any metropolitan city. It is a circumstance on roads where lots of vehicles are stuck, thus they are either moving very slowly or unable to make further movement for few minutes or may be hours. Underlying reasons of such traffic are due to the exponential rate increase in the number of vehicles, poor traffic manage-

DOI: 10.4018/978-1-5225-1022-2.ch011

ment, etc., whereas roads do not expand proportionally with the increased number of cars. Governments and corporations try to work on the threats of traffic congestion. However, many obstacles including poor public transport (i.e., absence of enough number of buses, trains, trams, etc., which leads to the necessity of traveling by individuals own cars. Also, numbers of unlicensed or fake licensed drivers are increasing daily. They know a little about traffic rules. Therefore, the tendency of overtaking and parking anywhere makes the situation of traffic a disaster. As a result, it is impossible for people to reach their destinations in time as well as leads to increase the number of accidents. Cities become inappropriate for people to live because of the increasing rate of pollution. Therefore, solving such problems requires monitoring the road network resources.

Most of the existing systems follow simple round robin algorithm to assign traffic lights on the lanes which is based on a fixed time quantum. Therefore, all the lanes in a junction are treated with equal priority irrespective of the number of vehicles present in each lane. Consequently, lanes with less or more traffic have to wait for the same time span. This useless waiting destroys ones valuable time. In order to control the lanes manually, it is necessary for the traffic police to have the entire view of the junction. However, it is quite difficult for the police who are standing in the middle of the junction. So, achieving the real time view is unfeasible in the existing systems. Additionally, due to the absence of traffic police at night it is impossible to supervise roads manually.

Currently, in order to improve this scenario many researches are interested to develop an Intelligent Traffic System (ITS) that involved in a much closer interaction with all the components of a traffic including vehicles, drivers and even pedestrian. It provides safety at intersections, and prevents traffic jam as well as manages the traffic as a whole.

Consequently, the proposed model follows an algorithm based on the length of traffic on each lane. Since, the length of traffic on the other lanes affects the time allotted to the current lane. Therefore, in the current work, proximity sensors are used instead of the wireless area network (WAN). These sensors are used to determine the length of the traffic. The proposed system can reduce the traffic in all lanes proportionately. In addition, the proposed system is quite cheap with respect to the systems used in developed countries where video recorders such as the closed-circuit television (CCTVs) that installed to monitor the lanes. Once the proposed system is implemented, it does not require any human assistance. However, the system can also be controlled manually in two ways:

1. By traffic police in the traffic control room by going through the sensor values in the computer and therefore gives the control to the respective lanes as required, or by
2. Controlling the system by using android application, which is loaded on the phone of the traffic police. All the communications is done via Bluetooth connection.

The operator first sets up the connection with the microcontroller by authenticating himself, then he can manually control the system. There are some cases when manual control is required like road constructions, and arrival of any minister's convoy that the proposed system can manage.

RELATED WORK

Developed countries have already implemented traffic systems on their roads and still many researches are going on to make traffic systems more advanced and suitable. Traffic systems are being researched and implemented through various means such as the use of wireless sensor networks and Radio frequency identification (RFID) that apply various concepts of graph theory to find the shortest path. In some cases, traffic managing system take the input of the current situation through video surveillance or WSNs and deal with the situation. The traffic signals are controlled according to the presence of vehicles and are operated automatically in real time. Some systems reduce the traffic congestion by the realization of the traffic density at a particular intersection for a given time. This data can be analysed to determine several factors such as the green light length, and the traffic at the particular time. Also, finding the best and shortest path to the destination can be used as a tool to minimize the traffic along a path. The traffic along the road can be sent to the incoming vehicle proving them the idea about the traffic and thus they can take an alternative path to the destination. Some systems take the present or statistical data and process them in the processor and then act according to a predefined algorithm. Some systems uses the data collected at one junction and are able to send to the other junctions informing them about the situation and allowing them to take measures. The same can be used in case of cars, ambulance and other vehicles. Several attempts were applied to optimise the length of the green light as it is the main causes of traffic congestion and leads to large red light delays.

Dotolie *et al.* (2003) investigated the traffic control issue in the urban areas. The model considered the traffic scenarios which also include pedestrians. This technique was applied for analysing real case studies. Wenjie *et al.* (2005) concentrated on calculating the time that a vehicle requires to reach the intersection from a particular point dynamically by the use of sensors. Various calculations were used to find the green light length. Queen and Albers (2007) used real time data to monitor current traffic flows in a junction, so that the traffic could be controlled in a convenient way. Reliable short-term forecasting video captured in a recorder plays an important role in monitoring the traffic management system. The data required can be easily provided by the CCTV cameras that can be beside the roads as per requirement. Tubaishat *et al.* (2007) described the traffic control on a real time basis using the traffic lights. Wireless sensors are deployed on each of the lanes that are able to detect number of vehicles passing and also the awaited vehicles and convey the information to the nearest control station. Chen *et al.* (2007) proposed a solution for minimising waiting time of vehicles by testing the setting problems of traffic light. The graph model was used to represent the traffic network. Van Daniker (2009) visualized the use of Transportation Incident Management Explorer (TIME) for calculating real time data. Ozkurt and Camci (2009) have proposed the use of video surveillance and neural network to reduce the traffic stress across the network.

Yousef *et al.* (2010) suggested a scheme of solving traffic congestion in terms of the average waiting time and length of the queue at the isolated intersection and provide efficient flow in global traffic control on multiple intersections with the accordance of real time data. Thus the data collected can be used in various ways depending perspective of the user.

Binbin Zhou *et al.* (2010) used the concept of adaptive traffic light control algorithm which manipulates both the sequence and length of traffic lights in accordance with the detected traffic. The algorithm used real time data like the waiting time of vehicles, and volume of traffic in each lane to determine the traffic light sequence and optimal length of green light. The algorithm used to lower the vehicle's average waiting time, thus providing much higher throughput. Challal *et al.* (2011) proposed a distributed wireless network of vehicular sensors to get a view of the actual scenario and used its various sectors to lower the congestion without taking decisions in real time. The use of two types of sensor network was proposed, namely Vehicular Sensor Network and Wireless Sensor Network. The combination of these two types permits the monitoring as well as managing of the traffic.

The system proposed by Sinhmar (2012) used infra-red (IR) sensors to determine the density of traffic based on which the traffic signals were updated to provide a smooth flow of vehicles. Kafi *et al.* (2012) used video surveillance. It was used with decreasing response time of the emergency cars by establishing communication between emergency cars and traffic lights. The data collected in real time can be used to determine the traffic density and also based on the traffic present. Srivastava *et al.* (2012) suggested ways to determine the number of vehicles using weight sensors, then with the use of a programmable logic controller to analyse the data and then park in automated parking or has diverge them accordingly. Hussain *et al.* (2013) proposed a system that uses a central microcontroller at every junction which received data wireless sensor placed along the road that determines the traffic density. The microcontroller used this data to control the traffic using the programmed algorithm to manage the traffic in an efficient manner. Kale and Dhok (2013) designed a system that uses the traffic information and sends it to the incoming ambulance by allowing it take way according to the situation. The various performance evaluation criteria are used such as average waiting time, the average distance travelled by vehicles and switching frequency of green light at a junction.

An optimized path for transportation using the concept of Ant Colony Optimization was proposed by Bertelle *et al.* (2003) and Di Caro *et al.* (2005). Once an optimized path is determined, several other features can be added to make the system more convenient and avoid traffic jams.

In order to achieve optimal solution, optimization algorithms such as Particle Swarm Optimisation, Ant colony Optimization and the Genetic Algorithms that used in several applications (Dey *et al.*, 2013a; 2013b; Samanta *et al.*, 2013a; 2013b; Chakraborty *et al.*, 2013; Dey *et al.*, 2014; Chakrabarty *et al.*, 2014; Kaliannan *et al.*, 2015) can be used to solve the traffic problem. Soh *et al.* (2010) presented a MATLAB simulation of fuzzy traffic controller for controlling traffic flow in the multilane isolated signalized intersection. The controller controls the traffic light timings and phase sequence to ensure smooth flow of traffic with minimal waiting time, queue length & delay time.

Kareem and Jantan (2011) proposed monitoring system in addition to the traffic light system to determine different street cases (e.g. empty, normal, crowded) with different weather conditions by using small associative memory depending on the stream of images, which are extracted from the streets video recorders. It provided a high flexibility to learn different street cases using different training images. Płaczek (2011) described a method which is designed to be implemented in an on-line simulation environment that enables optimization of adaptive traffic control strategies. Performance measures are computed using a fuzzy cellular traffic model, formulated as a hybrid system combining both cellular automata and fuzzy calculus. Dakhole and Moon (2013) used ARM7 based traffic control system that proposes a multiple traffic light control and monitoring system that reduce the possibilities of traffic jams, caused by traffic lights. This system uses ATmega16 and ARM7 for its processing.

MAIN FOCUS OF THE CHAPTER

System Requirements

There are various system requirements including resistors, breadboard, Vero board, connecting wires, and led lights. In addition, the following requirements are employed.

IR Proximity Sensors

The IR proximity sensors are used in the ITMS to detect the point up to which vehicles are present in a road. The sensors consist of a photodiode, an infrared emitter, LM358 op-amp chip, variable resistor, resistors and connecting wires. Once any obstacle (vehicle) comes in front of the sensors, the infrared waves emitted by the infrared led is reflected back, which causes the resistance of the photodiode, connected in reverse bias reach a finite value. This acts as a threshold value of the LM358 chip, which acts as a comparator circuit, and compares it with the input from the variable resistor. Whenever, the drop across the sensor is low than the variable resistor the output goes high indicating the presence of an obstacle near the sensor. The IR sensor module is operated with 5V DC supply.

HC05 Bluetooth Module

The HC05 is a Bluetooth serial port protocol module used to setup transparent and wireless communication. The ITMS use the HC05 to setup connection with the ITMS android application and the Arduino. The HC05 operates on 3.3V DC 50 mA input. It has Bluetooth V2.0+EDR (Enhanced Data Rate) 3Mbps modulation with complete 2.4GHz radio transceiver and baseband and uses CSR Blue core 04-External single chip Bluetooth system with AFH (Adaptive Frequency Hopping Feature) and CMOS technology. TX pin is connected to the RX pin of the Arduino and vice versa.

Arduino – Microcontroller Board

Arduino is an open source hardware/software company that designs kits for manufacturing various digital devices. It is mainly used by students for prototyping and enables the creation of devices that interacts with the environment with sensors and actuators. The board contains a centrally located microcontroller, an USB (Universal Serial Bus) are provided to load programs to the microcontroller. The board provides various digital and analog pins to interface with the external devices. There are various Arduino boards that are available in the market each having its own unique features and use. The ITMS uses Arduino Mega 2560. It contains the Atmega 2560 microcontroller with 16 analog pins and 54 digital pins. Among 54 digital pins 15 can be used as PWM pins. A range between 5V to 20V 1A DC supply 9V is recommended for optimal use. It has a Flash Memory of 256 KB of which 8 KB used by boot loader, SRAM of 8 KB, EEPROM of 4 KB and has a clock speed of 16 MHz. The ITMS prototype, containing 2 sensors in each road, to avoid interference of the sensors as the area is less for the prototype, uses 8 analog pins to receive the output from the sensors and 12 digital pins to control the traffic led. The TX digital pin 1 and RX pin 0 is used for connecting with the HC05 Bluetooth module.

Eclipse

The Android application of ITMS is designed using Eclipse. Eclipse is an IDE used for developing applications using various languages like JAVA, Android, J2EE, PHP and Perl. It was released under the terms of the Eclipse Public License, Eclipse SDK is free and open source software The graphical representation of the ITMS app, how the app will work is developed using Eclipse. Establishing Bluetooth connection, authentication and serial communication with the Arduino and all the other controls is developed by Eclipse.

Processing

Processing is an open source programming language and IDE used for developing graphical user interface for various electronics contents. It was developed by Casey Reas and Benjamin Fry in 2001. The stable version of it is Processing 2.2.1. The language builds on JAVA and uses .pde filename extensions. It can be downloaded free from the official website processing.org. The ITMS PC software has been developed using Processing. The whole graphical user interface (GUI) of the software is developed. It is used to communicate with the Arduino from the GUI, better to say used to "talk" with the Arduino.

Arduino – IDE

The programs which are uploaded to the Arduino microcontroller are written in C or C++ in the Arduino IDE, which is a cross-platform application written in Java, and derives from the IDE for the Processing programming language and the Wiring projects. It provides an interactive text editor for writing programs, compiling and uploading it to the board. A program code written in Arduino is called a sketch and saved with the filename extension .ino. The programmers need to define two functions setup() and loop() for continuous execution of any program. The function setup() is used to initialize settings and runs once at the start of the program and the function loop() is called repeatedly until the board powers off. The algorithm defined in ITMS is loaded into the board after translating into code, the Firmata library is used to enable the controlling through the ITMS PC software.

Proposed Methodology

The proposed model has two modes:

1.   Auto Control, and
2.   Manual Control.

Auto Control

The auto controlling module acts as the core of the ITMS. It enables the system to manage the traffic without any human support in an efficient manner. The infrared proximity sensors are used not as counters but as indicators that up to which point of a lane the vehicles are presented. There are basically four sensors placed along each lane. The first two are placed at the near end of the junction the other two at far from junction. Figure 1 illustrates the sensors location.

In Figure 1, A, B, C and D indicate the sensors. Four sensors are used to provide a clear idea about the traffic in the lane; more sensors can be placed to get a more accurate data. However, for the sake of simplicity, only four sensors are used in the proposed system. The distance between the same side sensors A and B depends upon the road, where the system is being implemented, and the distance between two oppositely A and C placed sensors should be such that their area of coverage do not intersect with each other resulting in erroneous data. The schematic view of the proposed model is demonstrated in Figure 2.

The on/off states of the sensors are used to mark the presence of the vehicles along the road. The indications of the sensors are interpreted by the microcontroller and are used as the sole parameter in making the decision that when a lane will be executed and for how long. If the sensors A and B are

Figure 1. Location of the sensors

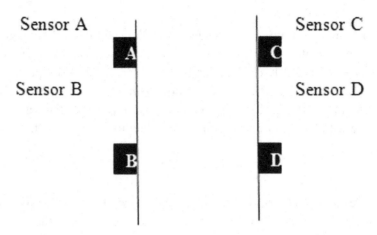

Figure 2. Schematic View of the proposed model

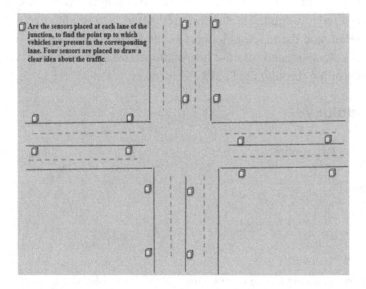

marked high, then it indicates that vehicles are present only along one side of the road and are intended to go either straight or in left direction. The same is applied for C and D in the right direction. When A and C are marked high, it indicated that vehicles are present only at the initial point and intend to go straight or in either direction, only B and D or B or D alone cannot be high as it will block the road, it is a unreal condition.

This data is used to determine the execution time of a lane as well as the way of execution of the time slot provided to that lane. The readings of the sensors are analysed by the microcontroller. Afterward, the priority is assigned to each lane, based on which the further decision making takes place. Table 1 lists how the priorities are set based on the readings of the sensors. In the case of all the sensors are high, it indicated that the road is full and is marked as with the highest priority. If the sensors of only one side and the initial of the opposite side are high, it is marked with second highest priority. If only the initial

Table 1. Priorities based on Sensors values of the lane

Sensor A	Sensor B	Sensor C	Sensor D	Priority
High	High	High	High	6 (HGHEST)
High	High	High	Low	5
High	Low	High	High	5
High	High	Low	Low	4
Low	Low	High	High	4
High	Low	High	Low	3
High	Low	Low	Low	2
Low	Low	High	Low	2
Low	High	Low	High	1
Low	High	Low	Low	1
Low	Low	Low	High	1
Low	Low	Low	Low	0 (LOWEST)

sensors are high, it is marked with a priority less than the previous case. If only one initial sensor is high, it is assigned with a priority less than others. Lastly, if none of the sensors is high priority, it is assigned to zero. The cases where only the sensors at final end are high are provided with some priority less than others but greater than zero, considering an incoming vehicle in the road.

Based on this data, the defined algorithm operates the traffic. The algorithm operates in three steps, namely Sense_and_Sort, Sense_Again, and Execute. The algorithms used in the proposed system are as follows.

Algorithm for the Control of Traffic Lights

The proposed algorithm initially senses the vehicular length of each lane and sets its priority and stores it into the array. The sequence in which the lanes are sorted, will be executed in this sequence only. Sense_and_Sort step checks the length of the vehicles and sets their time accordingly, also keeps a check that the next lane with lower priority initially may have acquired a higher priority than its preceding lane. In such case, the green light duration Ti that is provided to the present lane is decreased. The position of the array is decreased after execution of each lane. Once the pointer reach at the end of the array, the lanes are once again stored into the array according to priority and executed accordingly. (See Box 1)

In the first step, each lane meeting in the junction is sensed; their priorities are assigned to each of them based on the signal received from the sensors. Then, they are sorted according to their priority. Currently, according to the sorted sequence, the lanes will be processed. It is significant to note that no change will occur in this sequence till all the lanes are executed, no matter how the traffic changes in a lane while execution, the sequence of execution remains the same. This prevents the starvation of any lane due to high traffic in other lanes. Once the cycle is complete again the lanes are sensed and sorted. The sorted sequence acts as an input to the next step.

Afterward, the Sense_Again is performed. This step works in two sub parts as follows. The first part takes the lanes one by one in the decreasing order of their priority and once again sense the lane, now

Box 1.

Control Algorithm

Create an array: P_ARRAY [4]: Array to store the lane according to priority
Assign: Green Time Ti to the lane
Assign Priority: P_i, P_{i-1} to the current lane and the next lane
Sensing and sorting the P_ARRAY:
While (true) repeat Sense_and_Sort ()
　　While (Length.P_ARRAY not equal to 0) repeat
　　Sense_Again (P_ARRAY[i]): sense the length of the lane and set the priority for the lane in the P_ARRAY[i] and setting the green light time of the lane according to the priority assign to the lane in the P_ARRAY.
Execute the green Light: Execute (P_ARRAY[i], T_i) of the lane according to the time assign for each lane in the P_ARRAY[i]
End

Sense_and_Sort Algorithm

Sense each lane and prioritize them
Sort the lane according to their priority into P_ARRAY, the lane at the end of the P_ARRAY has maximum priority
End
Sense_Again (P_ARRAY[i])
Sense the priority of the current lane and the next lane
P_i= Priority of the current lane.
If (i=0)
P_{i-1} =0.
Else
P_{i-1} = Priority of the next lane.
Set T_i according to P_i
If ($P_i < P_{i-1}$)
Indicate that the vehicle length has increased after setting the P_ARRAY[i]
Update T_i
End

Execution Algorithm
Execute (P_ARRAY[i], T_i) *Set* the green light for the lane of the P_ARRAY[i] for time T_i *Set* the yellow light for the lane next of the P_ARRAY for time T_i indicating that it will be executed next and red to the other two lanes. *End*

the priority is judged again and based on the current priority, time is assigned to the lane. It may so happen that during the execution of the previous lanes the traffic in a lane with initial low priority has increased, but the sequence is to be maintained, to counter such situation the second part of this step sense the lane that is to be executed next and checks its current priority, if the priority of the lane which is to be executed is less than the next lane, the time assigned to the current lane is decreased to lower the waiting time of the next lane.

The Execute step sets the green light of the lane to be executed and others red. It also set the yellow light on for the vehicles to be executed next, as an indication. At the end of the cycle there exists no "next lane" in that cycle; red light is shown to all the lanes except from the executing lane. In this way, the whole algorithm works, after a cycle is completed another cycle is formed, this way of execution guarantees the execution of all lanes and allocation of time according to the traffic present in the lane. The order of execution depends only on the initial traffic, while the time of execution depends on the current traffic on the lane and the traffic of the lane to be executed next.

Manual Control

Auto control of ITMS uses the vehicular length in any lane as the primary parameter to determine the order and execution time of any lane. Situations may arise where the traffic could not be managed on the basis of vehicular length in each lane only. Since, the algorithm defined for automation does not take into consideration the occurrence of "Priority Vehicles" such as ambulance, fire brigade or any kind of unwanted situations. Sometimes the roads needs to be blocked for any construction works or some accident occurred at any place. The traffic also needs to be handled manually when any VIP cars will pass by. For these type of circumstances the traffic monitoring system ITMS provides means of controlling the traffic in manually through two efficient user interfaces, namely the Control using PC, and the Control using Android smartphone.

Controlling the traffic from the junction is a very hectic job especially in extreme weather conditions. The feature of controlling the traffic via PC not only eliminates this factor but also other issued faced in controlling the traffic standing at the junction. The ITMS PC software enables the traffic constable to control the traffic by sitting in a favourable place and still controlling the traffic in an efficient way. The software provides a simple and interactive user interface for the required purpose. It provides button for each lane at a junction, on pressing a button the green light of the corresponding lane goes high and red is shown to the other lanes, a timer is also provided to show the time for which the lane is on (i.e. the green light is on).

Once, a button is clicked, the timer is set, and an indication through a text "THIS LANE IS ON" is flashed in green on the screen till the lane is in on state. The information about the traffic in a lane is indicated by green and red marked text boxes in the screen corresponding to the sensors present at the road. Whenever, a sensor is high (i.e. there is vehicle up to that point in the lane) the text box corresponding to that goes green and the opposite is indicated by red mark. This provides the controller (Traffic police) by an abstract view of the traffic condition at the junction. Now, the controller can control the traffic just by a click on the computer screen comfortably sitting at a suitable place under the fan or AC.

The ITMS software is written in JAVA using Processing 2.2.1 as the platform. The details of Processing 2.2.1 is discussed in section 2.2. Two packages org.firmata.* and cc.arduino.* are used while writing the software to connect it with the Arduino, which uses the StandardInputOutput Firmata to translate the instruction sent from the software through serial communication. All the instructions sent to the Arduino from the software are sent through the Arduino class, where all the controls required for controlling Arduino is predefined. Like to make a pin high, suppose 13, the code is written as ard.digitalWrite(13,Arduino. HIGH); where "ard " is the object of the class Arduino defined in the program, similarly for analog pins we write ard.analogRead(13). It is remembered during coding that right port selection is also necessary to establish the connection, the port is mentioned while the object ard is initialized.

On spot control of the traffic is enabled through the ITMS Android Application that developed using Eclipse platform. The application (app) communicates with the Arduino through serial communication over Bluetooth. The app is password protected and the password is stored in Arduino instead of being in the phone memory. This is done to prevent mishandling of the application. The password is known only to the traffic authority and cannot be changed by any unauthorized person. The app provides similar controls as the ITMS software. After successful login, the main screen appears that contains the various controls required for controlling the traffic signals. The sensor data is omitted over here as the traffic authority needs to be present at the junction for controlling. Bluetooth module HC05 is used to establish the communication between the app and Arduino. When a button is pressed on the app sends a unique value to the Arduino, which then executes the code defined under that value.

The values sent to the Arduino is received by Serial.read() command and for receiving any string the Serial.readString() command is used. Similarly, to write any value back to the serial port Serial.write() command is used. The user interface (UI) of the android is kept simple, so that it can be easily handled. In order to start any connections whether with PC or mobile, the user needs to set the switch corresponding to the control, which indicates the microcontroller which part of the code is to be executed.

Proper synchronization is required for smooth running of the programs. Whenever, any transitions takes place all the red lights in the lane goes high waiting till the proper instruction is received. If both the switches are set low then the auto controlling part is executed. The users must turn on the switches while controlling and off while leaving the control to avoid any type malfunctioning of the system. It is to be noted that in case of manual control the yellow light always remains off, as it only depends upon the controller which lane to be executed next.

Results and Discussion

Figures 3 and 4 depict the prototype of the ITMS in different angles.

Figures 3 and 4 illustrate that in the prototype there are only two sensors present along each road, in place of four as proposed. This is because as the width of the roads in the prototype is small placing sensors in opposite directions will lead to coinciding of the area of coverage of the two sensors. This results in HIGH value even when there is no vehicle present along the road. The prototype represents a four way junction, which is sufficient to test the various situations encountered in real world.

Execution in Auto Control Mode

Figures 5 through 9 demonstrate the Auto Controlling of the model.

In Figure 5, a probable situation is shown where the east lane contains maximum numbers of vehicle. Both the sensors are high thus has maximum priority should be executed first (Figure 6), west lane which contains vehicles at the initial point has second highest priority and is to be executed next indicated by

Figure 3. Prototype (a)   *Figure 4. Prototype (b)*

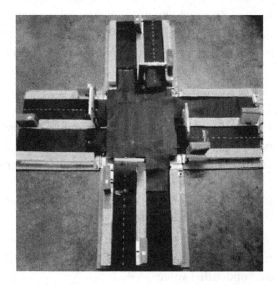 

Figure 5. Execution in auto control mode (a)

Figure 6. Execution in auto control: East lane mode

the glowing of the yellow light (Figure 7). Afterward, the south lane is executed as it has a vehicle at the far end; presently red light is shown (Figure 8). Finally, the north lane will be allowed green light for few seconds as it contains no vehicles. This is how the whole auto controlling takes place in the ITMS.

The ITMS PC Software

Figure 10 shows the starting screen of the ITMS PC Software.

In Figure 10, none of the lanes are set; the red and green boxes represent the low and high of the sensors; respectively. It is obvious that the boxes corresponding to west are both high indicating the

Figure 7. Execution in auto control: South lane mode

Figure 8. Execution in auto control: North lane mode

Figure 9. Execution in auto control: West lane mode

Figure 10. Starting screen of ITMS software PC

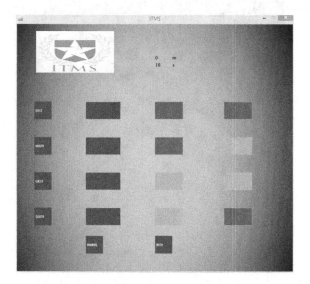

lane is full. The first box corresponding to north and the second box corresponding to south is high i.e. is green in color indicating that vehicles are present at initial and final points of the lanes; respectively.

In Figure 11, the north is set indicating that green light is shown to the north lane (Figure 12) and "North is on" is flashed on the screen. Figures 13 and 14 depict the same for west lane. The button "Manual" should be placed to indicate the microcontroller that the traffic will be now controlled manually. The button "Auto" indicates the opposite.

Figure 11. Setting North on in ITMS PC software

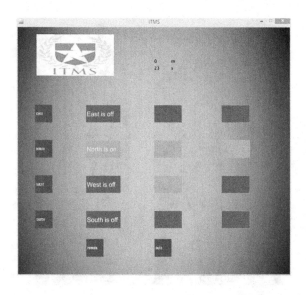

Figure 12. North is on in the prototype

Figure 13. Setting West on in ITMS PC software

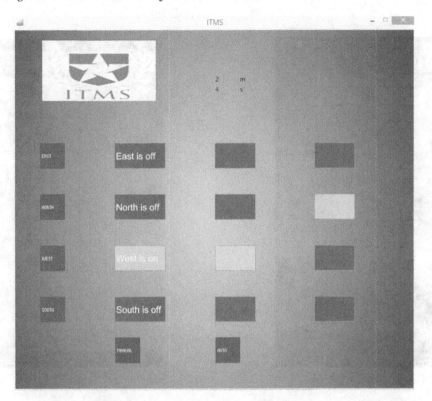

Figure 14. Setting West on in ITMS PC Software

The ITMS Android Application

Figure 15 through 18 represent the ITMS Android Application.

Figure 15 shows the connecting screen, the button "Connect" is used to connect to the HC05 Bluetooth module. Figure 16 represents the login screen that appears on successful connection, whereas Figure

Figure 15. Connecting screen of the Android application

Figure 16. Login screen of the ITMS Android application

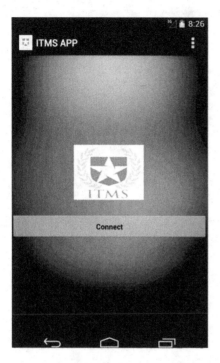

Figure 17. Control screen of the ITMS: Android application (a)

Figure 18. Control screen of the ITMS: Android application (b)

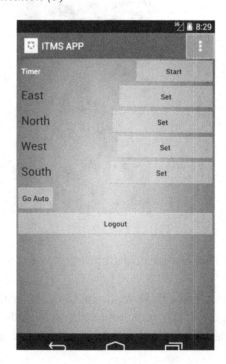

17appears on successful login, it provides various controls to handle the traffic, whenever the button "Set" corresponding to any lane is pressed, green light for the lane is set, on pressing the other buttons, the red LED is set for the other lanes except the set lane. The button "Go manual" should be placed to indicate the microcontroller that the traffic will be now controlled manually. The button "Go Auto" (Figure 18) indicates the opposite.

CONCLUSION

Road traffic is a critical problem that requires implementing an Intelligent Traffic System (ITS) prevent the traffic jam and manage the overall traffic problem. Such systems necessitate in time closer interaction with all the traffic modules including vehicles, drivers and even pedestrian.

The proposed model established its efficiency as a model to reduce various traffic problems. The automatic and manual controlling of the proposed traffic monitoring system can play a major role. Manually controlling the system using application can be positive point of this system. Furthermore, this traffic controlling system can be easily installed and attached to the existing traffic road infrastructure at a low cost and within a reasonable time. In addition, no traffic disruption will be necessary when a new traffic sensor is to be installed.

REFERENCES

Bertelle, C., Dutot, A., Lerebourg, S., Olivier, D., & du Havre, L. (2003). Road traffic management based on ant system and regulation model. In *Proc. of the Int. Workshop on Modeling and Applied Simulation* (pp. 35-43).

Chakrabarty, S., Pal, A. K., Dey, N., Das, D., & Acharjee, S. (2014, January). Foliage area computation using Monarch butterfly algorithm. In *Non Conventional Energy (ICONCE), 2014 1st International Conference on* (pp. 249-253). IEEE. doi:10.1109/ICONCE.2014.6808740

Chakraborty, S., Samanta, S., Biswas, D., Dey, N., & Chaudhuri, S. S. (2013, December). Particle swarm optimization based parameter optimization technique in medical information hiding. In *Computational Intelligence and Computing Research (ICCIC), 2013 IEEE International Conference on* (pp. 1-6). IEEE. doi:10.1109/ICCIC.2013.6724173

Challal, Y., Ouadjaout, A., Lasla, N., Bagaa, M., & Hadjidj, A. (2011). Secure and efficient disjoint multipath construction for fault tolerant routing in wireless sensor networks. *Journal of Network and Computer Applications*, *34*(4), 1380–1397. doi:10.1016/j.jnca.2011.03.022

Chen, S. W., Yang, C. B., & Peng, Y. H. (2007). Algorithms for the traffic light setting problem on the graph model. Taiwanese Association for Artificial Intelligence.

Dakhole, A. Y., & Moon, M. P. (2013). Design of intelligent traffic control system based on ARM. *International Journal (Toronto, Ont.)*, *1*(6).

Dey, N., Chakraborty, S., & Samanta, S. (2013b). Optimization of watermarking in biomedical signal. *Lambert Publication. Heinrich-Böcking-Straße, 6,* 66121.

Dey, N., Samanta, S., Chakraborty, S., Das, A., Chaudhuri, S. S., & Suri, J. S. (2014). Firefly algorithm for optimization of scaling factors during embedding of manifold medical information: An application in ophthalmology imaging. *Journal of Medical Imaging and Health Informatics, 4*(3), 384–394. doi:10.1166/jmihi.2014.1265

Dey, N., Samanta, S., Yang, X. S., Das, A., & Chaudhuri, S. S. (2013a). Optimisation of scaling factors in electrocardiogram signal watermarking using cuckoo search. *International Journal of Bio-inspired Computation, 5*(5), 315–326. doi:10.1504/IJBIC.2013.057193

Di Caro, G., Ducatelle, F., & Gambardella, L. M. (2005). AntHocNet: An adaptive nature-inspired algorithm for routing in mobile ad hoc networks. *European Transactions on Telecommunications, 16*(5), 443–455. doi:10.1002/ett.1062

Dotoli, M., Fanti, M. P., & Melon, C. (2003, October). Real time optimization of traffic signal control: application to coordinated intersections. In *Systems, Man and Cybernetics, 2003. IEEE International Conference on* (Vol. 4, pp. 3288-3295). IEEE. doi:10.1109/ICSMC.2003.1244397

Hussian, R., Sharma, S., Sharma, V., & Sharma, S. (2013). WSN applications: automated intelligent traffic control system using sensors. *International Journal of Soft Computing and Engineering*.

Kafi, M. A., Challal, Y., Djenouri, D., Bouabdallah, A., Khelladi, L., & Badache, N. (2012). A study of wireless sensor network architectures and projects for traffic light monitoring. *Procedia Computer Science, 10,* 543-552.

Kale, S. B., & Dhok, G. P. (2013). Design of intelligent ambulance and traffic control. *Int. J. Comput. Electron. Res, 2*(2).

Kaliannan, J., Baskaran, A., & Dey, N. (2015). Automatic generation control of thermal-thermal-hydro power systems with PID controller using ant colony optimization. *International Journal of Service Science, Management, Engineering, and Technology, 6*(2), 18–34. doi:10.4018/ijssmet.2015040102

Kareem, E. I. A., & Jantan, A. (2011). An intelligent traffic light monitor system using an adaptive associative memory. *IJIPM: International Journal of Information Processing and Management, 2*(2), 23–39. doi:10.4156/ijipm.vol2.issue2.4

Ozkurt, C., & Camci, F. (2009). Automatic traffic density estimation and vehicle classification for traffic surveillance systems using neural networks. *Mathematical and Computational Applications, 14*(3), 187–196. doi:10.3390/mca14030187

Płaczek, B. (2011). Performance evaluation of road traffic control using a fuzzy cellular model. In *Hybrid Artificial Intelligent Systems* (pp. 59–66). Springer Berlin Heidelberg. doi:10.1007/978-3-642-21222-2_8

Queen, C. M., & Albers, C. J. (2008, July). Forecasting traffic flows in road networks: A graphical dynamic model approach. In *Proceedings of the 28th International Symposium of Forecasting*. International Institute of Forecasters.

Samanta, S., Acharjee, S., Mukherjee, A., Das, D., & Dey, N. (2013a, December). Ant weight lifting algorithm for image segmentation. In *Computational Intelligence and Computing Research (ICCIC), 2013 IEEE International Conference on* (pp. 1-5). IEEE. doi:10.1109/ICCIC.2013.6724160

Samanta, S., Chakraborty, S., Acharjee, S., Mukherjee, A., & Dey, N. (2013b, December). Solving 0/1 knapsack problem using ant weight lifting algorithm. In *Computational Intelligence and Computing Research (ICCIC), 2013 IEEE International Conference on* (pp. 1-5). IEEE.

Sinhmar, P. (2012). Intelligent traffic light and density control using IR sensors and microcontroller. *International Journal of Advanced Technology & Engineering Research*, 2(2), 30–35.

Soh, A. C., Rhung, L. G., & Sarkan, H. M. (2010). MATLAB simulation of fuzzy traffic controller for multilane isolated intersection. *International Journal on Computer Science and Engineering*, 2(4), 924–933.

Srivastava, P. M. D., Sachin, S., Sharma, S., & Tyagi, U. (2012). Smart traffic control system using PLC and SCADA. *International Journal of Innovative Research in Science, Engineering and Technology*, 1(2).

Tubaishat, M., Shang, Y., & Shi, H. (2007, January). Adaptive traffic light control with wireless sensor networks. In *Proceedings of IEEE consumer communications and networking conference* (pp. 187-191). doi:10.1109/CCNC.2007.44

VanDaniker, M. (2009, August). Visualizing real-time and archived traffic incident data. In *Information Reuse & Integration, 2009. IRI'09. IEEE International Conference on* (pp. 206-211). IEEE. doi:10.1109/IRI.2009.5211552

Wenjie, C., Lifeng, C., Zhanglong, C., & Shiliang, T. (2005, June). A realtime dynamic traffic control system based on wireless sensor network. In *Parallel Processing, 2005. ICPP 2005 Workshops.International Conference Workshops on* (pp. 258-264). IEEE. doi:10.1109/ICPPW.2005.16

Yousef, K. M., Al-Karaki, M. N., & Shatnawi, A. M. (2010). Intelligent Traffic Light Flow Control System Using Wireless Sensors Networks. *J. Inf. Sci. Eng.*, 26(3), 753–768.

Zhou, B., Cao, J., Zeng, X., & Wu, H. (2010, September). Adaptive traffic light control in wireless sensor network-based intelligent transportation system. In *Vehicular technology conference fall (VTC 2010-Fall), 2010 IEEE 72nd* (pp. 1-5). IEEE. doi:10.1109/VETECF.2010.5594435

KEY TERMS AND DEFINITIONS

Arduino: Is an open-source electronics board (platform) that requires software to program it.

Automatic Control: Is the use of control theory to regulate any process without direct human intervention.

Bluetooth: Is a short-range radio (wireless) technology that simplify the communication among internet devices as well as between devices and the internet. Moreover, it facilitates data synchronization between computers and internet devices.

Intelligent Traffic Monitoring System (ITMS): A system that embodies a traffic knowledge model that provides information about the behavior at a strategic level.

Intelligent Transportation Systems (ITS): Are innovative applications that aim to provide advanced services related to different modes of traffic and transport management. It enables several users to be coordinated simply with more smart use of the transport networks.

Sensor: Is an object that detects changes in its environment, and then deliver a corresponding output.

Chapter 12
Reducing False Alarms in Vision–Based Fire Detection

Neethidevan Veerapathiran
Mepco Schlenk Engineering College, India

Anand S.
Mepco Schlenk Engineering College, India

ABSTRACT

Computer vision techniques are mainly used now a days to detect the fire. There are also many challenges in trying whether the region detected as fire is actually a fire this is perhaps mainly because the color of fire can range from red yellow to almost white. So fire region cannot be detected only by a single feature and many other features (i.e.) color have to be taken into consideration. Early warning and instantaneous responses are the preventing ideas to avoid losses affecting environment as well as human causalities. Conventional fire detection systems use physical sensors to detect fire. Chemical properties of particles in the air are acquired by sensors and are used by conventional fire detection systems to raise an alarm. However, this can also cause false alarms. In order to reduce false alarms of conventional fire detection systems, system make use of vision based fire detection system. This chapter discuss about the fundamentals of videos, various issues in processing video signals, various algorithms for video processing using vision techniques.

INTRODUCTION

What Is Video?

It combines a sequence of images to form a moving picture (Abidha T E, Paul P Matha). It transmits a signal to a screen and processes the order in which the screen captures should be shown. It usually have audio components that correspond with the pictures being shown on the screen.

The Video is a series of still images displayed many times per second to give the illusion of motion. A frame is a single still picture that fills the display screen.

DOI: 10.4018/978-1-5225-1022-2.ch012

The frame rate for video is nothing but, how many full screens are displayed per second. It's measured in fps (Frames per second). Even though analog video isn't progressive, for purposes of comparison it's useful to use the same units. Interlaced video does also have a field rate, which will always be twice the frame rate. Film has a frame rate of 24fps, PAL uses 25fps, and NTSC 30/1.001fps. The strange number for NTSC is because of changes that were made from the original 30fps black and white signal so color could be added. It's commonly noted as 29.97fps, or occasionally 30fps. Sometimes NTSC and PAL signals are identified by a combination of their frame rate and type (interlaced or progressive) or field rate and type. 25i or 50i refers to PAL while 30i or 60i is generally used for NTSC. This is generally in reference to a digital signal with either NTSC or PAL characteristics. Film can be referred to as 24p.

Video Surveillance System

The main components of an automatic video surveillance system are shown in Fig. 1: video cameras are connected to a video processing unit (either a general purpose PC or a dedicated computer equipment), for instance, to extract high-level information identified with alert situation; this processing unit could be connected throughout a network to a control and visualization center that manages, for example, alerts; another important component is a video database and retrieval tool where selected video segments, video objects, and related contents can be stored and inquired (Aishy Amer, Concordia University, Montre´al, Que´bec, Canada, Carlo Regazzoni, University of Genoa).

A video sequence is composed typically of a set of many video shots. To facilitate the processing of video objects, a video sequence is first segmented into video shots. A shot is a (finite) sequence of frames recorded contiguously from the same camera (usually without viewpoint change). In video surveillance, a shot represents a continuous, in time and space, action or event driven by moving objects (e.g., an object stopping at a restricted site). Video processing for surveillance applications aims at describing the data in successive frames of a video in terms of what is in the real scene, where it is located, when

Figure 1. A generic block diagram of a video surveillance system

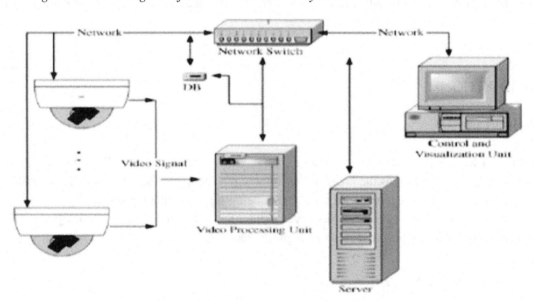

it occurred, and what are its features. It is the basic step towards an automated full understanding of the semantic contents in the input video. Semantically (perceptually) significant content, e.g., object activities, are generally related to moving objects. This is related to the human visual system (HVS) which is strongly attracted to moving objects creating luminance change.

A generic block diagram of a video object processing module for video surveillance is shown in Figure 1. As can be seen, several video processing steps are required to take out video objects and related high-level features: preprocessing (e.g., noise estimation or reduction), temporal segmentation (e.g., shot detection), video analysis (e.g., motion estimation, object segmentation, and tracking), object classification, video interpretation (i.e., extraction of high-level context-independent information), and video understanding (i.e., extraction of context-dependent semantically information). In a multi-level system such as in Figure 2 (or the sublevels of this figure such as video analysis with motion estimation, object segmentation, and tracking), the system architecture should be modular with special consideration to processing inaccuracies and errors of precedent levels. Processing errors need to be corrected or compensated at a higher level where more (higher level) information is available. Higher processing level can provide useful and more reliable information for the detection and correction of processing errors. Results of lower level processing are integrated to support higher processing. Higher levels may support lower levels through memory-based feedbacks.

Video Signal

It is basically any sequence of time varying images. A still image is a spatial distribution of intensities that remain constant with time, whereas a time varying image has a spatial intensity distribution that varies with time. Video signal is treated as a series of images called frames.

In a digital video, the picture information is digitized both spatially and temporally and the resultant pixel intensities are quantized. The block diagram depicting the process of obtaining digital video from continuous natural scene is shown in Figure 3.

Figure 2. Multi level system

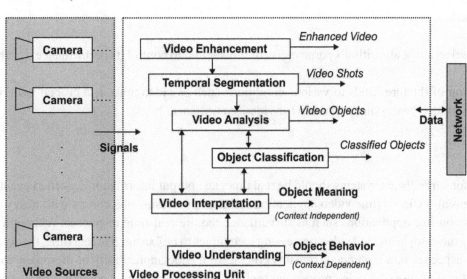

Figure 3. Digital video from natural scene

Fundamental Issues and Challenges

There are more issues and fundamental challenges still remains in video processing despite the many advancements in the field. It originated from several issues that can complicate the design and evaluation of processing algorithms for video surveillance.

Interpretation: The Object-oriented video processing aims at extracting video objects and their spatio-temporal features. To extract video objects, technically, the following definitions are given:

1. A video is a finite set of frames and each frame consists of an array of pixels;
2. The aim of analysis is to give each pixel a label based on some properties; and
3. An object consists of a connected group of pixels that share the same label.

Does interpretation consider a vehicle with a driver as one object or two objects? A person moving the body parts one or more objects? These questions indicate that it is difficult to design one single object-oriented video processing method that is valid for all applications. Object processing is subjective and can vary between observers and the evaluation of one observer can vary in time. This subjectivity cannot always be formulated by a precise mathematical definition.

The specific applications require, however, specific parameters to be fixed and even the designers of general systems can have difficulty adapting the system parameters to a specific application. Therefore, recent research efforts are increasingly dedicated to develop object processing methods that focus on a well-defined range of specific applications.

Automation

Automatic selection of algorithm's parameters (or thresholds) to control its performance under different conditions.

Adaptation of the thresholds to various conditions such as video noise is a crucial issue in the acceptance of a video processing algorithm for video surveillance.

Efficiency

Algorithm for surveillance systems should be real time, i.e., output information's, such as events, as they occur in the real scene. Offline video applications tolerate processing algorithms with heavy computation whereas on-line applications such as surveillance require real-time responses. Although there are increasingly more sophisticated video processing techniques to be executed in real time, the computation complexity aspect of new techniques remains crucial for a wide applicability of these new techniques. In general, the wide use of a video processing tool strongly depends on its computational efficiency.

Robustness

A video processing algorithm is required to be robust, i.e., deliver good or satisfactory performance for the given applications and conditions. Achieving robust algorithms is a challenge especially:

1. Under illumination variation due to weather conditions or indoor lighting changes, for example, in the presence of open doors or windows;
2. Under view changes;
3. In case of multiple objects with partial or complete occlusion or deformation;
4. In the presence of articulated or non-rigid objects;
5. In case of shadow, reflections, and clutter; and
6. With video noise (e.g., Gaussian white) and artifacts (e.g., due to compression).

Even though many advancements in solving these difficulties have been achieved but robustness remains a major issue to avoid system failure. Tradeoff: accuracy versus efficiency: Video and object changes complicates further two conflicting requirements: accuracy and efficiency. In a surveillance application, for instance, the emphasis is on the Robustness or stability of the video processing task (e.g., no system failure) rather than on accurate object extraction, e.g., pixel-precise object boundaries. On the other hand, algorithm complexity has an impact on the design and use of video processing modules. Therefore, the tradeoff between these conflicting requirements is still an ongoing video surveillance research.

Performance Evaluation

It is a crucial issue for the acceptance of video processing modules in real-world surveillance systems. Due to the difficulties of evaluating video processing in general, this is an open problem that has gained many research activities

BACKGROUND

Video Processing

To process the Video, there is a need for hardware, software, and combinations of the two for editing the images and sound recorded in video files. Extensive algorithms in the processing software and the peripheral equipment allow the user to perform editing functions using various filters. The desired effects can be produced by editing frame by frame or in larger batches.

Now-a-days most of the modern personal computers come with software that allows the user to compile images and videos, edit images, and create videos on a limited level. Storyboards allow the addition of audio files and the adjustment of visual images, transitions, and audio files, which, together, determine the overall length of the video. Lot of people including Videographers, electrical engineers, and computer science professionals use programs that are capable of a wider range of functions. The signal processing typically involves applying a combination of profilers, intra filters, and post filters.

What Is Video Analytics?

Video Analytics, also referred to as Video Content Analysis (VCA), is a common term used to define automated *processing and analysis of video streams [4]*. Computer analysis of video is currently implemented in many of the fields and industries, though the term "Video Analytics" is typically associated with analysis of video streams captured by surveillance systems. Its applications can perform a many tasks ranging from real-time analysis of live video for immediate detection of events of interest, to analysis of pre-recorded video.

Based on Video Analytics to automatically monitor cameras and alert for events of interest is in many cases much more effective than dependence on a human operator, which is a costly resource with limited vigilance and attention. Many surveys and research studies on real-life incidents indicate that an average human operator of a surveillance system, tasked with noting video screens, cannot remain alert and attentive for more than 20 minutes. Moreover, the operator's ability to monitor the video and effectively respond to events is considerably compromised as time goes by.

Also, there is a need to go through already recorded video and extract specific video slices containing an event of interest. This need is growing as the use of video surveillance becomes extensive use and the quantity of recorded video increases. Surveillance system users are also looking for additional ways to influence their recorded video, including by extracting statistical data for business intelligence purposes. Analyzing recorded video by human operators involved many manual tasks and time consuming process. it may have more issues and mistakes also.

While there is a necessity for this kind of system in modern research facilities in all the working places with various benefits, and the accompanying financial investment in deploying such surveillance system is significant, the actual benefit derived from a surveillance system is limited when relying on human operators alone. In contrast, by deploying video surveillance system with analytics, many more benefits are achieved.

Video Analytics is an ideal solution that meets the various requirements of surveillance system operators, security officers, and corporate managers, as they seek to make practical and effective use of their surveillance systems in everyday life.

What Is Video Analytics Used For?

Video surveillance systems are obviously installed to record video tape of areas of interest within sites of known interest, with a view to "catching" (allowing the user to be able to observe) and recording events related to security, safety, loss prevention, operational efficiency and even business intelligence.

It improves the system performance by examining real time video and detect various events of known interest, post-event analysis and removal of statistical data while saving manpower costs and increasing the effectiveness of the surveillance system operation.

Video Analytics for Real-Time Detections and Alerts

With various image processing algorithms already available in Literature, it can able to detect a variety of events, in real-time, such as:

- Penetration of unauthorized people / vehicles into restricted areas
- Tailgating of people / vehicles through secure checkpoints
- Traffic obstacles
- Unattended objects
- Vehicles stopped in no-parking zones, highways or roads
- Removal of assets
- Crowding or grouping
- Loitering
- Counting people
- Pointing criminals

And many more.

By defining the set of events that the surveillance system operator wants to be alerted to, the software constantly analyzes the video in real-time and provides an immediate alert upon detection of a appropriate event.

More for The "Discriminating" Business

With more sophisticated infrastructure in place and the potential expertise to use it, video analytics continues to evolve from the most basic application of motion detection, object detection and recognition to higher levels of discrimination.

Tripwire analysis can create intrusion parameters. For example, if someone or something enters by violating a security barrier, an event notification can be triggered. If, someone leaps over a counter into a bank teller's space, an event can be triggered. Taking it a step further can reveal how many people or objects (e.g., vehicles) are crossing a controlled boundary within a certain timeframe. This certainly has inferences for both commercial security and Homeland Security initiatives.

Table 1. List of known functionalities and a short description

Function	Description
Lively masking	Blocking a part of the video signal based on the signal itself, for example because of privacy concerns.
Ego motion estimation	Ego motion estimation is used to determine the location of a camera by analyzing its output signal.
Gesture detection	Gesture detection is used to determine the presence of relevant motion in the observed scene.
Shape recognition	Shape recognition is used to recognize shapes in the input video, for example circles or squares. This functionality is typically used in more advanced functionalities such as object detection.
Object detection	Object detection is used to determine the presence of a type of object or entity, for example a person or car. Other examples include fire and smoke detection.
Recognition	Face recognition and Automatic Number Plate Recognition are used to recognize, and therefore possibly identify, persons or cars.
Style detection	Style detection is used in settings where the video signal has been produced, for example for television broadcast. Style detection detects the style of the production process.
Tamper detection	Tamper detection is used to determine whether the camera or output signal is tampered with.
Video tracking	Video tracking is used to determine the location of persons or objects in the video signal, possibly with regard to an external reference grid

Directional analysis is the ability to distinguish performance by assigning specific values (low to high) to areas within a camera's field of view. The main objective is identifying the location, size and spread of a forest fire.

Objects-left-behind-or-removed analysis identifies that a threat can come from an inanimate object, such as an unattended bag or package in an airport terminal. On the other hand, the unauthorized removal of public or corporate property would trigger the dispatch of a security guard or investigator.

Behavioral analysis is mainly used for both safety and marketing purposes. A vehicle in a shopping center parking lot circles boundlessly, signaling a possible security situation. A person strolls aimlessly in that same parking lot, posturing a potential threat to others or him/herself – and potential accountability for management. Retail marketers are realizing tremendous benefit of analyzing store traffic patterns and the impact of product placement on product sales.

Removing the Human Equation through Automation

Some might say that video analytics removes the participation of human operators from the surveillance process. To some extent it does. It removes the deadlines involved in having one or more sets of eyes on a monitor for an extended period of time. In fact, the automation of video analytics permits the insertion of human judgment at the most critical time in the surveillance process.

Application of Video Analytics

- People Counting
- Loitering
- Forbidden Area Detect
- Scene Detection
- Object Direction
- Object Counting
- Unattended Object
- Missing Object
- Object Color
- Fire Detection
- Wrong Way
- Face Detection

Introduction to Video tracking

Video tracking is the process of tracing a moving object (or multiple objects) over time using a video camera[6,7]. It has a diversity of uses, some of which are: human-computer interaction, security and surveillance, video communication and compression, augmented reality, traffic control, number plate recognition, medical imaging and video editing. Video tracking can be a time consuming process due to the large amount of data that is contained in video and it has to be processed. Also to use object recognition techniques for tracking, a challenging problem is also there.

The objective of video tracking is to associate target objects in successive video frames. The association can be especially difficult when the objects are moving fast relative to the frame rate. The increased difficulty of the problem is when the tracked object changes its orientation overall time period. For this video tracking systems usually employ a motion model which describes how the image of the target might change for different possible motions of the object.

Problems Encountered in Video Tracking

The various problems encountered in people counting in video tracking are explained below.

Background Model

Background subtraction–Object detection can be easily achieved by building the background model, and then the decision deviates from the model for each incoming frame. Any important change in an image region from the background model signifies a moving object. The tracking process for one person in a stationary background may be relatively simple, the problem becomes very complicated with many people. They may be crossing in front of each other, behind occlusions, through different lighting, with shadows, and in groups.

Occlusion Problem

Among large number of objects, certain objects being fully visible and, many objects partially visible due to occlusions from other objects. The problem of detection and tracking partially visible or hidden objects becomes really challenging. The camera is not fixed and it is moving and there is a lot of camera shake, which further increases the difficulty of the problem. The poor resolution of the video also complexes the detection problem.

Blob Analysis

Blob tracing may be easy and fast, but it does not work normally, particularly with people moving in groups. A number of candidate algorithms claim to be capable of distinguishing people in clusters. People approaching from grocery stores or other similar shops often push trolleys ahead of them. These trolleys are typically of an identical size as a human being. This makes splitting blobs by height or width a hard task, as it must take movement into consideration; it is hard to see if the blob is real or collected from numerous people. It becomes impossible to detect that in hectic areas. Still, huge area blobs are expected to be groups of either people or something else that is moving. People standing in groups are tough to compact with, even for the human eye. When they stand close jointly or hold hands, they appear as one big blob. Blob detection (blob analysis) involves probing the blob image and finding each human being as a foreground object. This may, in some specific cases, become a rather difficult task, even for the human eye. When the moving persons have equivalent colors as the background, the blobs become rather unclear, and unusual situations appear when blobs contain slums. Dilation is not enough in this case to entirely dense the blob.

Lighting Conditions and Noise

The first frame on the video should have a static background; considering that frame as a base frame and judge against the real video frames against this base frame. It is hard to believe that stationary background frames with similar lighting conditions for day, night and noon might be obtained. Light is limited according to the seasons and non-natural lighting settings; so a more compound approach is needed. The image still contains a lot of noise and requires further processing. The subtracted resulting frame gives a attractive good approximation of the information. Noise should be eliminated in order to avoid its interference with the blob detection algorithm.

Color Classification

An object can be segmented based on its color in RGB or HSV space. This is perfect if the object color is distinct from the background. An object can also be recognized or classified based on its color. In the given video sequence, object colors are not distinct as they take a large range of colors. Object recognition based on color would work for some objects, like rickshaws having a dissimilar yellow color, but would fail for cars that can take numerous colors.

Object Boundaries

Edge-based– Object boundaries generally produce strong changes in image intensities. Edge detection is used to recognize these changes. A main property of edges is that, they are less sensitive to lighting changes compared to color features. Algorithms that follow the boundary of the things generally use edges as the representative feature. This technique is not suitable when there are numerous jumbled objects in a frame.

Different Gray Levels and Image Quality

These issues imposed problems especially for image segmentation and binarization techniques, where the foreground image is represented as black, and the background is white. They give high false alarms for images with high noise and patchy illumination. Although most images with a simple background and high contrast can be correctly localized and removed, images with low resolution and complex background will be difficult to extract. Most digital images and videos are generally stored, processed and transmitted in a compacted form.

Simple Algorithm for Video Surveillance

To perform video tracking an algorithm analyzes sequential video frames and outputs the movement of targets between the frames. There are a variety of algorithms, each having strengths and weaknesses. Considering the intended use is important when choosing which algorithm to use. There are two major components of a visual tracking system: target representation and localization, as well as filtering and data association

MAIN FOCUS OF THE CHAPTER

In this chapter, main focus will be discussing various video processing algorithms already available in the Literature. Also explain about its method of implementation, efficiency, merits and demerits etc.

Video Processing Algorithms

First, the valuation results depend on testing sequences. In other words, these results may change forcefully with a new set of video sequences. The reason is that each video sequence contains various problems at various difficulty levels and the final results are affected by all these factors. With a new video sequence, there is a new combination of problems. Thus, the algorithm performance on this sequence is changeable. Secondly, a video processing algorithm is usually designed to work in specific conditions (outdoor, indoor scene, containing fast/slow illumination changes etc.). However there is no quantitative measure to compute the difficulty level of a video sequence relatively to a given problem. Therefore it is difficult to know up to which difficulty level of the video, the algorithm can still achieve good result.

ETISEO, A Performance Evaluation Program

ETISEO, one of the latest evaluation programs, has tried to address these issues. One of the main objectives of ETISEO is to "acquire precise knowledge of vision algorithms". In other words, ETISEO tries to underline the "dependencies between algorithms and their conditions of use" [4]. ETISEO tries to address each video processing problem separately, by defining accurately the problem. For instance, algorithm should handle shadows within at least three different problems:

1. Shadows at different intensity levels (i.e. weakly or strongly contrasted shadows) with uniform non color background,
2. Shadows at the same intensity level with different types of background images in terms of color and texture, and
3. Shadows with different illumination sources in terms of source position and wavelengths.

Firstly, for each problem, it collects the video sequences demonstrating only the current problem. The video sequences should illustrate the problem at different difficulty levels. For instance, for the problem of shadows and intensity levels, algorithm should select video sequences containing shadows at different intensity levels (more or less contrasted). On these selected sequences, the suitable part of the ground truth is filtered and extracted to isolate video processing problems. For instance, for the detection task, evaluate the algorithm performance relatively to the problem of handling occluded objects by considering only the ground truth related to the occluded objects. Secondly, for a given task (object detection, tracking, object classification and event recognition) ETISEO defines sufficient number of metrics to measure and describe the algorithm performance on various aspects. For instance, in ETISEO there are 7 metrics for the task of object detection. Thirdly, ETISEO computes the reference data which resembles to the expected output of the algorithm to be evaluated relatively to a given video processing task. The reference data are computed from the ground truth provided by human operators and can be improved to better correspond to the expected results. For instance, instead of evaluating the mobile object positions from the ground truth (2Dpoints), make use of 3D-point reference data to measure the computation of

3D object position. Finally, ETISEO provides a unique automatic evaluation tool to accurately examine how a given algorithm address a given problem.

A new evaluation methodology to better evaluate video processing performance. An evaluation methodology that help to reuse the evaluation result and also try to isolate each video processing problem and define quantitative measures to compute the difficulty level of a video relatively to the given problem(A.T. Nghiem, F. Bremond, M. Thonnat, and R. Ma). The maximum difficulty level of the videos at which the algorithm is performing well enough is defined as the upper bound of the algorithm capacity for handling the problem.

Detection of Pedestrians

For pedestrian detection, a more generalized video processing technique is presented in Figure 4. (A. T. Nghiem, F. Bremond, M. Thonnat, V. Valentin ETISEO). The first stage is the image acquisition. The respective signal processing procedures are typically built in hardware, i.e. in the camera, and do not require additional resources

First the image acquisition and the image preprocessing stage is performed. This stage reduces noise of the image sensor and removes the intertwined scanning effect (Kumarguru Poobalan and Siau Chuin Liew) . The next step shown in Figure 4 formulates the so called region of interest (ROI), which is a particular part of the image for further processing. The properly selected ROI should consist of all

Figure 4. Fire detection techniques in video using image processing

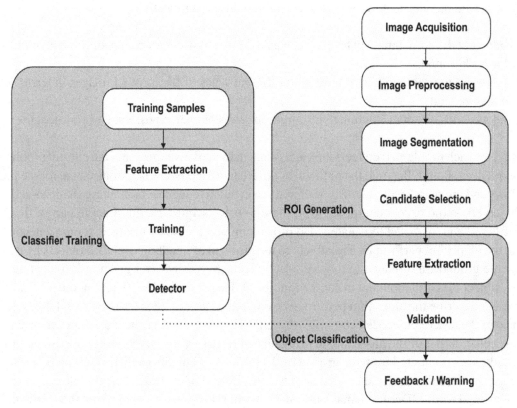

objects which even potentially are in scope of the interest (the pedestrian candidates), and cannot miss significant regions. In the presented case, the well generated ROI covers pedestrians to be detected, but at the same time significantly reduces the size of the analyzed image part, i.e. the amount of data which is moved to the next stages. Thus it can substantially speed-up the data processing. The first step of the ROI generation is image segmentation (Figure 4). When worked with 2D images, decided to use the threshold technique. The algorithm translates the input gray scale image to the binary image, while white objects are the potential candidates to be detected as pedestrians and the background is black.

Various Algorithms for Fire Detection

RGB Color Model

A fire image can be described by using its color properties. (John Adedapo Ojo Jamiu Alabi Oladosu). There are three different element of color pixel: R, G and B. The color pixel can be extracted into these three individual elements R, G and B, which is used for color detection.

RGB color model is used to detect red color information in image. In terms of RGB values, the corresponding inter-relation between R, G and B color channels: R>G and G>B. The combined condition for the captured image can be written as: R>G>B. In fire color detection R should be more stressed than the other component, and hence R becomes the domination color channel in an RGB image for fire. The flow chart is shown in Figure 5.

Figure 5. Flow chart of proposed algorithm for fire detection

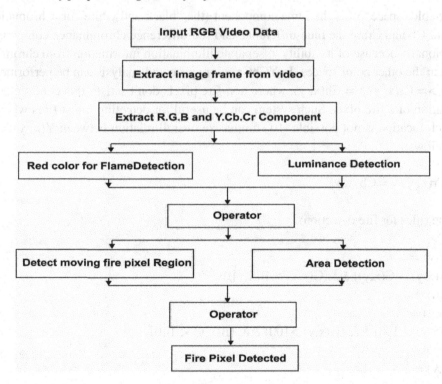

This imposes the condition for R as to be over some pre-determined threshold value RTH.

All of these conditions for fire color in image are summarized as following:

Condition1: R > RTH
Condition2: R > G > B. where RTH is the Red color threshold value for fire.

YCbCr Color Model

YCbCr color space is used in the model rather than other color spaces because of its ability to distinguish luminance information from chrominance information more effectively than other color model[10]. In order to create Y, Cb, Cr components from obtained RGB Image, the color space transformation equation is used to transform each RGB pixel in corresponding Y Channel, Cb Channel, Cr Channel pixel to form a corresponding Y, Cb, Cr image.

When the image is converted from RGB to YCbCr color space, intensity and chrominance is easily discriminated.

YCbCr color space can be easily modeled as following for the fire:

$$Y = 16 + R * 65.481 + G * 128.553 + B * 24.996;$$

$$Cb = 128 + R * -37.797 - G * 74.203 + B * 112.0;$$

$$Cr = 128 + R * 112.00 + G * -93.7864 + B * -18.214;$$

In YCbCr color space, Y′ is the luma component (the "black and white" or achromatic portion of the image) and Cb and Crare the blue-difference and red-difference chrominance components, will be chosen intentionally because of its ability to separate illumination information from chrominance more effectively than the other color spaces. In YCbCr color space and analysis can be performed. For a fire pixel $Y(x, y) >= Cr(x, y) >= Cb(x, y)$, where non-fire pixels don't satisfy this condition, where (x,y) is spatial location of a fire pixel. Such system can be useful for detecting forest fires where putting of sensors at each location is not possible. To summarize overall relation between Y(x, y), Cb(x, y) and Cr(x, y) as follows:

$$Y(x, y) >= Cr(x, y) >= Cb(x, y)$$

Now, some rules for fire detection:

Rule 1: R1(x,y) =
    1, if ((R(x,y) > G(x,y)) &&(G(x,y) > B(x,y)))
    0, otherwise
Rule 2: R2(x,y) =
    1, if (R(x,y) > 190) && (G(x,y) >100) && (B(x,y) < 140)
    0, otherwise
Rule3: R3(x,y) =
    1, if Y(x,y) >= Cb(x,y)

0, otherwise

Rule4: R4(x,y) =

    1, if (Cr(x,y) >= Cb(x,y)

    0, otherwise

Motion Detection in Video

Identifying moving objects from a video sequence is a fundamental and critical task in many computer-vision applications (Abhilash Nunes, Leroy Dias, Shalem Pereira, Meena Ugale). A common method is to perform background subtraction, which is used to identify any moving objects from the portion of a video frame that differs significantly from a background model. There are basically three approaches used in background subtraction for motion detection in a continuous video stream. One of the most common approaches is to compare the current frame with the previous one or with something called as background. This algorithm uses Frame differencing method to detect the moving pixels in a image. Frame differencing uses the video frame at time t-1 as the background for the frame at time t. The binary background difference is generated by comparing the current frame with the background frame. The flowchart for motion detection is shown in Figure 6.

This method has two major advantages. One obvious advantage is the modest computational load. Another is that the background model is highly adaptive. Since the background is based solely on the previous frame, it can adapt to changes in the background faster than any other method (at 1/fps to be precise). The frame difference method subtracts out unimportant background noise (such as flapping trees), much better than the more complex approximate median and mixture of Gaussians methods.

For noise removal a Gaussian filter can be used but need to be cautious about blurring.

Figure 6. Flowchart for motion detection

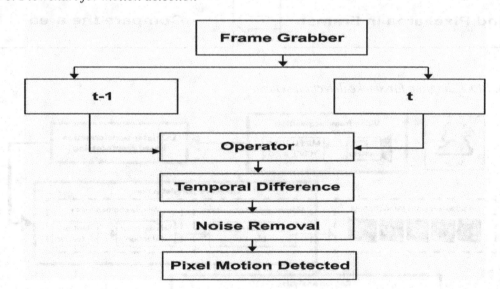

Area Detection

This method is used to detect distribution of fire pixel area in the sequential frames. Area counts the number of pixels in an object (http://in.mathworks.com/help/vision/ug/multipleobjecttracking. html?requestedDomain=www.mathworks.com). In area detection method, two sequential frames which comes out from color detector are compared and then check dispersion in minimum and maximum co-ordinate of X and Y axis. Area detection is modeled in Figure 7.

Block Diagram of the Smoke Detection Scheme

The block diagram of the smoke detection scheme is shown in Figure 8, which is composed of four stages:

1. Video frames acquisition stage,
2. DCT inter-transformation based preprocessing stage,
3. Smoke region detection stage and
4. Region analysis stage (Abhilash Nunes, Leroy Dias, Shalem Pereira, Meena Ugale Optimized Flame Detection,).

Figure 7. Area Detection

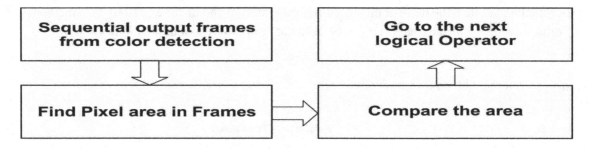

Figure 8. Block diagram for smoke detection scheme

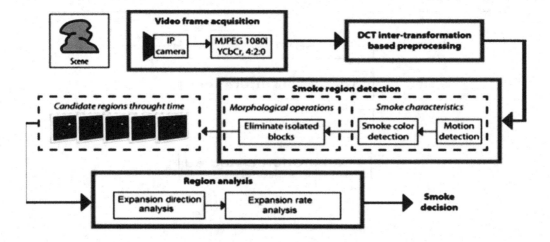

In the video frames acquisition stage, each frame of size $1,920 \times 1,080$ pixels is captured by an IP camera and encoded using an standard JPEG codec, in which bi-dimensional DCT is applied to non-overlapped blocks of 8×8 pixels of each frame. In the preprocessing stage, the DCT inter-transformation is applied to all DCT blocks of 8×8 coefficients of each frame to get DCT blocks of 4×4 coefficients without using the inverse DCT (IDCT).

In the smoke region detection stage, using the DC values of each DCT block of the 4×4 coefficients of several consecutive frames, motion and color properties of smoke are analyzed to get the smoke region candidates. The candidate regions are processed using morphological operations to eliminate isolated blocks

DCT Inter-Transformation Based Preprocessing

As mentioned before, an IP camera module provides DCT blocks of 8×8 coefficients of each frame; however this block size is too large for accurate analysis of smoke features and it is necessary to use a smaller block size. Traditionally if a DCT block with a size different from a current block size is required, the IDCT must be computed and then a new DCT with the required block size is re-calculated. These processes are highly time consuming operations. In ((Turgay,Kalirajan, K., and M. Sudha)), inter-transformation of DCT coefficients is proposed, in which the relationship between DCT coefficients of different block sizes is established.

Smoke Region Detection Stage

In the smoke region detection stage, some smoke block candidates are estimated using the motion and color properties of smoke. This stage receives DCT blocks of $Sb \times Sb$ coefficients previously calculated by the preprocessing stage of each frame, which is composed of three channels: luminance channel (Y) and two chrominance channels (Cb and Cr). The motion property of smoke is analyzed using only the luminance channel Y, and the smoke color property is analyzed using two chrominance channels Cb and Cr

Elimination of Isolated Blocks

Due to Illumination variations and motion caused by wind, that are considered to be the principal factors of erroneous block detection; however these erroneous blocks can be detected easily because these blocks are generally isolated. By taking in to account the expansion property of smoke, which occupies several connected blocks, the isolated blocks can be considered as erroneous blocks.

Region Analysis Stage

Once the smoke candidate regions are detected, the behavior of these regions through the several frames must be analyzed, because some objects hold similar properties to smoke.

Computer Vision-Based Fire Detection Algorithm

This algorithm consists of two main parts:

1. Fire color modeling, and
2. Motion detection (Vinayak G Ukinkar, Makrand Samvatsar).

The algorithm can be used in parallel with conventional fire detection systems to reduce false alarms. It can also be deployed as a stand-alone system to detect fire by using video frames acquired through a video acquisition device. A novel fire color model is developed in CIE $L*a*b*$ color space to identify fire pixels. The new fire color model is tested with ten diverse video sequences including different types of fire. The experimental results are quite inspiring in terms of correctly classifying fire pixels according to color information only. The Figure 9 shows flow chart of algorithm for fire detection in image sequences.

It is assumed that the image attainment device produces its output in RGB format. The algorithm consists of three main stages: fire pixel detection using color information, detecting moving pixels, and analyzing dynamics of moving fire pixels in consecutive frames.

RGB to CIE L*a*b* Color Space Conversion

The first stage of the algorithm is the conversion from RGB to CIE $L*a*b*$ color space. Most of the existing CCTV video cameras provide output in RGB color space, but there are also other color spaces used for data output representation.

Color Modeling for Fire Detection

A fire in an image can be described by using its visual properties. These visual properties can be stated using simple mathematical formulations. The Figure 10 shows sample images which contain fire and their CIE $L*a*b*$ color channels ($L*$, $a*$, $b*$). Figure 2 gives some clues about the way CIE $L*a*b*$ color channel values characterize fire pixels. Using such visual properties, rules were developed to detect fire using CIE $L*a*b*$ color space.

Figure 9. Flow chart of algorithm for fire detection in image sequences

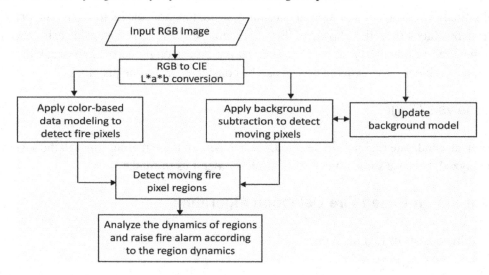

Figure 10. Sample RGB images containing fire and their CIE L*a*b* color channels: (a) RGB image, (b) L* color channel, (c) a* color

Channel, and (d) b* color channel. For visualization purposes, responses in different color channels are normalized into interval[0, 1].

Moving Pixel Detection

In moving pixel detection, it is assumed that the video camera is stable, that is, the camera is still, and there is no movement in spatial location of the video camera. There are three main parts in moving pixel detection: frame/background subtraction, background registration, and moving pixel detection

The first step is to compute the binary frame difference map by thresholding the difference between two consecutive input frames. At the same time, the binary background difference map is generated by comparing the current input frame with the background frame stored in the background buffer. The binary background difference map is used as primary information for moving pixel detection. In the second step, according to the frame difference map of past several frames, pixels which are not moving for a long time are considered as reliable background in the background registration. This step maintains an updated background buffer as well as a background registration map indicating whether the background information of a pixel is available or not. In the third step, the binary background difference map and the binary frame difference map are used together to create the binary moving pixel map. If the background registration map indicates that the background information of a pixel is available, the background difference map is used as the initial binary moving pixel map. Otherwise, the value in the binary frame difference map is copied to binary moving pixel map. The intensity channel $L*$ is used in moving pixel detection.

Moving Object Detection for Video Surveillance

The various steps are shown in Figure 11.

Figure 11. The flow diagram of the framework

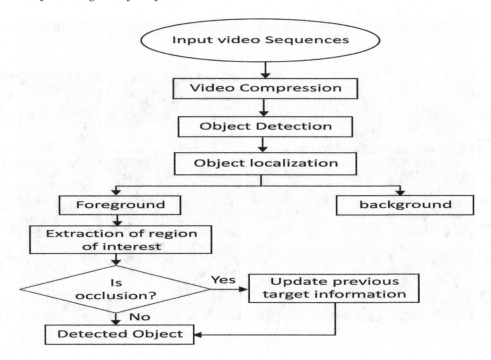

Video Compression

In the first phase of proposed framework, the input video frames are compressed using block processing algorithm called 2D discrete cosine transform (DCT).(Millan-Garcia, L.; Sanchez-Perez, G.; Nakano, M.; Toscano-Medina, K.; Perez-Meana, H.; Rojas-Cardenas, L). Let (u, V) be the transformed frame and let (x, y) be the original frame. Consider an image frame with dimensions of $(.\ N)$, where M and N are the rows and columns involved in each image frame

Object Detection

Object detection is mainly concentrated to detect the target position in each frame with coordinates, scale, and orientation. In object detection phase, the feature vectors are derived using 2D correlation. Correlation is one of the statistical approaches which provide a direct measure of the similarity between two video frames and it will not be influenced by illumination variation and object translations. However, it cannot cope with image rotation and scaling. The proposed model can further be extended to deal with an image rotation and scaling by incorporating the sophisticated object detection algorithm such as multi resolution analysis.

Object Localization

In this phase, an effective classifier is constructed to classify the matched features points into foreground and background using Bayesian rule (A. T. Nghiem, F. Bremond, M. Thonnat, V. Valentin ETISEO). Let (k, t) be the input image frame at time t in the position k,

(k, t) be the template frame, and fV, be the feature vector of target in the position k at time t. The posterior probability of feature vector that appears from the background at position k is calculated as follows:

$$P\left(\frac{f_{back}}{f_{v,t}}\right) = \frac{P(f_{v,t} \,/\, f_{back})P\left(f_{back}\right)}{P\left(f_{v,t}\right)}$$

where $f$back is the background and $P(f$V,t/ $f$back) is the probability of feature vector fV,t being observed as background. The prior probability of feature vector being identified at the position k is denoted by (fV,) and ($f$back) is the prior probability of feature vector belonging to background

A Fire-Alarming Method Based on Video Processing

In the proposed system, a Gaussian-smoothed color histogram is first created to detect fire-colored pixels, and then a temporal variation is assumed to determine the final fire pixels from the candidate pixels. Unfortunately, the algorithm fails for cases that lack temporal variation.

Any algorithm of fire detection may firstly require a segmentation of fire region from an image for analysis, so engaging suitable color model is vital to fire detection. To simulate the color sensing properties of the human visual system, RGB color information is usually transformed into a mathematical space that decouples the brightness (or luminance) information from the color information. Among these models, HIS (hue/saturation/intensity) color model is very suitable for providing a more people-oriented way of describing the colors, because the hue and saturation components are closely related to the way in which human beings perceive color. YCbCr is a model derived from YUV color model, the essential characteristic of which is to separating the brightness and the color information. I1I2I3 is a feature space produced by pattern recognition theory for classification. Unfortunately, all the color models above may ignore slight anomalies for not considering the types of burning materials. Moreover, each of them is only suitable for fire detection in specific scenes.

In spite of the varieties of flame colors, the initial fires often display red-to yellow color. In order to reduce the computational complexity, the RGB color model is employed in the algorithm proposed by (A. T. Nghiem, F. Bremond, M. Thonnat, V. Valentin ETISEO) for its simplicity. The first rule in detecting fire colors is defined as R≥G≥B. Also, there should be a stronger R in the captured fire image due to the fact that R becomes the *m*ajor component in a colorful image of fire. This is because that fire is also a light source and the video camera needs adequate brightness during the night to capture the useful video sequences. Hence, the second rule adds a constraint that the value of R component should exceed a threshold, RT. However, *the* background lighting may affect the saturation of flames or generate a fire alias, which leads to false detections. To avoid being affected by the background illumination, the saturation value of the extracted flame needs to be larger than a specified threshold, ST. Accordingly, the saturation S should be in inverse quantity to the R component. To sum up, the three decision rules for extracting fire regions from an image can be described as follows:

Rule 1: $R > RT$
Rule 2: $R \geq G \geq B$
Rule 3: $S \geq (255 - R)$i$ST \,/\, RT$

In *the* decision rules, both RT and ST *are* defined according to various experimental outcomes, and typical values range from 55 to 65 and 115 to 135 for ST and RT, *respec*tively.

Algorithm

Color alone is not sufficient to identify fire. There are many non-fire objects that share the same color as fire, such as a desert sun and red leaves. The important key is to distinguish the fire and the fire-colored objects is their motion properties. Flame moves significantly between consecutive frames (at 30 frames/s), as shown in Figure 12. Unfortunately, some fire-similar regions in an image may have the same color as fire, which are usually detected as the real fire from the image. Those fire aliases are produced by two cases:

- Non-fire objects with the same color as fire and
- Background with illumination of fire-similar light sources.

In the first case, the object with reddish colors may cause a false alarm of fire. The second case is typically caused by solar reflections and artificial lights which bring negative influences to fire detection, and it makes the process complex and unreliable. In addition to chromatic features, dynamic features are typically adopted to distinguish fire aliases, including sudden movements of flames, changeable shapes,

Figure 12. Flow chart of growth criterion

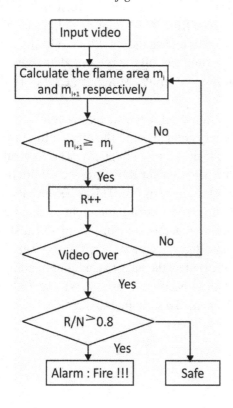

Figure 13. The flow chart of the detection criterion based on the invariability of centroid

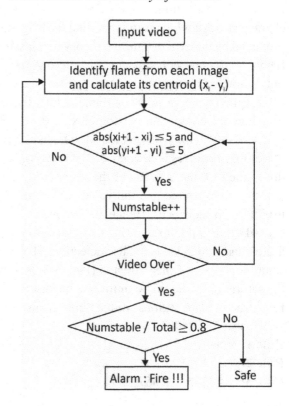

growing rate, and oscillation (or vibrations) in the infrared response. For improving the consistency of detection, algorithm utilize both the growth of fire pixels and the invariability of the centroid of flame to check the correctness of the fire regions detected by the chromatic features. Generally, the increasing of a burning fire will be mainly conquered by the air-flow and fuel type. The flame size keeps changing in an increasing trend due to air flowing, especially for initial burning flame. To obtain the growing feature of fire, algorithm calculate the number of fire-pixels in each frame and then compare the numbers of every two continuous frames. Let mi and $mi+1$ denote the number of fire-pixels in the current frame and next one, respectively. If the occurrence of $mi+1>mi$ is more than g times at intervals of tF during a time period T, it reveals that there is a growing trend of the fire which increases the reliability of the detection process, where g, tF and T rely on statistical data of experiments.

USE OF NAÏVE BAYES IN THE DECISION FUSION FRAMEWORK

Compound algorithm which contains of several sub-algorithms. Each sub-algorithm yields its own decision as a real number centered on zero, representing the level of confidence of that particular sub-algorithm. Each sub-algorithm has a weight related with it and the weights are updated online according to an active fusion method in accordance with the decisions made by a security guard. So, the weight of a sub-algorithm with poor performance declines during the training phase and the importance of that sub-algorithm becomes low in the decision making process. Decision values got by the sub-algorithms are linearly combined with these weights. The proposed automatic video-based wildfire detection algorithm is based on five sub-algorithms: (Kalirajan, K., and M. Sudha.)

Figure 14. Fire-alarming algorithm

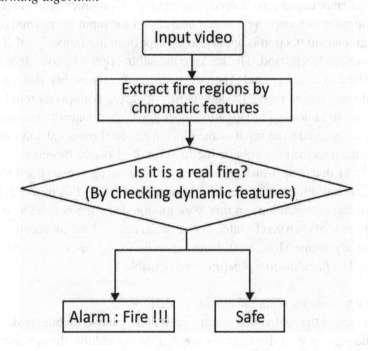

1. Slow moving video object detection;
2. Smoke-colored region detection;
3. Wavelet-transform-based region smoothness detection
4. Shadow detection and elimination; and
5. Covariance-matrix-based classification using Naïve Bayes classifier.

In this algorithm, for the purpose of final classification, Naive Bayes classifier is used. Various studies in image processing show that Naive Bayes classifier outdoes all other sophisticated algorithms such as SVM and is a best tool for image classification (Abidha T.E., Mathai P.P., Divya M). A Naive Bayes classifier is a simple probabilistic classifier based on applying Bayes' theorem with strong individuality assumptions. If these independence assumptions actually hold, a Naive Bayes classifier will converge earlier than other classifiers. Even if the NB assumption doesn't hold, a NB classifier still performs amazingly well in practice and it need less training data. Naive Bayes classifier is Fast to train (single scan), fast to classify, and not sensitive to irrelevant features. It can handle real, discrete and streaming data well. Naive Bayes classifiers have worked quite well in many complex real-world situations. There are sound theoretical reasons for the seemingly unlikely efficacy of naive Bayes classifier smoke detection algorithm for intelligent video surveillance system.

Algorithm uses motion and dissimilarity as the two key features for smoke detection. Motion is a primary sign and is used at the beginning for extraction from a current frame of candidate areas. In addition to consider a way of smoke distribution the movement estimation based on the optical ∞ow is applied. The relation of smoke Intensity to background intensity above than at objects with similar behavior, such as a fog, shadows from slowly moving objects and patches of light. Therefore contrast calculated with Weber formula is a good characteristic sign for a smoke. The algorithm is a group of the following modules as it is shown in Figure 15. A successive framesI_{t-2}, I_{t-1} and I_t and obtained from the stationary video surveillance camera are entered to an input of the preprocessing block. This block carries out some transformations which improve contrast qualities of the input frames and reduce calculations. Then adaptive background subtraction is applied to excerpt from the frame I_{t+1} of slowly moving areas and pixels of the so-called foreground. The background subtraction adaptive algorithm considers that a smoke slowly is mixed to a background. Then the connected components analysis is used in order to clear the foreground noise and to merge the slowly moving areas with pixels into blobs. The received connected blobs are moved into the classification block for Weber contrast analysis. At the same time the connected blobs are entered to an input of the block for optical ∞ow calculation Finally the classified block processes the information to obtain the final result of smoke detection.

The method starts by detecting which regions of the input image correspond to objects in motion many well-known techniques for motion detection can be applied for this purpose. Due to its simplicity -which is vital for fast computation - in this work motion detection is done by retaining a dynamic threshold to the magnitude of each pixel's intensity variation across three consecutive frames . The result of this process is a binary image $M(n)$. Next, focusing on these regions, the processing is split into two processing pipelines. The *fire detection pipeline* is responsible for:

1. Segmenting fire regions according to a color model;
2. Determining which of the segmented regions present a dynamic texture; and
3. filtering out the regions with dynamic texture that do not exhibit the spatio-temporal frequency signature of typical fires.

Figure 15. The method's processing pipeline

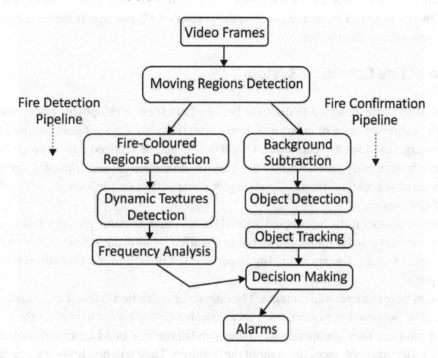

Despite the pipeline's strength, the presence of challenging fire-colored moving objects may still induce false fire alarms. To reduce the fire false alarm rate in these situations, knowledge about the location and category of the moving objects in the scene is used. This processing is the responsibility of the *fire confirmation pipeline*, which:

1. Detects foreground objects invariant to the presence of shadows;
2. Tracks the objects across frames;
3. Recognizes the objects' categories; and
4. Fuses the objects' awareness with the putative fire alarms to decide whether these should actually be issued.

A by-product of this pipeline is the possibility of generating alarms due to the presence of the specified categories of objects being detected and tracked. Ideally, to foster situational alertness in humans, fire alarms should be geo-referenced. For this purpose, the events natively described in the camera-frame need to be described in the world-frame, that is, they need to be mapped from pixel coordinates to GPS coordinates. Equally, the inverse mapping allows the object detection and tracking process to reject distracters based on the expected size of the objects' bounding boxes. One possibility to solve the mapping problem would be to know beforehand the GPS position of the camera, to make a few assumptions regarding the planarity of the environment, and to employ a camera calibration procedure. Nonetheless, here, calibration is done by learning from observing a moving person in the environment. This method avoids the hard planar assumption and makes the standardization procedure intuitive and, thus, easily deployable.

To accumulate learning data, a human equipped with a GPS-enabled PDA moves in the scene while being tracked by the system. During the process, the person's GPS position is stored alongside with the bounding box reported by the tracker

Importance of Fire Detection System

Fire detection systems are designed to discover fires early in their development when time will still be available for the safe evacuation of occupants. Early detection also plays a significant role in protecting the safety of emergency response personnel. Property loss can be reduced and stoppage for the operation minimized through early detection because control efforts are started while the fire is still small. Most alarm systems provide information to emergency responders on the location of the fire, speeding the process of fire control.

To be useful, detectors must be coupled with alarms. Alarm systems provide notice to at least the building occupants and usually transmit a signal to a staffed monitoring station either on or off site. In some cases, alarms may go directly to the fire department, although in most locations this is no longer the typical approach.

These systems have numerous advantages. The one major restriction is that they do nothing to contain or control the fire. Suppression systems such as automatic sprinklers act to control the fire. They also provide notification that they are operating, so they can fill the role of a heat detection-based system if connected to notification appliances throughout the building. They will not, however, operate as quickly as a smoke detection system. This is why facilities where rapid notice is essential, even when equipped with sprinklers, still need detection and alarm systems.

The most basic alarm system does not include detection. It has manual pull stations and sounds only a local alarm. This level of system is not what is typically used; it relies on an occupant to discover the fire, which can cause an important delay. The more quickly to be notified of the fire, the more costly the system you must install. Speed of detection is expensive. The slowest system to detect a fire is a heat detector, which is also the least expensive. An air-aspirating smoke detection system provides the most rapid indication of fire, but these systems are five to 10 times as expensive

CONCLUSION

One possible direction to improve the segmentation and recognition algorithm is to make use of flame characteristics such as flame color, intensity of flame brightness, shape, height, and flicker frequency. This is to augment the accuracy of video fire detection recognition. Another direction lies in the tracking and prediction of the direction of flame spread. A video based system which could do this, would be of assistance to those seeking refuge. It should be noted that experimental work is necessary for the testing and validation of different algorithms.

Little false detections are observed in the presence of significant noise, partial occlusions, and rapid angle change (J. H. Huang, J.-Y. Su, Z.-M. Lu, and J.-S. Pan). Future enhancement work includes the development of optical flow estimators with improved robustness to noise that take into account more than two frames at a time. While most of the challenges (such as variability of smoke density, lighting conditions, appearance that depends on the luminance conditions, colours of the scene, varying background, other moving objects and shadows; and unstable patterns) have been overcome using different novel techniques, many of the existing VFD approaches are still susceptible to false alarms (Phillips III, W., Shah, M., Lobo, N.V). The impact of the false fire alarms might include declaration of emergency situation that may eventually lead to general loss of confidence in the detection system and major economic losses. It is expected that future work in efficient dynamic texture segmentation will help in reducing this false alarm. Also more robust smoke detection with high detection accuracies and low false alarm rates can be obtained by combining multi-spectrum video information using various image fusion techniques (Arthur K. K. Wong, N. K. Fong, Elevier, R.Rathi, Dr. V. Ramalingam). Though such systems may be currently expensive, but they will be very useful with decrease in the costs of multi-spectrum image acquisition devices. To evaluate the video processing algorithms, new evaluation metrics on more video processing problems and tasks to validate the generalizing power of this evaluation methodology (A.T Nghiem, F. Bremond, M. Thonnat and R. Ma) . Also plant compute the dependencies between the parameter sets necessary for handling specific problems. Knowing these dependencies, future work will be able to estimate the reliability of the computation of the algorithm capacity upper bound.

REFERENCES

Abidha, T. E., & Matha, P. (2013). Reducing False Alarms in Vision Based Fire Detection with NB Classifier in EADF Framework. *International Journal of Scientific and Research Publications, 3*(8).

Abidha, T.E., Mathai, P.P., & Divya, M. (2013). Vision Based Wildfire Detection Using Bayesian Decision Fusion Framework. *International Journal of Advanced Research in Computer and Communication Engineering, 2*(12).

Adedapo, Jamiu, & Oladosu. (2014). Video-based Smoke Detection Algorithms: *A Chronological Survey. Computer Engineering and Intelligent Systems.*

Amer & Regazzoni. (2005). Introduction to the special issue on video object processing for surveillance applications. *Real-Time Imaging, 11*(3).

Celik, T. (2010). Fast and Efficient Method for Fire Detection Using Image Processing. *ETRI Journal, 32*(6), 2010. doi:10.4218/etrij.10.0109.0695

Huang, H., Su, J.-Y., Lu, Z.-M., & Pan, J.-S. (2006). A fire-alarming method based on video processing. In *IEEE International Conference on Intelligent Information Hiding and Multimedia Signal Processing.* doi:10.1109/IIH-MSP.2006.265017

Kalirajan, K., & Sudha, M. (2015). Moving Object Detection for Video Surveillance. *TheScientificWorldJournal, 2015*, 1–10. doi:10.1155/2015/907469 PMID:25861686

Millan-Garcia, L., Sanchez-Perez, G., Nakano, M., Toscano-Medina, K., Perez-Meana, H., & Rojas-Cardenas, L. (2012). An Early Fire Detection Algorithm Using IP Cameras. *Sensors (Basel, Switzerland), 12*(12), 5670–5686. doi:10.3390/s120505670 PMID:22778607

Nghiem, Bremond, Thonnat, & Ma. (2007). A New Evaluation Approach for Video Processing Algorithms. Academic Press.

Nghiem, A. T., Bremond, F., & Thonnat, M. (2007). *V. Valentin ETISEO, performance evaluation for video surveillance systems*. AVSS.

Nghiem, A. T., Bremond, F., Thonnat, M., & Ma, R. (2007). A New Evaluation Approach for Video Processing Algorithms. In *Proc. IEEE Workshop Motion and Video Computing*. Retrieved from http://www.agentvi.com/20-Technology-56-What_is_Video_AnalyticsETISEO

Nunes, Dias, Pereira, & Ugale. (2015, March). Optimized Flame Detection. *International Journal of Computer Applications*.

Ojha, S., & Sakhare, S. (2015). Image processing techniques for object tracking in video surveillance-A survey. In *Pervasive Computing (ICPC),2015International Conference on*. doi:10.1109/PERVA-SIVE.2015.7087180

Phillips, W. III, Shah, M., & Lobo, N. V. (2012, May). Flame recognition in video. *Pattern Recognition Letters, 23*(1–3), 319–327.

Piniarski, Pawłowski, & Da. browski. (2015). Video Processing Algorithms for Detection of Pedestrians. *CMST, 21*(3).

Rathi & Ramalingam. (2014). Pattern Classification Techniques for Flame Detection in Videos Using Optical Flow Estimation. *International Journal of Science, Engineering and Technology Research, 3*(2).

Ukinkar & Samvatsar. (2012). Object detection in dynamic background using image segmentation: A review. *International Journal of Engineering Research and Applications, 2*(3), 232-236.

Wong & Fong. (2013). *Experimental study of video fire detection and its applications*. International Conference on Performance-based Fire and Fire Protection Engineering, Wuhan, China.

Compilation of References

Abidha, T. E., & Matha, P. (2013). Reducing False Alarms in Vision Based Fire Detection with NB Classifier in EADF Framework. *International Journal of Scientific and Research Publications, 3*(8).

Abidha, T.E., Mathai, P.P., & Divya, M. (2013). Vision Based Wildfire Detection Using Bayesian Decision Fusion Framework. *International Journal of Advanced Research in Computer and Communication Engineering, 2*(12).

Adedapo, Jamiu, & Oladosu. (2014). Video-based Smoke Detection Algorithms: *A Chronological Survey. Computer Engineering and Intelligent Systems.*

Adel, A. (2010). *Steganography-Based Secret and Reliable Communications: Improving Steganography Capacity and Imperceptibility* (Doctor of Philosophy Thesis). Department of Information Systems and Computing, Brunel University.

Aggarwal, J. K., & Ryoo, M. S. (2011). Human activity analysis: A review. *ACM Computing Surveys, 43*(3), 16. doi:10.1145/1922649.1922653

Agrawal, D.D., Dubey, S.R., & Jalal, A.S. (2014). Emotion Recognition from Facial Expressions based on Multi-level Classification. *International Journal of Computational Vision and Robotics.*

Akita, K. (1984). Image Sequence Analysis of Real World Human Motion. *Pattern Recognition, 17*(1), 73–83. doi:10.1016/0031-3203(84)90036-0

Al-Azzawi, N. A., Sakim, H. A. M., & Abdullah, W. A. K. W. (2010). MR image monomodal registration based on the nonsubsampled contourlet transform and mutual information.*2010 International Conference on Computer Applications and Industrial Electronics (ICCAIE)*, (pp. 481 - 485). doi:10.1109/ICCAIE.2010.5735128

Ali, S., & Shah, M. (2010). Human action recognition in videos using kinematic features and multiple instance learning. *Pattern Analysis and Machine Intelligence. IEEE Transactions on, 32*(2), 288–303.

Alshennawy, A. A., & Aly, A. A. (2009). *Edge Detection in Digital Images Using Fuzzy Logic Technique.* World Academy of Science, Engineering and Technology.

Amer & Regazzoni. (2005). Introduction to the special issue on video object processing for surveillance applications. *Real-Time Imaging, 11*(3).

Amirtharajan, R., & Rayappan, J. B. B. (2012). An Intelligent Chaotic Embedding Approach to Enhance Stego-image Quality. *Information Sciences, 193*, 115–124. doi:10.1016/j.ins.2012.01.010

Angulo de vision. (n.d.). In *Wikipedia.* Retrieved from https://es.wikipedia.org/wiki/%C3%81ngulo_de_visi%C3%B3n

Anthimopoulos, M. M., Gianola, L., Scarnato, L., Diem, P., & Mougiakakou, S. G. (2014). A food recognition system for diabetic patients based on an optimized bag-of-features model. *Biomedical and Health Informatics. IEEE Journal of, 18*(4), 1261–1271.

Araki, T., Ikeda, N., Dey, N., Chakraborty, S., Saba, L., Kumar, D., & Suri, J. S. et al. (2015). A comparative approach of four different image registration techniques for quantitative assessment of coronary artery calcium lesions using intravascular ultrasound. *Computer Methods and Programs in Biomedicine, 118*(2), 158–172. doi:10.1016/j.cmpb.2014.11.006 PMID:25523233

Archana, S., Judice, A. A., & Kaliyamurthie, K. P. (2013). A Novel Approach on Image Steganography Methods for Optimum Hiding Capacity. *International Journal of Engineering and Computer Science, 2*(2), 378 - 385.

Artilec. (n.d.). *Artilec Seguridad Electrónica.* Retrieved June 07, 2012, from http://www.artilec.cl/site/productos.php?r=01&s=5

Ashburner, J. T., & Friston, K. J. (2007). *Rigid body registration. In Statistical Parametric Mapping: The Analysis of Functional Brain Images* (pp. 49–62). Academic Press. doi:10.1016/B978-012372560-8/50004-8

Atallah, M. A. (2012). A New Method in Image Steganography with Improved Image Quality. *Applied Mathematical Sciences Journal, 6*(79), 3907–3915.

Babita & Ayushi. (2013). Secure Image Steganography Algorithm using RGB Image Format and Encryption Technique. *International Journal of Computer Science & Engineering Technology, 4*(06), 758–762.

Bai, W., & Brady, M. (2008). Regularized B-spline deformable registration for respiratory motion correction in PET images. *2008 IEEE Nuclear Science Symposium Conference Record.*

Barrett, D. M., Beaulieu, J. C., & Shewfelt, R. (2010). Color, flavor, texture, and nutritional quality of fresh-cut fruits and vegetables: Desirable levels, instrumental and sensory measurement, and the effects of processing. *Critical Reviews in Food Science and Nutrition, 50*(5), 369–389. doi:10.1080/10408391003626322 PMID:20373184

Bashir, F., & Porikli, F. (2006). Performance evaluation of object detection and tracking systems. *Proc. IEEE Int. Workshop on Performance Evaluation of Tracking Systems.*

Berret, D. (2013, July 10). Online surveillance camera for every 11 people in Britain, says CCTV survey. *The Telegraph.*

Bertelle, C., Dutot, A., Lerebourg, S., Olivier, D., & du Havre, L. (2003). Road traffic management based on ant system and regulation model. In *Proc. of the Int. Workshop on Modeling and Applied Simulation* (pp. 35-43).

Bhardwaj, S., & Mittal, A. (2012). A Survey on Various Edge Detector Techniques. *Procedia Technology, 4,* 220–226. doi:10.1016/j.protcy.2012.05.033

Bharti, C., & Shaily, J. (2016). Video Steganography: A Survey. *IOSR Journal of Computer Engineering, 18*(1).

Bhavna, S., Shrikant, B., & Anant, G. K. (2013). Biometric Feature Based Steganography Scheme- An Approach Based On LSB Technique And Huffman Coding. *International Journal of Innovative Research and Studies, 2*(8), 247–257.

Bingham, G. P., Schmidt, R. C., & Rosenblum, L. D. (1995). Dynamics and the Orientation of Kinematic Forms in Visual Event Recognition. *Journal of Experimental Psychology. Human Perception and Performance, 21*(6), 1473–1493. doi:10.1037/0096-1523.21.6.1473 PMID:7490589

Biswas, R., & Sil, J. (2012). An Improved Canny Edge Detection Algorithm Based on Type-2 Fuzzy Sets. *Procedia Technology, 4,* 820–824. doi:10.1016/j.protcy.2012.05.134

Blake, R., & Shiffrar, M. (2007). Perception of human motion. *Annual Review of Psychology, 58*(1), 47–73. doi:10.1146/annurev.psych.57.102904.190152 PMID:16903802

Blank, M., Gorelick, L., Shechtman, E., Irani, M., & Basri, R. (2005). Actions as space-time shapes. In *Computer Vision, 2005. ICCV 2005. Tenth IEEE International Conference on* (Vol. 2, pp. 1395-1402).

Blasco, J., Aleixos, N., Gómez-Sanchis, J., & Moltó, E. (2009). Recognition and classification of external skin damage in citrus fruits using multispectral data and morphological features. *Biosystems Engineering, 103*(2), 137–145. doi:10.1016/j.biosystemseng.2009.03.009

Bouchrika, I. (2008). *Gait Analysis and Recognition for Automated Visual Surveillance*. University of Southampton.

Bouchrika, I., Carter, J. N., Nixon, M. S., Morzinger, R., & Thallinger, G. (2010). Using gait features for improving walking people detection. *20th International Conference on Pattern Recognition (ICPR)* (pp. 3097-3100). doi:10.1109/ICPR.2010.758

Bouchrika, I., & Nixon, M. S. (2006). Markerless Feature Extraction for Gait Analysis. *IEEE SMC Chapter Conference on Advanced in Cybernetic Systems*.

Bouwmans, T. (2011). Recent advanced statistical background modeling for foreground detection: A systematic survey. *Recent Patents on Computer Science, 4*(3).

Bouwmans, T., Porikli, F., Hferlin, B., & Vacavant, A. (2014). *Background Modeling and Foreground Detection for Video Surveillance*. Chapman and Hall/CRC. doi:10.1201/b17223

Bovik, A. C. (2000). *Handbook of Image and Video Processing*. Academic Press.

Bowyer, K., Kranenburg, C., & Dougherty, S. (2001). Edge detector evaluation using empirical ROC curves. *Journal on Computer Vision and Understanding, 84*(1), 77–103. doi:10.1006/cviu.2001.0931

Brutzer, S., Hoferlin, B., & Heideman, G. (2011). Evaluation of background subtraction techniques for video surveillance. *Proc. IEEE Conf. Computer Vision Pat. Recog.* (pp. 1937–1944). doi:10.1109/CVPR.2011.5995508

Burton, A., & Radford, J. (1978). *Thinking in perspective: critical essays in the study of thought processes*. Methuen.

Cámara inteligente. (n.d.). In *Wikipedia*. Retrieved from https://es.wikipedia.org/wiki/C%C3%A1mara_inteligente

CámarasIP. (n.d.). *CámarasIP*. Retrieved June 07, 2012, from http://www.camarasip.cl/es/

Canny, J. (1986). A computational approach to edge detection. *Pattern Analysis and Machine Intelligence, IEEE Transactions on*, (6), 679-698.

Canny, J. (1986). A computational approach to edge detection. *IEEE Transactions on Pattern Analysis and Machine Intelligence, 6*(6), 679–698. doi:10.1109/TPAMI.1986.4767851 PMID:21869365

Carnegie Mellon University. (2007). *Embedded vision processor, CMUcam3 datasheet* [Data file]. Retrieved from www.cmucam.org/.../48/CMUcam3_datasheet.pdf

Castejón, L., Miravete, A., & Cuartero, J. (2006). Composite bus rollover simulation and testing. *International Journal of Heavy Vehicle Systems, 13*(4), 281–297. doi:10.1504/IJHVS.2006.010584

Cedras, C., & Shah, M. (1995). Motion-based Recognition: A survey. *Image and Vision Computing, 13*(2), 129–155. doi:10.1016/0262-8856(95)93154-K

Celik, T. (2010). Fast and Efficient Method for Fire Detection Using Image Processing. *ETRI Journal, 32*(6), 2010. doi:10.4218/etrij.10.0109.0695

Chakrabarty, S., Pal, A. K., Dey, N., Das, D., & Acharjee, S. (2014, January). Foliage area computation using Monarch butterfly algorithm. In *Non Conventional Energy (ICONCE), 2014 1st International Conference on* (pp. 249-253). IEEE. doi:10.1109/ICONCE.2014.6808740

Chakraborty, S., Samanta, S., Biswas, D., Dey, N., & Chaudhuri, S. S. (2013, December). Particle swarm optimization based parameter optimization technique in medical information hiding. In *Computational Intelligence and Computing Research (ICCIC), 2013 IEEE International Conference on* (pp. 1-6). IEEE. doi:10.1109/ICCIC.2013.6724173

Challal, Y., Ouadjaout, A., Lasla, N., Bagaa, M., & Hadjidj, A. (2011). Secure and efficient disjoint multipath construction for fault tolerant routing in wireless sensor networks. *Journal of Network and Computer Applications, 34*(4), 1380–1397. doi:10.1016/j.jnca.2011.03.022

Chandrakant, B., Ashish, K. D., Bhupesh, K. P., Keerti, Y., & Kaushal, K. S. (2012). A new steganography technique: Image hiding in Mobile application. *International Journal of Advanced Computer and Mathematical Sciences, 3*(4), 556–562.

Chang, C. C., Lin, C. C., Tseng, C. S., & Tai, W. L. (2007). Reversible Hiding in DCT-based Compressed Images. *Information Sciences, 177*(13), 2768–2786. doi:10.1016/j.ins.2007.02.019

Chat, I. R. C. (n.d.). In *Wikipedia*. Retrieved from http://es.wikipedia.org/wiki/C%C3%A1mara_inteligente

Chaudhry, R., Ravichandran, A., Hager, G., & Vidal, R. (2009). *Histograms of oriented optical flow and binet-cauchy kernels on nonlinear dynamical systems for the recognition of human actions.* Paper presented at the Computer Vision and Pattern Recognition, 2009. CVPR 2009. IEEE Conference on. doi:10.1109/CVPR.2009.5206821

Chen, Feris, Zhai, Brown, & Hampapur. (2012). *An Integrated System for Moving Object Classification in Surveillance Videos.* Academic Press.

Chen, S. W., Yang, C. B., & Peng, Y. H. (2007). Algorithms for the traffic light setting problem on the graph model. Taiwanese Association for Artificial Intelligence.

Chen, Y. B., & Chen, O. T. C. (2009). Image segmentation method using thresholds automatically determined from picture contents. *EURASIP Journal on Image and Video Processing, 2009*, 1–15. doi:10.1155/2009/140492

Chinchu, E. A., & Iwin, Th. J. (2013). An Analysis of Various Steganography Algorithms. *International Journal of Advanced Research in Electronics and Communication Engineering, 2*(2), 116–123.

Chou, C. L., Lin, C. H., Chiang, T. H., Chen, H. T., & Lee, S. Y. (2015, June). Coherent event-based surveillance video synopsis using trajectory clustering. In *Multimedia & Expo Workshops (ICMEW), 2015 IEEE International Conference on*, (pp. 1-6).

Christensen, G. E., & He, J. (2001). Consistent nonlinear elastic image registration. *MM-BI, A01*, 1–5.

Chung, W. Y., & Oh, S. J. (2006). Remote monitoring system with wireless sensors module for room environment. *Sensors and Actuators. B, Chemical, 113*(1), 64–70. doi:10.1016/j.snb.2005.02.023

Comaniciu, D., & Meer, P. (1997, June). Robust analysis of feature spaces: color image segmentation. In *Computer Vision and Pattern Recognition, 1997. Proceedings., 1997 IEEE Computer Society Conference o*n (pp. 750-755). IEEE. doi:10.1109/CVPR.1997.609410

Comaniciu, D., Ramesh, V., & Meer, P. (2000). Real-time Tracking of Non-Rigid Objects using Mean Shift.*Proceedings. IEEE Conference on Computer Vision and Pattern Recognition, 2.* doi:10.1109/CVPR.2000.854761

Crandall, R. (1998). *Some Notes on Steganography.* Posted on Steganography Mailing List.

Cucchiara, R., Grana, C., Piccardi, M., & Prati, A. (2003). Detecting moving objects, ghosts, and shadows in video streams. *Pattern Analysis and Machine Intelligence. IEEE Transactions on, 25*(10), 1337–1342.

Cuevas, E. V., Zaldivar, D., & Rojas, R. (2005). *Kalman filter for vision tracking.* Academic Press.

Dakhole, A. Y., & Moon, M. P. (2013). Design of intelligent traffic control system based on ARM. *International Journal (Toronto, Ont.), 1*(6).

Dalal, N., & Triggs, B. (2005). Histograms of oriented gradients for human detection. *Computer Vision and Pattern Recognition, 2005. CVPR 2005. IEEE Computer Society Conference on.* doi:10.1109/CVPR.2005.177

Danese, G., Giachero, M., Leporati, F., Nazzicari, N., & Nobis, M. (2008). *An embedded acquisition system for remote monitoring of tire status in F1 race cars through thermal images.* Paper presented at 11thEuromicro Conference on Digital System Design, Parma, Italy. doi:10.1109/DSD.2008.55

Daniels, B., & Daniels, A. (n.d.). *Beaudaniels-illustration.* Retrieved from http://www.beaudaniels-illustration.com/technical-drawing-site-2/Cutaway.html

Dass, R., & Devi, P. S. (2012). Image Segmentation Techniques. *International Journal of Electronics & Communication Technology, 3*(1), 66–70.

Deepa, S., & Umarani, R. (2015). A Prototype for Secure Information using Video Steganography. *International Journal of Advanced Research in Computer and Communication Engineering, 4*(8).

Deepa, S., & Umarani, R. (2013). A Study on Digital Image Steganography. *International Journal of Advanced Research in Computer Science and Software Engineering, 3*(1), 54–57.

Dehariya, V. K., Shrivastava, S. K., & Jain, R. C. (2010, November). Clustering of image data set using K-means and fuzzy K-means algorithms. In *Computational Intelligence and Communication Networks (CICN), 2010 International Conference on* (pp. 386-391). IEEE.

Derrington, A. M., Allen, H. A., & Delicato, L. S. (2004). Visual mechanisms of motion analysis and motion perception. *Annual Review of Psychology, 55*(1), 181–205. doi:10.1146/annurev.psych.55.090902.141903 PMID:14744214

Deshayes, R., Mens, T., & Palanque, P. (2013). *A generic framework for executable gestural interaction models.* Paper presented at the Visual Languages and Human-Centric Computing (VL/HCC), 2013 IEEE Symposium on. doi:10.1109/VLHCC.2013.6645240

Dey, N., Roy, A. B., Pal, M., & Das, A. (2012). *FCM based blood vessel segmentation method for retinal images.* arXiv preprint arXiv: 1209.1181.

Dey, N., Chakraborty, S., & Samanta, S. (2013b). Optimization of watermarking in biomedical signal. *Lambert Publication. Heinrich-Böcking-Straße, 6,* 66121.

Dey, N., Samanta, S., Chakraborty, S., Das, A., Chaudhuri, S. S., & Suri, J. S. (2014). Firefly algorithm for optimization of scaling factors during embedding of manifold medical information: An application in ophthalmology imaging. *Journal of Medical Imaging and Health Informatics, 4*(3), 384–394. doi:10.1166/jmihi.2014.1265

Dey, N., Samanta, S., Yang, X. S., Das, A., & Chaudhuri, S. S. (2013a). Optimisation of scaling factors in electrocardiogram signal watermarking using cuckoo search. *International Journal of Bio-inspired Computation, 5*(5), 315–326. doi:10.1504/IJBIC.2013.057193

Di Caro, G., Ducatelle, F., & Gambardella, L. M. (2005). AntHocNet: An adaptive nature-inspired algorithm for routing in mobile ad hoc networks. *European Transactions on Telecommunications, 16*(5), 443–455. doi:10.1002/ett.1062

Dittrich, W. H. (1993). Action Categories and the Perception of Biological Motion. *Perception, 22*(1), 15–22. doi:10.1068/p220015 PMID:8474831

Dotoli, M., Fanti, M. P., & Melon, C. (2003, October). Real time optimization of traffic signal control: application to coordinated intersections. In *Systems, Man and Cybernetics, 2003. IEEE International Conference on* (Vol. 4, pp. 3288-3295). IEEE. doi:10.1109/ICSMC.2003.1244397

Dubey, S. R., & Jalal, A. S. (2014). *Fusing Color and Texture Cues to Categorize the Fruit Diseases from Images*. Academic Press.

Electrónica. (n.d.). In *Encyclopedia Libre Wikipedia*. Retrieved from http://es.wikipedia.org/wiki/Electr%C3%B3nica

Elhoseiny, Bakry, & Elgammal. (2013). Multi Class Object Classification in Video Surveillance Systems. In *Proceedings of CVPR 2013*, (pp. 788-793).

Erdem, C. E., Sankur, B., & Tekalp, A. M. (2004). Performance Measures for Video Object Segmentation and Tracking. *IEEE Transactions on Image Processing*, *13*(7), 937–951. doi:10.1109/TIP.2004.828427 PMID:15648860

Fan, L., Gao, T., & Chang, C. C. (2013, May). Mathematical Analysis of Extended Matrix Coding for Steganography. In *Sensor Network Security Technology and Privacy Communication System (SNS & PCS), 2013 International Conference on* (pp. 156-160). IEEE. doi:10.1109/SNS-PCS.2013.6553856

Fan, L., Gao, T., & Cao, Y. (2013). Improving the Embedding Efficiency of Weight Matrix-based Steganography for Grayscale Images. *Computers & Electrical Engineering*, *39*(3), 873–881. doi:10.1016/j.compeleceng.2012.06.014

Fan, L., Gao, T., Yang, Q., & Cao, Y. (2011). An Extended Matrix Encoding Algorithm for Steganography of High Embedding Efficiency. *Computers & Electrical Engineering*, *37*(6), 973–981. doi:10.1016/j.compeleceng.2011.08.006

Farbman, Z., Hoffer, G., Lipman, Y., Cohen-Or, D., & Lischinski, D. (2009, July). Coordinates for instant image cloning. *ACM Transactions on Graphics*, *28*(3), 67. doi:10.1145/1531326.1531373

Fawcett, T. (2006). An Introduction to ROC Analysis. *Pattern Recognition Letters*, *27*(8), 861–874. doi:10.1016/j.patrec.2005.10.010

Felzenszwalb, P. F., & Huttenlocher, D. P. (2004). Graph-based image segmentation. *International Journal of Computer Vision*, *59*(2), 167–181. doi:10.1023/B:VISI.0000022288.19776.77

Feng, S., Liao, S., Yuan, Z., & Li, S. Z. (2010, August). Online principal background selection for video synopsis. In *Pattern Recognition (ICPR), 2010 20th International Conference on*, (pp. 17-20). doi:10.1109/ICPR.2010.13

Firas, A. J. (2013). A Novel Steganography Algorithm for Hiding Text in Image using Five Modulus Method. *International Journal of Computers and Applications*, *72*(17), 39–44.

Fortun, D., Bouthemy, P., & Kervrann, C. (2015). Optical flow modeling and computation: A survey. *Computer Vision and Image Understanding*, *134*, 1–21. doi:10.1016/j.cviu.2015.02.008

Fragkiadaki, K., & Shi, J. (2012). Video segmentation by tracing discontinuities in a trajectory embedding. *Proceedings of the 2012 IEEE Conference on Computer Vision and Pattern Recognition* (pp. 1946-1853). doi:10.1109/CVPR.2012.6247883

Frame rate#Visible frame rate. (n.d.). In *Wikipedia*. Retrieved from https://en.wikipedia.org/wiki/Frame_rate#Visible_frame_rate

Fram, J. R., & Deutsch, E. S. (1975). On the Quantitative Evaluation of Edge Detection Schemes and Their Comparison with Human Performance. *IEEE Transactions on Computers*, *100*(6), 616–628. doi:10.1109/T-C.1975.224274

Freiman, M., Voss, S. D., & Warfield, S. K. (2011). Demons registration with local affine adaptive regularization: application to registration of abdominal structures. *2011 IEEE International Symposium on Biomedical Imaging: From Nano to Macro*. doi:10.1109/ISBI.2011.5872621

Galasso, F., Shankar Nagaraja, N., Jimenez Cardenas, T., Brox, T., & Schiele, B. (2013). A unified video segmentation benchmark: Annotation, metrics and analysis.*Proceedings of the IEEE International Conference on Computer Vision* (pp. 3527-3534). doi:10.1109/ICCV.2013.438

Gao, W., Zhang, X., Yang, L., & Liu, H. (2010) An Improved Sobel Edge Detection.*Proceedings of the 2010 3rd IEEE International Conference on Computer Science and Information Technology (ICCSIT)* (Vol. 5, pp. 67-71).

Gavrila, D., & Davis, L. (1995). *Towards 3-d model-based tracking and recognition of human movement: a multi-view approach.* International workshop on automatic face-and gesture-recognition.

Gayathri, C., & Kalpana, V. (2013). Study on Image Steganography Techniques. *International Journal of Engineering and Technology, 5*(2), 572–577.

Genovese, M., Bifulco, P., De Caro, D., Napoli, E., Petra, N., Romano, M., & Strollo, A. G. (2015). Hardware Implementation of a Spatio-temporal Average Filter for Real-time Denoising of Fluoroscopic Images. *Integration, the VLSI Journal, 49*, 114-124.

Ghaffari, A., & Fatemizadeh, E. (2013). Mono-modal image registration via correntropy measure. *2013 8th Iranian Conference on Machine Vision and Image Processing* (MVIP), (pp. 223 - 226). doi:10.1109/IranianMVIP.2013.6779983

Ghaffari, A., & Fatemizadeh, E. (2013). Sparse based similarity measure for mono-modal image registration. *2013 8th Iranian Conference on Machine Vision and Image Processing* (MVIP), (pp. 462 - 466).

Giacomo, C. (2009). *New techniques for steganography and steganalysis in the pixel domain* (Ph.D. Thesis). Dipartimento di Ingegneria dell'Informazione, Universita Degli Studi di Siena.

Glasgow, H. B., Burkholder, J. M., Reed, R. E., Lewitus, A. J., & Kleinman, J. E. (2004). Real-time remote monitoring of water quality: A review of current applications, and advancements in sensor, telemetry, and computing technologies. *Journal of Experimental Marine Biology and Ecology, 300*(1–2), 409–448. doi:10.1016/j.jembe.2004.02.022

Gobar, L. T. D. A. (n.d.). *Gobar, Seguridad - Electronica Electricidad.* Retrieved June 07, 2012, from http://www.gobar.cl/?ver=detalle&id=1142

Goddard, N. H. (1992). *The Perception of Articulated Motion: Recognizing Moving Light Displays.* University of Rochester.

Gómez-Hernández, E., Feregrino-Uribe, C., & Cumplido, R. (2008, March). FPGA Hardware Architecture of the Steganographic Context Technique. In *Electronics, Communications and Computers, 2008. CONIELECOMP 2008, 18*[th] *International Conference on* (pp. 123-128). IEEE doi:10.1109/CONIELECOMP.2008.24

Gómez-Moreno, H., Maldonado-Bascón, S., & López-Ferreras, F. (2001). Edge detection in noisy images using the support vector machines. In Connectionist Models of Neurons, Learning Processes, and Artificial Intelligence (pp. 685-692). Springer Berlin Heidelberg. doi:10.1007/3-540-45720-8_82

Gonzalez, R. C., & Woods, R. E. (2002). *Digital image processing* (2nd ed.). Prentice Hall.

Goyette, N., Jodoin, P.-M., Porikli, F., Konrad, J., & Ishwar, P. (2012). Changedetection.net: A new change detection benchmark dataset.*Proceedings of the IEEE CVPR change detection workshop.* doi:10.1109/CVPRW.2012.6238919

Goyette, N., Jodoin, P.-M., Porikli, F., Konrad, J., & Ishwar, P. (2014). A Novel Video Dataset for Change Detection Benchmarking. *IEEE Transactions on Image Processing, 23*(11), 4663–4679. doi:10.1109/TIP.2014.2346013 PMID:25122568

Greg, K. (2004). *Investigator's Guide to Steganography.* Auerbach Publications.

Guglielmo Marconi. (n.d.). In *Encyclopedia Asifunciona online.* Retrieved from http://www.asifunciona.com/biografias/marconi/marconi.htm

Guo, Y., Xu, G., & Tsuji, S. (1994). Understanding Human Motion Patterns. *Pattern Recognition, Conference B: Computer Vision & Image Processing., Proceedings of the 12th IAPR International. Conference on, 2.*

Han & Kamber. (2006). *Data Mining: Concepts and Techniques* (2nd ed.). Morgan Kaufmann Publishers.

Hansen, M. S., Larsen, R., Glocker, B., & Navab, R. (2008). Adaptive parametrization of multivariate B-splines for image registration. *IEEE Conference on Computer Vision and Pattern Recognition.* doi:10.1109/CVPR.2008.4587760

Hao, L., Cao, J., & Li, C. (2013, June). Research of GrabCut algorithm for single camera video synopsis. In *Intelligent Control and Information Processing (ICICIP), 2013 Fourth International Conference on,* (pp. 632-637). doi:10.1109/ICICIP.2013.6568151

Haralick, R. M., & Shapiro, L. G. (1985). Image segmentation techniques. *CVGIP, 29,* 100–132.

Harsh, P., Tushar, S., Gyanendra, O., & Sunil, C. (2012). Information Hiding in an Image File: Steganography. *International Journal of Computer Science and Information Technologies, 3*(3), 4216–4217.

Heath, M., Sarkar, S., Sanocki, T., & Bowyer, K. (1998). Comparison of Edge Detectors. *Computer Vision and Image Understanding, 69*(1), 38–54. doi:10.1006/cviu.1997.0587

Hemalatha, R., Santhiyakumari, N., & Suresh, S. (2015, January). Implementation of Medical Image Segmentation using Virtex FPGA kit. In *Signal Processing and Communication Engineering Systems (SPACES), 2015 International Conference on* (pp. 358-362). IEEE. doi:10.1109/SPACES.2015.7058283

Hemalatha, S., Dinesh, U. A., Renuka, A., & Priya, R. K. (2013). A Secure and High Capacity Image Steganography Technique. *Signal & Image Processing: An International Journal, 4*(1), 83 - 89.

Hernandez, M., Medioni, G., Hu, Z., & Sadda, H. (2015). Multimodal Registration of Multiple Retinal Images Based on Line Structures. *2015 IEEE Winter Conference on Applications of Computer Vision,* (pp. 907 - 914). doi:10.1109/WACV.2015.125

Hines, G. D., Rahman, Z. U., Jobson, D. J., & Woodell, G. A. (2004, July). DSP Implementation of the Retinex Image Enhancement Algorithm. In *Defense and Security* (pp. 13–24). International Society for Optics and Photonics.

Ho, J., Yang, M. H., Rangarajan, A., & Vemuri, B. (2007). A New Affine Registration Algorithm for Matching 2D Point Sets. *IEEE Workshop on Applications of Computer Vision.* doi:10.1109/WACV.2007.6

Höferlin, B., Höferlin, M., Weiskopf, D., & Heidemann, G. (2011). Information-based adaptive fast-forward for visual surveillance. *Multimedia Tools and Applications, 55*(1), 127–150. doi:10.1007/s11042-010-0606-z

Hong, G., & Zhang, Y. (2005). The Image Registration Technique for High Resolution Remote Sensing Image in Hilly Area. *International Society of Photogrammetry and Remote Sensing Symposium.*

Hong, Z. Q. (1991). Algebraic feature extraction of image for recognition. *Pattern Recognition, 24*(3), 211–219. doi:10.1016/0031-3203(91)90063-B

Hoover, A., Jean-Baptiste, G., Jiang, X., Flynn, P. J., Bunke, H., Goldgof, D. B., & Fisher, R. B. et al. (1996). An experimental comparison of range image segmentation algorithms. *IEEE Transactions on Pattern Analysis and Machine Intelligence, 18*(7), 673–689. doi:10.1109/34.506791

Hore, S., Chakroborty, S., Ashour, A. S., Dey, N., Ashour, A. S., Sifaki-Pistolla, D., & Chowdhury, S. R. (2015). Finding Contours of Hippocampus Brain Cell Using Microscopic Image Analysis. *Journal of Advanced Microscopy Research, 10*(2), 93–103. doi:10.1166/jamr.2015.1245

Hsia, C. H., Chiang, J. S., Hsieh, C. F., & Hu, L. C. (2013, November). A complexity reduction method for video synopsis system. In *Intelligent Signal Processing and Communications Systems (ISPACS), 2013 International Symposium on*, (pp. 163-168). doi:10.1109/ISPACS.2013.6704540

Huang, H., Su, J.-Y., Lu, Z.-M., & Pan, J.-S. (2006). A fire-alarming method based on video processing. In *IEEE International Conference on Intelligent Information Hiding and Multimedia Signal Processing*. doi:10.1109/IIH-MSP.2006.265017

Huang, C. R., Chung, P. C. J., Yang, D. K., Chen, H. C., & Huang, G. J. (2014). Maximum a Posteriori Probability Estimation for Online Surveillance Video Synopsis. *Circuits and Systems for Video Technology. IEEE Transactions on*, *24*(8), 1417–1429.

Hung, M. H., Pan, J. S., & Hsieh, C. H. (2014). A fast algorithm of temporal median filter for background subtraction. *Journal of Information Hiding and Multimedia Signal Processing*, *5*(1), 33–40.

Hussain, M., & Hussain, M. (2013). *A Survey of Image Steganography Techniques*. Academic Press.

Hussian, R., Sharma, S., Sharma, V., & Sharma, S. (2013). WSN applications: automated intelligent traffic control system using sensors. *International Journal of Soft Computing and Engineering*.

Ikizler, N., Cinbis, R. G., & Duygulu, P. (2008). *Human action recognition with line and flow histograms*. Paper presented at the Pattern Recognition, 2008. ICPR 2008. 19th International Conference on. doi:10.1109/ICPR.2008.4761434

Indradip, B., Souvik, B., & Gautam, S. (2013). Study and Analysis of Steganography with Pixel Factor Mapping (PFM) Method. *International Journal of Application or Innovation in Engineering & Management*, *2*(8), 268–266.

Irani, M., & Peleg, S. (1991). Improving resolution by image registration. *CVGIP: Graphical Models and Image Proc.*, *53*, 231–239.

Isard, M. C., & Blake, A. C. (1998). CONDENSATION: Conditional Density Propagation for Visual Tracking. *International Journal of Computer Vision*, *29*(1), 5–28. doi:10.1023/A:1008078328650

Jack, W., & Russ, R. (2007). *Techno Security's Guide to Managing Risks for IT Managers, Auditors, and Investigators*. Elsevier Inc.

Jain, A. K. (1989). *Fundamentals of Digital Image Processing*. Prentice-Hall, Inc.

Jain, A. K., & Dubes, R. C. (1988). *Algorithms for clustering data*. Prentice-Hall, Inc.

Jain, R., & Binford, T. (1991). Ignorance, myopia, and naivete in computer vision systems. *CVGIP. Image Understanding*, *53*(1), 112–117. doi:10.1016/1049-9660(91)90009-E

Janiczek, R. L., Gilliam, A. D., Antkowiak, P., Acton, S. T., & Epstein, F. H. (2005). Automated Affine Registration of First-Pass Magnetic Resonance Images. *Conference Record of the Thirty-Ninth Asilomar Conference on Signals, Systems and Computers*. doi:10.1109/ACSSC.2005.1599747

Jeeshna, P. V., & Kuttymalu, V. K. (2015). A Technique for Object Movement Based Video Synopsis. IEEE Transaction on International Journal of Engineering and Advanced Technology, 20(9), 1303-1315.

Jiménez, A. R., Jain, A. K., Ceres, R., & Pons, J. L. (1999). Automatic fruit recognition: A survey and new results using range/attenuation images. *Pattern Recognition*, *32*(10), 1719–1736. doi:10.1016/S0031-3203(98)00170-8

Jiri, H. (2011). *Artificial Intelligence Applied on Cryptoanalysis Aimed on Revealing Weaknesses of Modern Cryptology and Computer Security* (Dissertation thesis). Department of Informatics and Artificial Intelligence, Zlin Faculty of Applied Informatics, Tomas Bata University.

Johansson, G. (1973). Visual Perception of Biological Motion and a Model for its Analysis. *Perception & Psychophysics, 14*(2), 201–211. doi:10.3758/BF03212378

Kaehler, A., & Bradski, G. (2013). *Learning OpenCV – Computer Vision in C++ with OpenCV Library* (1st ed.). O'Reilly Media.

Kafi, M. A., Challal, Y., Djenouri, D., Bouabdallah, A., Khelladi, L., & Badache, N. (2012). A study of wireless sensor network architectures and projects for traffic light monitoring. *Procedia Computer Science, 10*, 543-552.

Kaganami, H. G., & Beiji, Z. (2009, September). Region-based segmentation versus edge detection. In *Intelligent Information Hiding and Multimedia Signal Processing, 2009. IIH-MSP'09. Fifth International Conference on* (pp. 1217-1221). IEEE. doi:10.1109/IIH-MSP.2009.13

Kale, S. B., & Dhok, G. P. (2013). Design of intelligent ambulance and traffic control. *Int. J. Comput. Electron. Res, 2*(2).

Kaliannan, J., Baskaran, A., & Dey, N. (2015). Automatic generation control of thermal-thermal-hydro power systems with PID controller using ant colony optimization. *International Journal of Service Science, Management, Engineering, and Technology, 6*(2), 18–34. doi:10.4018/ijssmet.2015040102

Kalirajan, K., & Sudha, M. (2015). Moving Object Detection for Video Surveillance. *TheScientificWorldJournal, 2015*, 1–10. doi:10.1155/2015/907469 PMID:25861686

Kang, H. W., Matsushita, Y., Tang, X., & Chen, X. Q. (2006, June). Space-time video montage. In computer vision and pattern recognition, 2006 IEEE computer society conference on, (vol. 2, pp. 1331-1338).

Kang, W. X., Yang, Q. Q., & Liang, R. P. (2009, March). The comparative research on image segmentation algorithms. In *2009 First International Workshop on Education Technology and Computer Science* (pp. 703-707). IEEE. doi:10.1109/ETCS.2009.417

Kantharia, P., Patel, T., & Thakker, M. (2014). Design of sensor fault detection and remote monitoring system for temperature measurement. *International Journal of Current Engineering and Technology, 4*(2), 504–508.

Karaulova, I. A., Hall, P. M., & Marshall, A. D. (2000). A Hierarchical Model of Dynamics for Tracking People with a Single Video Camera. In *Proceedings of the 11th British Machine Vision Conference, 1*, 352-361. doi:10.5244/C.14.36

Kareem, E. I. A., & Jantan, A. (2011). An intelligent traffic light monitor system using an adaptive associative memory. *IJIPM: International Journal of Information Processing and Management, 2*(2), 23–39. doi:10.4156/ijipm.vol2.issue2.4

Karthikeyan, B., Vaithiyanathan, V., Venkatraman, B., & Menaka, M. (2012). Analysis of image segmentation for radiographic images. *Indian Journal of Science and Technology, 5*(11), 3660–3664.

Kaustubh, C. (2012). Image Steganography and Global Terrorism. *Global Security Studies, 3*(4), 115–135.

Kedar, N. C., & Aakash, W. (2015). A Survey Paper on Video Steganography. *International Journal of Computer Science and Information Technologies, 6*(3).

Khaire, P. A., & Thakur, N. V. (2012). A Fuzzy Set Approach for Edge Detection. *International Journal of Image Processing, 6*(6), 403–412.

Khaire, P. A., & Thakur, N. V. (2012). An Overview of Image Segmentation Algorithms. *International Journal of Image Processing and Vision Sciences, 1*(2), 62–68.

Khalifa, A. R. (2010). Evaluating the effectiveness of region growing and edge detection segmentation algorithms. *J. Am. Sci, 6*(10).

Khan, J. F., Bhuiyan, S., & Adhami, R. R. (2011). Image segmentation and shape analysis for road-sign detection. *Intelligent Transportation Systems. IEEE Transactions on, 12*(1), 83–96.

Kliper-Gross, O., Gurovich, Y., Hassner, T., & Wolf, L. (2012). Motion interchange patterns for action recognition in unconstrained videos. *European Conference on Computer Vision*, (pp. 256-269). doi:10.1007/978-3-642-33783-3_19

Ko, T. (2008). *A survey on behavior analysis in video surveillance for homeland security applications.* Paper presented at the Applied Imagery Pattern Recognition Workshop, 2008. AIPR'08. 37th IEEE. doi:10.1109/AIPR.2008.4906450

Koprinska, I., & Carrato, S. (2001). Temporal video segmentation: A survey. *Signal Processing Image Communication, 16*(5), 477–500. doi:10.1016/S0923-5965(00)00011-4

Kozlowski, L. T., & Cutting, J. E. (1978). Recognizing the Gender of Walkers from Point-Lights Mounted on Ankles: Some Second Thoughts. *Perception & Psychophysics, 23*(5), 459. doi:10.3758/BF03204150

Kuehne, H., Jhuang, H., Garrote, E., Poggio, T., & Serre, T. (2011). *HMDB: a large video database for human motion recognition.* Paper presented at the Computer Vision (ICCV), 2011 IEEE International Conference on. doi:10.1109/ICCV.2011.6126543

Kühnel, C., Westermann, T., Hemmert, F., Kratz, S., Müller, A., & Möller, S. (2011). I'm home: Defining and evaluating a gesture set for smart-home control. *International Journal of Human-Computer Studies, 69*(11), 693–704. doi:10.1016/j.ijhcs.2011.04.005

Ladjailia, A., Bouchrika, I., Merouani, H. F., & Harrati, N. (2015a). Automated Detection of Similar Human Actions using Motion Descriptors. *16th international conference on Sciences and Techniques of Automatic control and computer engineering (STA).* IEEE.

Ladjailia, A., Bouchrika, I., Merouani, H. F., & Harrati, N. (2015b). On the Use of Local Motion Information for Human Action Recognition via Feature Selection.*4th IEEE International Conference on Electrical Engineering (ICEE).* doi:10.1109/INTEE.2015.7416792

Lakshmanan, A. G., Swarnambiga, A., Vasuki, S., & Raja, A. A. (2013). Affine based image registration applied to MRI brain.*2013 International Conference on Information Communication and Embedded Systems (ICICES).* doi:10.1109/ICICES.2013.6508186

Langote, V. B., & Chaudhari, D. D. (2012). Segmentation Techniques for Image Analysis. *International Journal of Advanced Engineering Research and Studies, 1.*

Laptev, I. (2005). On space-time interest points. *International Journal of Computer Vision, 64*(2-3), 107–123. doi:10.1007/s11263-005-1838-7

Lara, O. D., & Labrador, M. A. (2013). A survey on human activity recognition using wearable sensors. *IEEE Communications Surveys and Tutorials, 15*(3), 1192–1209. doi:10.1109/SURV.2012.110112.00192

Lee, B. Y., View, L. H., Cheah, W. S., & Wang, Y. C. (2014). *Occlusion Handling in Videos Object Tracking: A Survey.* IOP Conference.

Leezaben, A. P. (2010). *Steganography Using Cylinder Insertion Algorithm and Mobile Based Stealth Steganography* (Master of Science Thesis). San Diego State University, San Diego, CA.

Lehmann, F. (2011). Turbo segmentation of textured images. Pattern Analysis and Machine Intelligence. *IEEE Transactions on, 33*(1), 16–29.

Li, Y., Zhang, T., & Tretter, D. (2001). *An overview of video abstraction techniques*. Technical Report HPL-2001-191. HP Laboratory.

Li, H., Manjunath, B. S., & Mitra, S. K. (1995). A contour based approach to multisensor image registration. *IEEE Transactions on Image Processing, 4*(3), 320–334. doi:10.1109/83.366480 PMID:18289982

Li, H., & Ngan, K. N. (2007). Automatic video segmentation and tracking for content-based applications. *IEEE Communications Magazine, 45*(1), 27–33. doi:10.1109/MCOM.2007.284535

Lijuan, Z., Dongming, L., Junnan, W., & Hui, Z. (2012). High-accuracy image registration algorithm using B-splines. *2012 2nd International Conference on Computer Science and Network Technology* (ICCSNT), (pp. 279 - 283). doi:10.1109/ICCSNT.2012.6525938

Li, L., Huang, W., Gu, I. Y. H., Leman, K., & Tian, Q. (2003, October 5-8). Principal color representation for tracking persons.*Proceedings of the IEEE International Conference on Systems, Man and Cybernetics*, (pp. 1007-1012).

Lim, Y. W., & Lee, S. U. (1990). On the color image segmentation algorithm based on the thresholding and the fuzzy c-means techniques. *Pattern Recognition, 23*(9), 935–952. doi:10.1016/0031-3203(90)90103-R

Lino, A. C. L., Sanches, J., & Fabbro, I. M. D. (2008). Image processing techniques for lemons and tomatoes classification. *Bragantia, 67*(3), 785–789. doi:10.1590/S0006-87052008000300029

Li, Q., Sato, I., & Murakami, I. (2007). Affine registration of multimodality images by optimization of mutual information using a stochastic gradient approximation technique.*2007 IEEE International Geoscience and Remote Sensing Symposium*. doi:10.1109/IGARSS.2007.4422814

Li, S., Tian, J., Yang, Z., & Qiao, F. (2013). Research and implement of remote vehicle monitoring and early-warning system based on GPS/GPRS. In *Proceedings of SPIE 8768- International Conference on Graphic and Image Processing (ICGIP 2012)*. Singapore: SPIE. doi:10.1117/12.2010751

Liu, A., & Yang, Z. (2010, October). An interactive method for dynamic video synopsis generation. In *Computer Application and System Modeling (ICCASM), 2010 International Conference on*.

Liu, W., Pokharel, P. P., & Principe, J. C. (2008). The kernel least-mean-square algorithm. *Signal Processing. IEEE Transactions on, 56*(2), 543–554. doi:10.1109/TSP.2007.907881

Lo, B. P. L., & Velastin, S. A. (2001). Automatic congestion detection system for underground platforms. In *Intelligent Multimedia, Video and Speech Processing, 2001.Proceedings of 2001 International Symposium on*, (pp. 158-161). doi:10.1109/ISIMP.2001.925356

Lu, M., Wang, Y., & Pan, G. (2013, May). Generating fluent tubes in video synopsis. In *Acoustics, Speech and Signal Processing (ICASSP), 2013 IEEE International Conference on*, (pp. 2292-2296). doi:10.1109/ICASSP.2013.6638063

Lucas, B. D., & Kanade, T. et al. (1981). An iterative image registration technique with an application to stereo vision. *IJCAI, 81*, 674–679.

Lucas, B., & Kanade, T. (1981). An iterative image registration technique with an application to stereo vision.*Proc. DARPA Image Understanding Workshop*, (pp. 121–130).

Lu, H., Reyes, M., Šerifović, A., Weber, S., Sakurai, Y., Yamagata, H., & Cattin, P. C. (2010). Multi-modal diffeomorphic demons registration based on point-wise mutual information.*2010 IEEE International Symposium on Biomedical Imaging: From Nano to Macro*, (pp. 372 – 375). doi:10.1109/ISBI.2010.5490333

Luo, J., Gray, R. T., & Lee, H. C. (1998, October). Incorporation of derivative priors in adaptive Bayesian color image segmentation. In *Image Processing, 1998. ICIP 98. Proceedings. 1998 International Conference on* (pp. 780-784). IEEE.

Ma, W. Y., & Manjunath, B. S. (1997, June). Edge flow: a framework of boundary detection and image segmentation. In *Computer Vision and Pattern Recognition, 1997. Proceedings., 1997 IEEE Computer Society Conference on* (pp. 744-749). IEEE. doi:10.1109/CVPR.1997.609409

Mainville Electric Motor Co. (n.d.). *Motores de carro de golf*. Retrieved from http://www.manvillemotor.com/golf_cart_motors_spanish.htm

Makrogiannis, S., Wellen, J., Wu, Y., Bloy, L., & Sarkar, S. K. (2007). A Multimodal Image Registration and Fusion Methodology Applied to Drug Discovery Research. *IEEE 9th Workshop on Multimedia Signal Processing*. doi:10.1109/MMSP.2007.4412883

Mamta, J., & Parvinder, S. S. (2013). Information Hiding using Improved LSB Steganography and Feature Detection Technique. *International Journal of Engineering and Advanced Technology, 2*(4), 275 - 279.

Manoj, G., Senthur, T., Sivasankaran, M., Vikram, M., & Bharatha, S. G. (2013). AES Based Steganography. *International Journal of Application or Innovation in Engineering & Management, 2*(1), 382 - 389.

Marszalek, M., Laptev, I., & Schmid, C. (2009). *Actions in context*. Paper presented at the Computer Vision and Pattern Recognition, 2009. CVPR 2009. IEEE Conference on. doi:10.1109/CVPR.2009.5206557

Martin, D., Fowlkes, C., Tal, D., & Malik, J. (2001). A database of human segmented natural images and its application to evaluating segmentation algorithms and measuring ecological statistics.*Proceedings of the 8th International Conference Computer Vision* (pp. 416-423). doi:10.1109/ICCV.2001.937655

Martínez, F., Manzanera, A., & Romero, E. (2012). *A motion descriptor based on statistics of optical flow orientations for action classification in video-surveillance. In Multimedia and Signal Processing* (pp. 267–274). Springer.

Materka & Strzelecki. (1998). Texture Analysis Methods – A Review. Institute of Electronics, COST B11 report.

Meier, T., & Ngan, K. N. (1998). Automatic segmentation of moving objects for video object plane generation. *IEEE Transactions on Circuits and Systems for Video Technology, 8*(5), 525–538. doi:10.1109/76.718500

Mendhurwar, K., Patil, S., Sundani, H., Aggarwal, P., & Devabhaktuni, V. (2011). Edge-Detection in Noisy Images Using Independent Component Analysis. *ISRN Signal Processing*.

Mennatallah, M. S., Amal, S. K., & Mostafa, G. M. (2014). *Video steganography: A comprehensive review*. Springer.

Mente, R., Dhandra, B. V., & Mukarambi, G. (2014). Color Image Segmentation and Recognition based on Shape and Color Features. *International Journal on Computer Science and Engineering, 3*(01).

Millan-Garcia, L., Sanchez-Perez, G., Nakano, M., Toscano-Medina, K., Perez-Meana, H., & Rojas-Cardenas, L. (2012). An Early Fire Detection Algorithm Using IP Cameras. *Sensors (Basel, Switzerland), 12*(12), 5670–5686. doi:10.3390/s120505670 PMID:22778607

Mishra, A., Mondal, P., & Banerjee, S. (2012). Modified Demons deformation algorithm for non-rigid image registration. *2012 4th International Conference on Intelligent Human Computer Interaction* (IHCI), (pp. 1 - 5). doi:10.1109/IHCI.2012.6481800

Mishra, A., Mondal, P., & Banerjee, S. (2015). VLSI-Assisted Nonrigid Registration Using Modified Demons Algorithm. *IEEE Transactions on Very Large Scale Integration (VLSI) Systems, 23*(12), 2913–2921.

Mishra, B., Pati, U. C., & Sinha, U. (2015). Modified demons registration for highly deformed medical images.*2015 Third International Conference on Image Information Processing (ICIIP)*, (pp. 152 - 156). doi:10.1109/ICIIP.2015.7414757

Moeslund, T. B., Hilton, A., & Krüger, V. (2006). A survey of advances in vision-based human motion capture and analysis. *Computer Vision and Image Understanding, 104*(2), 90–126. doi:10.1016/j.cviu.2006.08.002

Mohammed, N. H. A. (2012). Text Realization Image Steganography. *International Journal of Engineering, 6*(1), 1–9.

Mohanty & Sethi. (2013). A Frame Based Decision Pooling Method for Video Classification. *Annual IEEE India Conference (INDICON)*.

Mohd, B. J., Abed, S., Al-Hayajneh, T., & Alouneh, S. (2012, May). FPGA Hardware of the LSB Steganography Method. In *Computer, Information and Telecommunication Systems (CITS), 2012 International Conference on* (pp. 1-4). IEEE. doi:10.1109/CITS.2012.6220393

Monitorización. (n.d.). In *Encyclopedia Libre Wikipedia*. Retrieved June 09, 2011, from http://es.wikipedia.org/wiki/Monitorizaci%C3%B3n

Motorman, S. A. (n. d.). *Motorman*. Retrieved April 01, 2016, from http://www.motorman.cl/

Motwani, M. C., Gadiya, M. C., Motwani, R. C., & Harris, F. C. (2004, September). Survey of image denoising techniques. In *Proceedings of GSPX* (pp. 27-30).

Mushrif, M. M., & Ray, A. K. (2008). Color image segmentation: Rough-set theoretic approach. *Pattern Recognition Letters, 29*(4), 483–493. doi:10.1016/j.patrec.2007.10.026

Mustafa, A. E., ElGamal, A.M.F., ElAlmi, M.E., & Ahmed, B. D. (2011). A Proposed Algorithm For Steganography In Digital Image Based on Least Significant Bit. *Research Journal Specific Education,* (21), 752 - 767.

Nadernejad, E., Sharifzadeh, S., & Hassanpour, H. (2008). Edge detection techniques: Evaluations and comparison. *Applied Mathematical Sciences, 2*(31), 1507–1520.

Nagham, H., Abid, Y., Badlishah, A., & Osamah, M. (2012). Image Steganography Techniques: An Overview. *International Journal of Computer Science and Security, 6*(3), 168–187.

Nagham, H., Abid, Y., Badlishah, R., Dheiaa, N., & Lubna, K. (2013). Steganography in image files: A survey. *Australian Journal of Basic and Applied Sciences, 7*(1), 35–55.

Naidele, K. M. (2012). *Stealthy Plaintext. Master's Projects, Master's Theses and Graduate Research. San Jose State (SJSU)* University.

Nascimento, J., & Marques, J. (2006). Performance evaluation of object detection algorithms for video surveillance. *IEEE Transactions on Multimedia, 8*(8), 761–774. doi:10.1109/TMM.2006.876287

Naz, S., Majeed, H., & Irshad, H. (2010, October). Image segmentation using fuzzy clustering: A survey. In *Emerging Technologies (ICET), 2010 6th International Conference on* (pp. 181-186). IEEE. doi:10.1109/ICET.2010.5638492

Nelson, R. C., & Polana, R. (1992). Qualitative recognition of motion using temporal texture. *CVGIP. Image Understanding, 56*(1), 78–89. doi:10.1016/1049-9660(92)90087-J

Nghiem, A. T., Bremond, F., Thonnat, M., & Ma, R. (2007). A New Evaluation Approach for Video Processing Algorithms. In *Proc. IEEE Workshop Motion and Video Computing*. Retrieved from http://www.agentvi.com/20-Technology-56-What_is_Video_AnalyticsETISEO

Nghiem, Bremond, Thonnat, & Ma. (2007). A New Evaluation Approach for Video Processing Algorithms. Academic Press.

Nghiem, A. T., Bremond, F., & Thonnat, M. (2007). *V. Valentin ETISEO, performance evaluation for video surveillance systems*. AVSS.

Niebles, J. C., Wang, H., & Fei-Fei, L. (2008). Unsupervised learning of human action categories using spatial-temporal words. *International Journal of Computer Vision, 79*(3), 299–318. doi:10.1007/s11263-007-0122-4

Nie, Y., Sun, H., Li, P., Xiao, C., & Ma, K. L. (2014). Object Movements Synopsis via Part Assembling and Stitching. *Visualization and Computer Graphics. IEEE Transactions on, 20*(9), 1303–1315.

Nie, Y., Xiao, C., Sun, H., & Li, P. (2013). Compact video synopsis via global spatiotemporal optimization. *Visualization and Computer Graphics. IEEE Transactions on, 19*(10), 1664–1676.

Ninawe, P., & Pandey, M. S. (2014). A Completion on Fruit Recognition System Using K-Nearest Neighbors Algorithm. *International Journal of Advanced Research in Computer Engineering & Technology, 3*(7).

Nishi, K., & Kanchan, S. G. (2015). Video Steganography by Using Statistical Key Frame Extraction Method and LSB Technique. *International Journal of Innovative Research in Science, Engineering and Technology, 4*(10).

Nunes, Dias, Pereira, & Ugale. (2015, March). Optimized Flame Detection. *International Journal of Computer Applications*.

Ogale, A. S., Karapurkar, A., & Aloimonos, Y. (2007). *View-invariant modeling and recognition of human actions using grammars. In Dynamical vision* (pp. 115–126). Springer.

Oh, J., Wen, Q., Hwang, S., & Lee, J. (2004). Video abstraction. *Video data management and information retrieval*, 321-346.

Oikonomopoulos, A., Patras, I., & Pantic, M. (2005). Spatiotemporal salient points for visual recognition of human actions. *Systems, Man, and Cybernetics, Part B: Cybernetics. IEEE Transactions on, 36*(3), 710–719.

Ojha, S., & Sakhare, S. (2015). Image processing techniques for object tracking in video surveillance- A survey. In *Pervasive Computing (ICPC),2015International Conference on.* doi:10.1109/PERVASIVE.2015.7087180

Oshin, O., Gilbert, A., & Bowden, R. (2014). Capturing relative motion and finding modes for action recognition in the wild. *Computer Vision and Image Understanding, 125*, 155–171. doi:10.1016/j.cviu.2014.04.005

Ou, Y., & GuangZhi, D. (2001). Color Edge Detection Based on Data Fusion Technology in Presence of Gaussian Noise. *Procedia Engineering, 15*, 2439–2443. doi:10.1016/j.proeng.2011.08.458

Ozkurt, C., & Camci, F. (2009). Automatic traffic density estimation and vehicle classification for traffic surveillance systems using neural networks. *Mathematical and Computational Applications, 14*(3), 187–196. doi:10.3390/mca14030187

Padhiyar, S., & Joshi, D. (2015). *A Survey. International Journal of Engineering Sciences & Research Technology*.

Pal, G., Acharjee, S., Rudrapaul, D., Ashour, A. S., & Dey, N. (2015). Video segmentation using minimum ratio similarity measurement. *International Journal of Image Mining, 1*(1), 87–110. doi:10.1504/IJIM.2015.070027

Pal, N. R., & Pal, S. K. (1993). A review on image segmentation techniques. *Pattern Recognition, 26*(9), 1277–1294. doi:10.1016/0031-3203(93)90135-J

Pan, J., Li, S., & Zhang, Y. (2000). Automatic extraction of moving objects using multiple features and multiple frames. *Proceedings of IEEE Symposium Circuits and Systems*, (Vol. 1, pp. 36-39).

Papari, G., & Petkov, N. (2011). Edge and line oriented contour detection: State of the art. In Image and Vision Computing (pp. 79–103).

Patel & Patel. (2013). Illumination Invariant Moving Object Detection. *International Journal of Computer and Electrical Engineering, 5*(1).

Patel, J., Patwardhan, J., Sankhe, K., & Kumbhare, R. (2011). Fuzzy Inference based Edge Detection System using Sobel and Laplacian of Gaussian Operators.*Proceedings of the International Conference & Workshop on Emerging Trends in Technology ICWET'11* (pp. 694-697). doi:10.1145/1980022.1980171

Pauwels, E. J., & Frederix, G. (1999). Finding salient regions in images-non-parametric clustering for image segmentation and grouping. *Computer Vision and Image Understanding*, 75(1-2), 73–85. doi:10.1006/cviu.1999.0763

Pepik, Stark, Gehler, & Schiele. (2013). *Occlusion Patterns for Object Detection*. CVPR 2013.

Pérez, P., Gangnet, M., & Blake, A. (2003, July). Poisson image editing. *ACM Transactions on Graphics*, 22(3), 313–318. doi:10.1145/882262.882269

Petrovic, N., Jojic, N., & Huang, T. S. (2005). Adaptive video fast forward. *Multimedia Tools and Applications*, 26(3), 327–344. doi:10.1007/s11042-005-0895-9

Phillips, W. III, Shah, M., & Lobo, N. V. (2012, May). Flame recognition in video. *Pattern Recognition Letters*, 23(1–3), 319–327.

Piccardi, M. (2004). *Background subtraction techniques: a review*. Academic Press.

Piccardi, M. (2004, October). Background subtraction techniques: a review. In Systems, man and cybernetics, 2004 IEEE international conference on, (vol. 4, pp. 3099-3104). doi:10.1109/ICSMC.2004.1400815

Pietikainen, Maenpaa, & Viertola. (2002). *Color Texture Classification with Color Histograms and Local Binary Patterns*. Machine Vision Group.

Piniarski, Pawłowski, & Da̧browski. (2015). Video Processing Algorithms for Detection of Pedestrians. *CMST, 21*(3).

Płaczek, B. (2011). Performance evaluation of road traffic control using a fuzzy cellular model. In *Hybrid Artificial Intelligent Systems* (pp. 59–66). Springer Berlin Heidelberg. doi:10.1007/978-3-642-21222-2_8

Polana, R., & Nelson, R. (1994). *Low level recognition of human motion (or how to get your man without finding his body parts)*. Paper presented at the Motion of Non-Rigid and Articulated Objects. doi:10.1109/MNRAO.1994.346251

Poorani, M., Prathiba, T., & Ravindran, G. (2013). Integrated Feature Extraction for Image Retrieval. *IJCSMC*, 2(2), 28–35.

Poornima, R., & Iswarya, R. J. (2013). An Overview Of Digital Image Steganography. *International Journal of Computer Science & Engineering Survey*, 4(1), 23–31. doi:10.5121/ijcses.2013.4102

Poppe, R. (2010). A survey on vision-based human action recognition. *Image and Vision Computing*, 28(6), 976–990. doi:10.1016/j.imavis.2009.11.014

Pradeepa, P., & Vennila, I. (2012). A multimodal image registration using mutual information.*2012 International Conference on Advances in Engineering, Science and Management (ICAESM)*, (pp. 474 - 477).

Pratt, W. K. (2007). *Digital Image Processing* (4th ed.). Wiley & Sons, Inc. doi:10.1002/0470097434

Price, K. (1986). Anything You Can Do, I Can Do Better (No You Can't). *Computer Vision Graphics and Image Processing*, 36(2-3), 387–391. doi:10.1016/0734-189X(86)90083-6

Pritch, Y., Ratovitch, S., Hende, A., & Peleg, S. (2009, September). Clustered synopsis of surveillance video. In *Advanced Video and Signal Based Surveillance, 2009. AVSS'09. Sixth IEEE International Conference on*, (pp. 195-200). doi:10.1109/AVSS.2009.53

Pritch, Y., Rav-Acha, A., & Peleg, S. (2008). Nonchronological video synopsis and indexing. *Pattern Analysis and Machine Intelligence. IEEE Transactions on*, 30(11), 1971–1984.

Queen, C. M., & Albers, C. J. (2008, July). Forecasting traffic flows in road networks: A graphical dynamic model approach. In *Proceedings of the 28th International Symposium of Forecasting*. International Institute of Forecasters.

Radke, R., Andra, S., Al-Kofahi, O., & Roysam, B. (2005). Image change detection algorithms: A systematic survey. *IEEE Transactions on Image Processing, 14*(3), 294–307. doi:10.1109/TIP.2004.838698 PMID:15762326

Rahman, S., See, J., & Ho, C. C. (2015). *Action Recognition in Low Quality Videos by Jointly Using Shape, Motion and Texture Features*. Paper presented at the IEEE Int. Conf. on Signal and Image Processing Applications. doi:10.1109/ICSIPA.2015.7412168

Rajagopalan, S., Amirtharajan, R., Upadhyay, H. N., & Rayappan, J. B. B. (2012). Survey and Analysis of Hardware Cryptographic and Steganographic Systems on FPGA. *Journal of Applied Sciences, 12*(3), 201–210. doi:10.3923/jas.2012.201.210

Rajagopalan, S., Prabhakar, P. J., Kumar, M. S., Nikhil, N. V. M., Upadhyay, H. N., Rayappan, J. B. B., & Amirtharajan, R. (2014). MSB Based Embedding with Integrity: An Adaptive RGB Stego on FPGA Platform. *Information Technology Journal, 13*(12), 1945–1952. doi:10.3923/itj.2014.1945.1952

Rajanikanth, R. K. (2009). *A High Capacity Data-Hiding Scheme in LSB-Based Image Steganography* (Master of Science Thesis). Graduate Faculty of the University of Akron, Akron, OH.

Rathi & Ramalingam. (2014). Pattern Classification Techniques for Flame Detection in Videos Using Optical Flow Estimation. *International Journal of Science, Engineering and Technology Research, 3*(2).

Rav-Acha, A., Pritch, Y., & Peleg, S. (2006, June). Making a long video short: Dynamic video synopsis. In *Computer Vision and Pattern Recognition, 2006 IEEE Computer Society Conference on*, (vol. 1, pp. 435-441).

Reducindo, I., Arce-Santana, E. R., Campos-Delgado, D. U., & Alba, A. (2010). Evaluation of multimodal medical image registration based on Particle Filter. *2010 7th International Conference onElectrical Engineering Computing Science and Automatic Control* (CCE), (pp. 406 - 411). doi:10.1109/ICEEE.2010.5608648

Ren, Z., Meng, J., Yuan, J., & Zhang, Z. (2011). *Robust hand gesture recognition with kinect sensor*. Paper presented at the 19th ACM international conference on Multimedia. doi:10.1145/2072298.2072443

Rodríguez, S. (n.d.). *Masadelante.com, servicios y recursos para tener éxito en internet*. Retrieved from http://www.masadelante.com/faqs/resolucion

Rohit, G. (2012). Comparison Of Lsb & Msb Based Steganography In Gray-Scale Images. *International Journal of Engineering Research & Technology, 1*(8), 1–6.

Rohr, K. (1994). Towards Model-Based Recognition of Human Movements in Image Sequences. *CVGIP. Image Understanding, 59*(1), 94–115. doi:10.1006/ciun.1994.1006

Rotem, O., Greenspan, H., & Goldberger, J. (2007, June). Combining region and edge cues for image segmentation in a probabilistic gaussian mixture framework. In *Computer Vision and Pattern Recognition, 2007. CVPR'07. IEEE Conference on* (pp. 1-8). IEEE. doi:10.1109/CVPR.2007.383232

Roy, P., Goswami, S., Chakraborty, S., Azar, A. T., & Dey, N. (2014). Image segmentation using rough set theory: A review. *International Journal of Rough Sets and Data Analysis, 1*(2), 62–74. doi:10.4018/ijrsda.2014070105

Rui, Y., Huang, T. S., & Chang, S.-F. (1999). Image retrieval: Current techniques, promising directions, and open issues. *Journal of Visual Communication and Image Representation, 10*(1), 39–62. doi:10.1006/jvci.1999.0413

Saha, B. N., Ray, N., Greiner, R., Murtha, A., & Zhang, H. (2012). Quick detection of brain tumors and edemas: A bounding box method using symmetry. *Computerized Medical Imaging and Graphics, 36*(2), 95–107. doi:10.1016/j. compmedimag.2011.06.001 PMID:21719256

Samanta, S., Acharjee, S., Mukherjee, A., Das, D., & Dey, N. (2013a, December). Ant weight lifting algorithm for image segmentation. In *Computational Intelligence and Computing Research (ICCIC), 2013 IEEE International Conference on* (pp. 1-5). IEEE. doi:10.1109/ICCIC.2013.6724160

Samanta, S., Chakraborty, S., Acharjee, S., Mukherjee, A., & Dey, N. (2013b, December). Solving 0/1 knapsack problem using ant weight lifting algorithm. In *Computational Intelligence and Computing Research (ICCIC), 2013 IEEE International Conference on* (pp. 1-5). IEEE.

Samsung Techwin Co. Ltd. (2010). *High resolution day & night camera SCB-2000, specifications* [Data file]. Retrieved from http://www.hanwhatechwinamerica.com/SAMSUNG/upload/Product_Specifications/SCB-2000_Datasheet.pdf

San Biagio, Bazzani, Cristani, & Murino. (2014). *Weighted Bag of Visual Words for Object Recognition.* Academic Press.

Sapna Varshney, S., Rajpa, N., & Purwar, R. (2009, December). Comparative study of image segmentation techniques and object matching using segmentation. In *Methods and Models in Computer Science, 2009. ICM2CS 2009.Proceeding of International Conference on* (pp. 1-6). IEEE. doi:10.1109/ICM2CS.2009.5397985

Sassi, O. B., Delleji, T., Taleb-Ahmed, A., Feki, I., & Hamida, A. B. (2008). *MR Image Monomodal Registration Using Structure Similarity Index. 2008 First Workshops on Image Processing Theory* (pp. 1–5). Tools and Applications.

Satyanarayana, G. V., & Mazaruddin, S. D. (2013). *Wireless sensor based remote monitoring system for agriculture using ZigBee and GPS.* Paper presented at Conference on Advances in Communication and Control Systems 2013 (CAC2S 2013), India.

Schindler, K., & Van Gool, L. (2008). *Action snippets: How many frames does human action recognition require?* Paper presented at the Computer Vision and Pattern Recognition, 2008. CVPR 2008. IEEE Conference on. doi:10.1109/ CVPR.2008.4587730

Schüldt, C., Laptev, I., & Caputo, B. (2004). *Recognizing human actions: a local SVM approach.* Paper presented at the Pattern Recognition. doi:10.1109/ICPR.2004.1334462

Seifedine, K., & Sara, N. (2013). New Generating Technique for Image Steganography. *Lecture Notes on Software Engineering, 1*(2), 190–193.

Sekhar Panda, C., & Patnaik, S. (2010). Filtering Corrupted Image and Edge Detection in Restored Grayscale Image Using Derivative Filters. *International Journal of Image Processing, 3*(3), 105–119.

Setayesh, M., Zhang, M., & Johnston, M. (2011). Detection of Continuous, Smooth and Thin Edges in Noisy Images Using Constrained Particle Swarm Optimization.*Proceedings of the 13th annual conference on Genetic and evolutionary computation* (pp. 45-52).

Shafait, F., Keysers, D., & Breuel, T. M. (2008, January). Efficient implementation of local adaptive thresholding techniques using integral images. In *Electronic Imaging 2008* (p. 681510–681510). International Society for Optics and Photonics. doi:10.1117/12.767755

Shah, M., & Jain, R. (2013). *Motion-based recognition* (Vol. 9). Springer Science & Business Media.

Shah, N. N., & Dalal, U. D. (2016). Hardware Efficient Double Diamond Search Block Matching Algorithm for Fast Video Motion Estimation. *Journal of Signal Processing Systems for Signal, Image, and Video Technology, 82*(1), 115–135. doi:10.1007/s11265-015-0993-5

Shatha A. B., & Ahmed, S. N. (2013). Steganography in Mobile Phone over Bluetooth. *International Journal of Information Technology and Business Management, 16*(1), 111-117.

Shawn, D. D. (2007). *An Overview of Steganography.* James Madison University.

Shechtman, E., & Irani, M. (2007). *Matching local self-similarities across images and videos.* Paper presented at the Computer Vision and Pattern Recognition. doi:10.1109/CVPR.2007.383198

Shih, F. Y., & Cheng, S. (2004). Adaptive mathematical morphology for edge linking. *Information Sciences, 167*(1), 9–21. doi:10.1016/j.ins.2003.07.020

Shi, J., & Malik, J. (2000). Normalized cuts and image segmentation. *IEEE Transactions on Pattern Analysis and Machine Intelligence, 22*(8), 888–905. doi:10.1109/34.868688

Shikha, S., & Sumit, B. (2013). Image Steganography: A Review. *International Journal of Emerging Technology and Advanced Engineering, 3*(1), 707–710.

Shoushtarian, B. (2003). A practical approach to real-time dynamic background generation based on a temporal median filter. *Journal of Sciences, 14*(4), 351–362.

Sinhmar, P. (2012). Intelligent traffic light and density control using IR sensors and microcontroller. *International Journal of Advanced Technology & Engineering Research, 2*(2), 30–35.

Siu, A., & Lau, E. (2005). Image Registration for Image-Based Rendering. *IEEE Transactions on Image Processing, 14*(1), 241–252. doi:10.1109/TIP.2004.840690 PMID:15700529

Smeets, D., Keustermans, J., Hermans, J., Vandermeulen, D., & Suetens, P. (2012). Feature-based piecewise rigid registration in 2-D medical images. *2012 9th IEEE International Symposium on Biomedical Imaging* (ISBI).

Sneha, A., & Sanyam, A. (2013). A New Approach for Image Steganography using Edge Detection Method. *International Journal of Innovative Research in Computer and Communication Engineering, 1*(3), 626–629.

Soh, A. C., Rhung, L. G., & Sarkan, H. M. (2010). MATLAB simulation of fuzzy traffic controller for multilane isolated intersection. *International Journal on Computer Science and Engineering, 2*(4), 924–933.

Somasundaram, S. K., & Alli, P. (2012). A Review on Recent Research and Implementation Methodologies on Medical Image Segmentation. *Journal of Computer Science, 8*(1), 170–174. doi:10.3844/jcssp.2012.170.174

Somol, P., Pudil, P., Novovičová, J., & Paclík, P. (1999). Adaptive floating search methods in feature selection. *Pattern Recognition Letters, 20*(11), 1157–1163. doi:10.1016/S0167-8655(99)00083-5

Song, Y., Glasbey, C. A., Horgan, G. W., Polder, G., Dieleman, J. A., & van der Heijden, G. W. A. M. (2014). Automatic fruit recognition and counting from multiple images. *Biosystems Engineering, 118*, 203–215. doi:10.1016/j.biosystemseng.2013.12.008

Sonka, M., Hlavac, V., & Boyle, R. (2014). *Image processing, analysis, and machine vision.* Cengage Learning.

Souvik, B., Indradip, B., & Gautam, S. (2011). A Survey of Steganography and Steganalysis Technique in Image, Text, Audio and Video as Cover Carrier. *Journal of Global Research in Computer Science, 2*(4), 1–16.

Sravanthi, G.S., Sunitha, D. B., Riyazoddin, S.M., & Janga, R. M. (2012). A Spatial Domain Image Steganography Technique Based on Plane Bit Substitution Method. *Global Journal of Computer Science and Technology Graphics & Vision, 12*(15).

Srivastava, P. M. D., Sachin, S., Sharma, S., & Tyagi, U. (2012). Smart traffic control system using PLC and SCADA. *International Journal of Innovative Research in Science, Engineering and Technology, 1*(2).

Sudipta, K. (2010). *An Enhanced, Secure and Comprehensive Data Hiding Approach Using 24 Bit Color Images* (Master of Technology thesis). Faculty Council for UG and PG Studies in Engineering & Technology, Jadavpur University, Kolkata, India.

Suma, E. A., Krum, D. M., Lange, B., Koenig, S., Rizzo, A., & Bolas, M. (2013). Adapting user interfaces for gestural interaction with the flexible action and articulated skeleton toolkit. *Computers & Graphics, 37*(3), 193–201. doi:10.1016/j.cag.2012.11.004

Sun, S., Haynor, D. R., & Kim, Y. (2003). Semiautomatic video object segmentation using Vsnakes. *IEEE Transactions on Circuits and Systems for Video Technology, 13*(1), 75–82. doi:10.1109/TCSVT.2002.808089

Sural, S., Qian, G., & Pramanik, S. (2002). Segmentation and histogram generation using the HSV color space for image retrieval. In *Image Processing. 2002. Proceedings. 2002 International Conference on* (Vol. 2, pp. II-589). IEEE. doi:10.1109/ICIP.2002.1040019

Tanaka, M., Kamio, R., & Okutomi, M. (2012, November). Seamless image cloning by a closed form solution of a modified poisson problem. In SIGGRAPH Asia 2012 Posters (p. 15). ACM. doi:10.1145/2407156.2407173

Tang, S., Andriluka, M., & Schiele, B. (2014). Detection and tracking of occluded people. *International Journal of Computer Vision, 110*(1), 58–69. doi:10.1007/s11263-013-0664-6

Tatiana, E. (2013). *Reversible Data Hiding in Digital Images* (Master's thesis). Tampere University of Technology.

Tayana, M. (2012). *Image Steganography Applications for Secure Communication* (Master of Science Thesis). Faculty of Engineering, Built Environment and Information Technology, University of Pretoria, Pretoria, South Africa.

Tekalp, A. M. (1995). *Digital Video Processing*. Prentice-Hall, Inc.

Thakkar, M., & Shah, H. (2010). *Automatic thresholding in edge detection using fuzzy approach*. IEEE. doi:10.1109/ICCIC.2010.5705868

The Berkeley Segmentation Dataset and Benchmark. (n.d.). Retrieved from https://www.eecs.berkeley.edu/Research/Projects/CS/vision/bsds/

Toklu, C., Tekalp, M., & Erdem, A. T. (2000). Semi-automatic video object segmentation in the presence of occlusion. *IEEE Transactions on Circuits and Systems for Video Technology, 10*(4), 624–629. doi:10.1109/76.845008

Toyama, K., Krumm, J., Brumitt, B., & Wallflower, B. M. (1999). Principles and practice of background maintenance. *Proc. IEEE Int. Conf. Computer Vision* (Vol. 1, pp. 255–261).

Trendnet. (2009). *ProView poe internet camera tv-ip501p, specifications* [Data file]. Retrieved from http://downloads.trendnet.com/tv-ip501p/datasheet/en_spec_tv-ip501p(v1.0r)-091009.pdf

Trendnet. (n.d.). *Trendnet*. Retrieved April 01, 2016, from https://www.trendnet.com/langsp/products/proddetail?prod=165_TV-IP501P

Trojan Battery Company. (2015). *T-605 datasheet* [Data file]. Retrieved from http://www.trojanbattery.com/pdf/datasheets/T605_Trojan_Data_Sheets.pdf

Troje, N. F., Westhoff, C., & Lavrov, M. (2005). Person Identification from Biological Motion: Effects of Structural and Kinematic Cues. *Perception & Psychophysics, 67*(4), 667–675. doi:10.3758/BF03193523 PMID:16134460

Tron, R., & Vidal, R. (2007). A benchmark for the comparison of 3-D motion segmentation algorithms.*Proceedings of the 2007 IEEE Conference on Computer Vision and Pattern Recognition.* doi:10.1109/CVPR.2007.382974

Truong, B. T., & Venkatesh, S. (2007). Video abstraction: A systematic review and classification. *ACM Transactions on Multimedia Computing, Communications, and Applications, 3*(1). doi:10.1145/1198302.1198305

Tseng, Y. C., Chen, Y. Y., & Pan, H. K. (2002). A Secure Data Hiding Scheme for Binary Images. *Communications. IEEE Transactions on, 50*(8), 1227–1231. doi:10.1109/TCOMM.2002.801488

Tubaishat, M., Shang, Y., & Shi, H. (2007, January). Adaptive traffic light control with wireless sensor networks. In *Proceedings of IEEE consumer communications and networking conference* (pp. 187-191). doi:10.1109/CCNC.2007.44

Turaga, P., Chellappa, R., Subrahmanian, V. S., & Udrea, O. (2008). Machine recognition of human activities: A survey. *Circuits and Systems for Video Technology. IEEE Transactions on, 18*(11), 1473–1488.

Tustison, N. J., Avants, B. A., & Gee, J. C. (2007). Improved FFD B-Spline Image Registration. *2007 IEEE 11th International Conference on Computer Vision.*

Ukinkar & Samvatsar. (2012). Object detection in dynamic background using image segmentation: A review. *International Journal of Engineering Research and Applications, 2*(3), 232-236.

Uma, S., & Saurabh, M. (n.d.). A Secure Data Hiding Technique Using Video Steganography. *International Journal of Computer Science & Communication Networks, 5*(5), 348-357.

Unnikrishnan, R., Pantofaru, C., & Hebert, M. (2007). Toward Objective Evaluation of Image Segmentation Algorithms. *IEEE Transactions on Pattern Analysis and Machine Intelligence, 29*(6), 929–944. doi:10.1109/TPAMI.2007.1046 PMID:17431294

VanDaniker, M. (2009, August). Visualizing real-time and archived traffic incident data. In *Information Reuse & Integration, 2009. IRI'09. IEEE International Conference on* (pp. 206-211). IEEE. doi:10.1109/IRI.2009.5211552

Vanmathi, C., & Prabu, S. (2013). A Survey of State of the Art techniques of Steganography. *IACSIT International Journal of Engineering and Technology, 5*(1), 376–379.

Vezzani, R., & Cucchiara, R. (2010). Video surveillance online repository (visor): An integrated framework. *Multimedia Tools and Applications, 50*(2), 359–380. doi:10.1007/s11042-009-0402-9

Viraj, S. G. (2010). *Steganography Using Cone Insertion Algorithm and Mobile Based Stealth Steganography* (Master of Science Thesis). San Diego State University, San Diego, CA.

Vishwakarma, S., & Agrawal, A. (2013). A survey on activity recognition and behavior understanding in video surveillance. *The Visual Computer, 29*(10), 983–1009. doi:10.1007/s00371-012-0752-6

Voravuthikunchai, W., Cremilleux, B., & Jurie, F. (2014). Histograms of Pattern Sets for Image Classification and Object Recognition. *CVP.*

Wahed, M., El-tawel, G. S., & El-karim, A. G. (2013). Automatic Image Registration Technique of Remote Sensing Images. *International Journal (Toronto, Ont.), 4*, 177–187.

Wang, W., Liu, L., Jiang, Y., & Kuang, G. (2011). Point-based rigid registration using Geometric Topological Inference algorithm. *2011 3rd International Asia-Pacific Conference on Synthetic Aperture Radar* (APSAR), (pp. 1-3).

Wang, Y., & Wang, Q. (2010, October). Image segmentation based on multi-scale local feature. In *Image and Signal Processing (CISP), 2010 3rd International Congress on* (Vol. 3, pp. 1406-1409). IEEE. doi:10.1109/CISP.2010.5648301

Wang, Y., Huang, K., & Tan, T. (2007). Human activity recognition based on r transform. *Computer Vision and Pattern Recognition, 2007. CVPR'07. IEEE Conference on* (pp. 1-8).

Wang, J., Thiesson, B., Xu, Y., & Cohen, M. (2004). Image and video segmentation by anisotropic kernel mean shift.*Proceedings of the European Conference of Computer Vision (ECCV '04)*, (LNCS), (Vol. 3022, pp. 238-249). doi:10.1007/978-3-540-24671-8_19

Wang, K., Lu, Z. M., & Hu, Y. J. (2013). A High Capacity Lossless Data Hiding Scheme for JPEG Images. *Journal of Systems and Software, 86*(7), 1965–1975. doi:10.1016/j.jss.2013.03.083

Wang, Y., Jodoin, P. M., Porikli, F., Konrad, J., Benezeth, Y., & Ishwar, P. (2014). CDnet 2014: An Expanded Change Detection Benchmark Dataset.*Proceedings of the IEEE Conference on Computer Vision and Pattern Recognition Workshops* (pp. 387-394). doi:10.1109/CVPRW.2014.126

Weinland, D., & Boyer, E. (2008). Action recognition using exemplar-based embedding. *Computer Vision and Pattern Recognition, 2008. CVPR 2008. IEEE Conference on* (pp. 1-7).

Weinland, D., Özuysal, M., & Fua, P. (2010). *Making action recognition robust to occlusions and viewpoint changes. In Computer Vision–ECCV 2010* (pp. 635–648). Springer.

Weinland, D., Ronfard, R., & Boyer, E. (2011). A survey of vision-based methods for action representation, segmentation and recognition. *Computer Vision and Image Understanding, 115*(2), 224–241. doi:10.1016/j.cviu.2010.10.002

Welch, G., & Bishop, G. (2001). An Introduction to the Kalman Filter. *ACM SIGGRAPH 2001 Course Notes.*

Wenjie, C., Lifeng, C., Zhanglong, C., & Shiliang, T. (2005, June). A realtime dynamic traffic control system based on wireless sensor network. In *Parallel Processing, 2005. ICPP 2005 Workshops.International Conference Workshops on* (pp. 258-264). IEEE. doi:10.1109/ICPPW.2005.16

Wikipedia. (n.d.). *F1 Score.* Retrieved from https://en.wikipedia.org/wiki/F1_score

Wireless Marketing Company Limited. (2011). *Linksys WRT54GL.* Retrieved from http://www.taladwireless.com/product_info.php?products_id=265

Wong & Fong. (2013). *Experimental study of video fire detection and its applications.* International Conference on Performance-based Fire and Fire Protection Engineering, Wuhan, China.

Wu, H., Liu, L., & Yuan, X. (2010). *Remote monitoring system of mine vehicle based on wireless sensor network.* Paper presented at 2010 International Conference on Intelligent Computation Technology and Automation, Changsha, China. doi:10.1109/ICICTA.2010.209

Xie, Z., & Farin, G. E. (2004). Image registration using hierarchical B-splines. *IEEE Transactions on Visualization and Computer Graphics, 10*(1), 85–94. doi:10.1109/TVCG.2004.1260760 PMID:15382700

Xuan, G., Shi, Y. Q., Ni, Z., Chai, P., Cui, X., & Tong, X. (2007). Reversible Data Hiding for JPEG Images based on Histogram Pairs. In *Image Analysis and Recognition* (pp. 715–727). Springer Berlin Heidelberg. doi:10.1007/978-3-540-74260-9_64

Yam, C.-Y., & Nixon, M. (2009). Gait Recognition, Model-Based. In Encyclopedia of Biometrics, (pp. 633-639). Academic Press.

Yamato, J., Ohya, J., & Ishii, K. (1992). Recognizing human action in time-sequential images using hidden markov model. *Proceedings CVPR, 92,* 1992.

Yang, G., & Xu, F. (2011). Research and analysis of Image edge detection algorithm Based on the MATLAB. *Procedia Engineering, 15*, 1313–1318. doi:10.1016/j.proeng.2011.08.243

Yanowitz, S. D., & Bruckstein, A. M. (1988, November). A new method for image segmentation. In *Pattern Recognition, 1988., 9th International Conference on* (pp. 270-275). IEEE.

Yeffet, L., & Wolf, L. (2009). Local trinary patterns for human action recognition. *Computer Vision, 2009 IEEE 12th International Conference on* (pp. 492-497).

Yildiz, A., Ozgur, A., & Akgul, Y. S. (2008, October). Fast non-linear video synopsis. In *Computer and Information Sciences, 2008. ISCIS'08. 23rd International Symposium on*, (pp. 1-6). doi:10.1109/ISCIS.2008.4717951

Yilmaz, A., Javed, O., & Shah, M. (2006). Object tracking: A survey. *ACM Computing Surveys, 38*(4). doi:10.1145/1177352.1177355

Yogameena, B., & Priya, K. S. (2015, January). Synoptic video based human crowd behavior analysis for forensic video surveillance. In *Advances in Pattern Recognition (ICAPR), 2015 Eighth International Conference on*, (pp. 1-6). doi:10.1109/ICAPR.2015.7050662

Young, D., & Ferryman, J. (2005). PETS metrics: Online performance evaluation service.*Proc. IEEE Int. Workshop on Performance Evaluation of Tracking Systems* (pp. 317–324).

Yousef, K. M., Al-Karaki, M. N., & Shatnawi, A. M. (2010). Intelligent Traffic Light Flow Control System Using Wireless Sensors Networks. *J. Inf. Sci. Eng., 26*(3), 753–768.

Ze-Nian, L., & Mark, S. D. (2004). *Fundamentals of Multimedia*. Pearson Education International.

Zhang, X., Zhang, Y., & Zheng, R. (2011). Image edge detection method of combining wavelet lift with Canny operator. *Procedia Engineering, 15*, 1335–1339. doi:10.1016/j.proeng.2011.08.247

Zhang, Y. J. (1997). Evaluation and comparison of different segmentation algorithms. *Pattern Recognition Letters, 18*(10), 963–974. doi:10.1016/S0167-8655(97)00083-4

Zhang, Y.-J. (2006).*Advances in Image and Video Segmentation*. Hershey, PA: IGI Global. doi:10.4018/978-1-59140-753-9

Zhao, J., Tow, J., & Katupitiya, J. (2005, August). On-tree fruit recognition using texture properties and color data. In *Intelligent Robots and Systems, 2005.(IROS 2005). 2005 IEEE/RSJ International Conference on* (pp. 263-268). IEEE. doi:10.1109/IROS.2005.1545592

Zhongzhi, H., Jing, L., Yougang, Z., & Yanzhao, L. (2012). Grading System of Pear's Appearance Quality Based on Computer Vision. *International Conference on System and Informatics (ICSAI-2012)*.

Zhou, B., Cao, J., Zeng, X., & Wu, H. (2010, September). Adaptive traffic light control in wireless sensor network-based intelligent transportation system. In Vehicular technology conference fall (VTC 2010-Fall), 2010 IEEE 72nd (pp. 1-5). IEEE. doi:10.1109/VETECF.2010.5594435

Zhu, J., Liao, S., & Li, S. Z. (2015). *Multi-Camera Joint Video Synopsis*. Academic Press.

Zhu, Y., Nayak, N. M., & Roy-Chowdhury, A. K. (2013). Context-aware activity recognition and anomaly detection in video. *Selected Topics in Signal Processing. IEEE Journal of, 7*(1), 91–101.

About the Contributors

Nilanjan Dey, PhD., is an Asst. Professor in the Department of Information Technology in Techno India College of Technology, Rajarhat, Kolkata, India. He holds an honorary position of Visiting Scientist at Global Biomedical Technologies Inc., CA, USA and Research Scientist of Laboratory of Applied Mathematical Modeling in Human Physiology, Territorial Organization of Scientific and Engineering Unions, Bulgaria, Associate Researcher of Laboratoire RIADI, University of Manouba, Tunisia. He is the Editor-in-Chief of International Journal of Ambient Computing and Intelligence (IGI Global), US, International Journal of Synthetic Emotions (IJSE) and the International Journal of Rough Sets and Data Analysis (IGI Global), US, Series Editor of Advances in Geospatial Technologies (AGT) Book Series, (IGI Global), US, Executive Editor of International Journal of Image Mining (IJIM), Inderscience, Regional Editor-Asia of International Journal of Intelligent Engineering Informatics (IJIEI), Inderscience and Associated Editor of International Journal of Service Science, Management, Engineering, and Technology, IGI Global. His research interests include: Medical Imaging, Soft computing, Data mining, Machine learning, Rough set, Mathematical Modeling and Computer Simulation, Modeling of Biomedical Systems, Robotics and Systems, Information Hiding, Security, Computer Aided Diagnosis, Atherosclerosis. He has 8 books and 160 international conferences and journal papers. He is a life member of IE, UACEE, ISOC, etc.

Amira S. Ashour is an Asst. Professor and Vice Chair of Computer Engineering Department of in Computers and Information Technology College, Taif University, KSA. Vice Chair of Computers Science Department of in Computers and Information Technology College, Taif University, KSA during 2006 till 2015. She is a Lecturer of Electronics and Electrical Communications Engineering, Faculty of Engg., Tanta University, Egypt. She got her Ph.D. in Smart Antenna (2005) in the Electronics and Electrical Communications Engineering, Tanta University, Egypt. She had her master in "Enhancement of Electromagnetic Non-Destructive Evaluation Performance Using Advanced Signal Processing Techniques," Faculty of Engineering, Egypt, 2000. Her research interests include Image Processing, Medical Imaging, Smart Antenna and Adaptive Antenna Arrays.

Suvojit Acharjee is an Assistant Professor, NIT Agartala. He was M-Tech in Intelligent Automation and Robotics, Jadavpur University, 2012. He worked as Assistant System Engineer in Tata Consultancy Services Ltd, from 2012 to 2014. He got Indian Patent for a work titled: An efficient transreceiver for Bio-Medical data using Ant Wight Lifting Algorithm on 2014. He has several publications in SCI journals and several conferences. His research interests include Bio-Medical Image Processing, Image Segmentation, Video Compression, Soft Computing.

* * *

Abdelmgeid Amin Ali is a Professor, Computer Science Dept. Faculty of Science, Minia University, Al Minia, Egypt, and Dean of Faculty of Computers & Information, Minia University, Al Minia, Egypt.

R. Aswath received his B. E, M. Tech, and Ph. D degrees in Low power VLSI Design from MGR University, Tamilnadu, India in 1991, 1994, and 2006, respectively. He is currently a Professor and Head of the Department of Telecommunication Engineering of Dayananda Sagar College of Engineering, Bengaluru, India. He has published 25 technical articles and proceedings in IEEE, Springer, Elsevier, etc. His areas of interest include Low power VLSI design and embedded systems.

Sayan Chakraborty is currently working as an asst. professor in Bengal College of Engineering & Technology. He has worked on multiple number of national and international papers.

Imed Bouchrika received his BSc and PhD degrees in Electronics and Computer Science from the University of Southampton (United Kingdom) in 2004 and 2008 respectively. Since 2008, he has worked as a research fellow at the Information: Signals, Images, Systems Research Group of the University of Southampton. He is now a lecturer of Computer Science at the University of Souk Ahras. His research areas are image processing and biometrics.

Shefali Gandhi works as an Assistant Professor in Department of Computer Engineering. The area of research/interest are image/video processing, Operating System, Data mining.

Xiao-Zhi Gao received B.Sc, M.Sc, and D.Sc degrees in Electrical Engineering from Helsinki University of Technology, Finland in 1993, 1996, and 1999, respectively. He is currently a Professor in the Department of Electrical Engineering and Automation, Aalto University School of Electrical Engineering, Finland. He has published 185 technical articles in refereed journals and proceedings such as IEEE, Springer, Elsevier, etc. His research interests include nature-inspired computing methods with applications in optimization, prediction, data mining, signal processing, and control.

Nouzha Harrati received her bachelor degree in computer science from the University of Annaba, Algeria, She obtained a Magister Degree in Computer Science from the University of Souk Ahras. Harrati is now working towards her PhD degree at the Image Processing Research Group at the University of Bejaia. She is an assistant lecturer of Computer Science. Her research includes usability analysis, affective computing and automated classification of facial expressions.

T. Naga Jyothi Completed her M.Tech (CSE) from VR Siddhartha Engineering College in the year 2015. She published one paper in IEEE IACC 2015. Presently working in TCS Chennai.

Pushpajit A. Khaire is currently an Assistant Professor, Department of Computer Science and Engineering, Shri Ramdeobaba College of Engineering and Management, Nagpur, Maharashtra, India. Completed Bachelor of Engineering degree from K.D.K College of Engineering, Nagpur, Maharashtra in 2010 and Master of Technology (M.Tech) in Computer Science and Engineering from SRCOEM, RTM Nagpur University, Nagpur in 2012. He has published many papers in International Journals and Conferences. His Research Interest includes Image Processing, Computer Vision, Pattern Recognition, Neural Networks, Cryptography and Machine Learning.

Ram Kumar obtained his B. Tech in Electronics and Communication Engineering from Rajasthan Institute of Engineering and Technology Jaipur in 2010. He obtained his M. Tech in microelectronics and VLSI from National Institute of Technology, Silchar (India) in 2013. Currently he is a Research Scholar in National Institute of Technology, Silchar. His area of interest is biomedical engineering and VLSI circuits. He has published around 20 research journal and conference. He is also the author and co-author of book.

Ammar Ladjailia obtained an engineering degree in computer science and a magister degree in 2000 and 2003 respectively from the University of Annaba, Algeria. He is now an assistant lecturer of Computer Science at the University of Souk Ahras. His research areas are image and video processing, human activities recognition and visual surveillance.

Zohra Mahfouf is a research fellow at the Image Processing research group at the University of Souk Ahras. She is working towards her PhD in the area of Biometrics and visual surveillance.

M. C. Hanumantharaju received his B. E, M. Tech, and Ph. D degrees in VLSI Signal and Image Processing from Visvesvaraya Technological University, Belgaum, Karnataka, India in 2001, 2004 and 2014, respectively. He is currently a Professor and Head of the Department of ECE at BMS Institute of Technology and Management, Bengaluru, India. He has published two books and 30 technical articles in refereed journals and proceedings such as IEEE, Intelligent Systems, Particle Swarm Optimization, etc. His research interests include Design of Hardware Architectures for Signal and Image Processing Algorithms, RTL Verilog Coding, FPGA/ASIC Design.

Anand S. is working as Associate Professor in ECE department of Mepco Schlenk Engineering College having 24 years of teaching experience and research area is Image processing. He is reviewer for many leading journals.

Abahan Sarkar is a Research Scholar in the department of Electrical Engineering in National Institute of Technology Silchar, Assam, India. Mr. Sarkar worked as Faculty member in Faculty of Science and Technology, The ICFAI University Tripura from 2007 to 2012. He has obtained his B.Tech degree in Electronics and Communication Engineering from North Eastern Regional Institute of Science and Technology, Nirjuli, Arunachal Pradesh, India in 2003. He has completed his Master of Technology in Computer Science and Technology from Tripura University, Suryamaninagar, Tripura, India in 2008. His research interest includes, Industrial automation, and Image processing and Machine vision.

K. Sathish Shet received his B. E degree in Electronics and Communication Engineering and M. Tech degree in VLSI Design and Embedded Systems from Visvesvaraya Technological University, Belgaum, India in 2000 and 2006, respectively. He is currently an assistant professor in the Department of Electronics and Communication Engineering (ECE) at JSS Academy of Technical Education and a research scholar in ECE Department of Dayananda Sagar College of Engineering, Bengaluru, India. He has published six technical articles in peer-reviewed journals and proceedings such as Springer, IEEE, etc. His research interests include VLSI and Image processing.

Claudio Urrea was born in Santiago, Chile. He received the M.Sc. Eng. and the Dr. degrees from Universidad de Santiago de Chile, Santiago, Chile in 1999, and 2003, respectively; and the Ph.D. degree from Institut National Polytechnique de Grenoble, France in 2003. Ph.D. Urrea is currently Professor at the Department of Electrical Engineering, Universidad de Santiago de Chile, from 1998. He has developed and implemented a Robotics Laboratory, where intelligent robotic systems are development and investigated. He is currently Director of the Doctorate in Engineering Sciences, Major in Automation, at the Universidad de Santiago de Chile.

S. Vasavi is working as a Professor in Computer Science & Engineering Department, VR Siddhartha Engineering College, Vijayawada, India. With 19 years of experience. She currently holds funded Research projects from University Grants Commission (UGC) and Advanced Data Research Institute (ADRIN, ISRO). Her research areas are Data mining and Image Classification. She is a life member of Computer Society of India. She published 35 papers in various International conferences and journals. She filed two patents.

Neethidevan Veerapathiran is working as Assistant Professor in MCA department of Mepco Schlenk Engineering College having 19 years of teaching experience and research area is image processing.

Index

user-friendly 220, 240, 243

V

Verilog 145, 147, 156, 166
VHDL 145
video 28-33, 35-38, 42-48, 52-53, 70-71, 74-75, 79, 87-89, 95, 97-99, 104, 109, 116, 125-127, 133-135, 140-141, 145-147, 171-174, 176, 180-181, 183-185, 193-198, 200-210, 212-216, 218-220, 224-230, 233, 236-237, 239-240, 245-247, 263-274, 277, 279-283, 286, 288-289
video abstraction 194-195, 219
Video Compression 282

Video Dataset 42, 53
video processing 31, 33, 37, 134, 219, 263-267, 273-274, 283, 289
video segmentation 28-33, 35-38, 42, 45-48, 53
video summarization 173
video synopsis 193-197, 204-206, 212-216, 218-219
VLSI 108, 146

W

wireless 194, 221-222, 224, 230-231, 233, 243, 245-248, 261

Printed in the United States
By Bookmasters